Lecture Notes in Mathematics

Edited by A. Dold and B. Eckmann

932

Analytic Theory of Continued Fractions

Proceedings of a Seminar-Workshop held at Loen, Norway, 1981

Edited by W. B. Jones, W. J. Thron, and H. Waadeland

Springer-Verlag
Berlin Heidelberg New York 1982

Editors
William B. Jones
W.J. Thron
Department of Mathematics, Campus Box 426
University of Colorado
Boulder, Colorado 80309, USA

Haakon Waadeland
Institutt for Matematikk og Statistikk
Universitetet i Trondheim
7055 Dragvoll, Norway

AMS Subject Classifications (1980): 30B70, 30E05, 30E10, 30E20, 40A15, 41A20, 41A21, 65B10, 65D20, 65G05

ISBN 3-540-11567-6 Springer-Verlag Berlin Heidelberg New York
ISBN 0-387-11567-6 Springer-Verlag New York Heidelberg Berlin

Printing and binding: Beltz Offsetdruck, Hemsbach/Bergstr.
2141/3140-543210

PREFACE

This volume of LECTURE NOTES IN MATHEMATICS contains the proceedings of a research Seminar-Workshop on recent progress in the Analytic Theory of Continued Fractions held at Loen, Norway from June 5 to June 30, 1981. In recent years there has been a renewed interest in the subject of continued fractions. This is due in part to the advent of computers and the resulting importance of the algorithmic character of continued fractions. It is also due to the close connection between continued fractions and Padé approximants and their application to theoretical physics. Primary emphasis at the Workshop was on the analytic aspects of the subject; however, considerable attention was also given to applied and computational problems. These interests are reflected in the Workshop proceedings.

The sessions at Loen were devoted not only to reports on recent work but also to the development of new results and the formulation of further problems. The authors whose papers appear in these proceedings either attended the Workshop or, if unable to attend, had their work presented and discussed at Loen.

The Seminar-Workshop was organized by Haakon Waadeland of the University of Trondheim and was made possible by grants from the Norwegian Research Council for Science and the Humanities (NAVF) and from the University of Trondheim. Support for travel expenses of the American participants came from the United States National Science Foundation, the University of Colorado at Boulder, Colorado State University, NAVF, and Fridtjof Nansen's and Affiliated Funds for the Advancement of Science and the Humanities. The latter also supported a visit to the University of Colorado for follow-up discussions of research topics. The University of Colorado provided a small grant for expenses related to publication of the proceedings. We gratefully acknowledge these contributions.

We also wish to thank the director and staff of the Alexandra Hotel in Loen for providing excellent working facilities and a cordial atmosphere for the Workshop. The professional assistance of the technical typists, Burt Rashbaum and Alexandra Hunt, at the Mathematics Department of the University of Colorado is greatly appreciated. Finally, we would like to thank Professor B. Eckmann, ETH Zürich, for accepting this volume for the Springer series of LECTURE NOTES IN MATHEMATICS.

CONTENTS

LIST OF CONTRIBUTORS AND PARTICIPANTS

JOHN GILL

Department of Mathematics, University of Southern Colorado, Pueblo, Colorado 81001 U.S.A.

LISA JACOBSEN

Institutt for Matematikk, Universitetet i Trondheim, 7034 Trondheim-NTH, Norway

WILLIAM B. JONES

Department of Mathematics, Campus Box 426, University of Colorado, Boulder, Colorado 80309 U.S.A.

L. J. LANGE

Department of Mathematics, University of Missouri, Columbia, Missouri 65201 U.S.A.

ARNE MAGNUS

Department of Mathematics, Colorado State University, Fort Collins, Colorado 80523 U.S.A.

MARIUS OVERHOLT

Matematikk, Rogaland Distriktshögskole, 4000 Stavanger, Norway

ELAINE PULÉO

Department of Mathematics, University of Southern Colorado, Pueblo, Colorado 81001 U.S.A.

WALTER M. REID

Department of Mathematics, University of Wisconsin-Eau Claire, Eau Claire, Wisconsin 54701 U.S.A.

ALLAN STEINHARDT

Department of Electrical Engineering, University of Colorado, Boulder, Colorado 80309 U.S.A.

W. J. THRON

Department of Mathematics, Campus Box 426, University of Colorado, Boulder, Colorado 80309 U.S.A.

HAAKON WAADELAND

Department of Mathematics and Statistics, University of Trondheim, N-7055 Dragvoll, Norway

BASIC DEFINITIONS AND NOTATION

To help unify the contributions to this volume we have asked all authors to use the same notation for certain basic concepts. These concepts and their definitions are listed here so as to avoid unnecessary duplication in the introduction to various articles. The complex plane shall be denoted by $\mathbb{C}$. For the extended complex plane, that is $\mathbb{C} \cup [\infty]$, we use the notation $\hat{\mathbb{C}}$.

The continued fraction algorithm is a function K that associates with ordered pairs of sequences $\langle\{a_n\},\{b_n\}\rangle$, with $a_n \in \mathbb{C}$, $a_n \neq 0$ for $n \geq 1$ and $b_n \in \mathbb{C}$ for $n \geq 0$, a third sequence $\{f_n\}$ with $f_n \in \hat{\mathbb{C}}$. A continued fraction is an ordered pair

$$\langle\langle\{a_n\}, \{b_n\}\rangle, \{f_n\}\rangle ,$$

where the sequence $\{f_n\}$ is defined as follows. Let $\{s_n\}$ and $\{S_n\}$ be sequences of linear fractional transformations (l.f.t.) defined by

(DN1a) $$s_0(w) = b_0 + w , \quad s_n(w) = \frac{a_n}{b_n + w} , \quad n = 1, 2, 3, \ldots$$

and

(DN1b) $$S_0(w) = s_0(w) , \quad S_n(w) = S_{n-1}(s_n(w)) , \quad n = 1, 2, 3, \ldots .$$

Then

(DN1c) $$f_n = S_n(0) , \quad n = 0, 1, 2, \ldots .$$

The numbers a_n and b_n are called the n th partial numerator and n th partial denominator, respectively, of the continued fraction $\langle\langle\{a_n\},\{b_n\}\rangle, \{f_n\}\rangle$; they are also called the elements. f_n is called the n th approximant of the continued fraction. If $f = \lim_{n\to\infty} f_n$ exists in $\hat{\mathbb{C}}$, we say that the continued fraction is convergent and that its value is f . If the limit does not exist we speak of a divergent continued fraction.

More generally we shall also be interested in sequences of l.f.t.'s $\{S_n^{(m)}\}$ defined as follows:

(DN2)
$$S_0^{(0)}(w) = w ,$$

$$S_n^{(m)}(w) = s_{m+1} \circ s_{m+2} \circ \cdots \circ s_{m+n}(w) , \quad n = 1, 2, 3, \ldots, \; m = 0, 1, 2, \ldots .$$

Here the symbol $\circ$ denotes functional composition; that is, for functions g and h , $g \circ h(w) = g(h(w))$, provided that the domain of g contains the range of h . It can be seen that

$$S_n^{(m)}(w) = \cfrac{a_{m+1}}{b_{m+1} + \cfrac{a_{m+2}}{b_{m+2} + \ddots + \cfrac{a_{m+n}}{b_{m+n} + w}}} .$$

One also obtains the following

(DN3) $$S_n(w) = s_0 \circ S_n^{(0)}(w) = s_0 \circ s_1 \circ \cdots \circ s_n(w) , \quad n = 0, 1, 2, \ldots,$$

so that

(DN4) $$f_n = S_n(0) = s_0 \circ S_n^{(0)}(0) , \quad n = 0, 1, 2, \ldots .$$

Thus when $b_0 = 0$ (so that $s_0(w) = w$), we have

$$S_n(w) = S_n^{(0)}(w) \quad \text{and} \quad f_n = S_n(0) = S_n^{(0)}(0) .$$

Since $S_n(w)$ is the composition of non-singular l.f.t.'s, it is itself a non-singular l.f.t. It is well known that $S_n(w)$ can be written in the form

$$S_n(w) = \frac{A_n + w A_{n-1}}{B_n + w B_{n-1}} , \quad n = 0, 1, 2, \ldots$$

where the A_n and B_n (called the n th <u>numerator</u> and n th <u>denominator</u>, respectively, of the continued fraction) are defined by the second order linear <u>difference equations</u>

(DN6a) $$A_{-1} = 1 , \; A_0 = b_0 , \; B_{-1} = 0 , \; B_0 = 1 ,$$

$$A_n = b_n A_{n-1} + a_n A_{n-2} , \quad n = 1, 2, 3, \ldots,$$

(DN6b)

$$B_n = b_n B_{n-1} + a_n B_{n-2} , \quad n = 1, 2, 3, \ldots .$$

For simplicity a continued fraction $\langle\langle \{a_n\} , \{b_n\} \rangle , \{f_n\} \rangle$ will be denoted by the symbol

(DN7) $$b_0 + \operatorname*{K}_{n=1}^{\infty} \left(\frac{a_n}{b_n} \right)$$

or

(DN8) $$b_0 + \frac{a_1}{b_1} + \frac{a_2}{b_2} + \frac{a_3}{b_3} + \cdots .$$

The symbols (DN7) and (DN8) may be used to denote both the continued fraction and its value when it is convergent.

The notation (DN8) shall also be used for finite combinations. Thus we write

$$S_n^{(m)}(w) = \frac{a_{m+1}}{b_{m+1}} + \frac{a_{m+2}}{b_{m+2}} + \cdots + \frac{a_{m+n}}{b_{m+n} + w} , \quad m \geq 0 .$$

We shall call the continued fraction

$$\frac{a_{m+1}}{b_{m+1}} + \frac{a_{m+2}}{b_{m+2}} + \frac{a_{m+3}}{b_{m+3}} + \cdots, \quad (m \geq 0)$$

the m th <u>tail</u> of (DN8) and we denote its n th approximant by $f_n^{(m)}$. Thus

$$f_n^{(m)} = S_n^{(m)}(0) \ , \ n = 1, 2, 3, \ldots, \quad m = 0, 1, 2, \ldots \ .$$

If $\lim_{n\to\infty} f_n^{(m)}$ exists in $\hat{\mathbb{C}}$, then the m th tail is convergent and its value is denoted by $f^{(m)}$. One thus has, for $m \geq 0$,

$$\text{(DN9)} \qquad f^{(m)} = \frac{a_{m+1}}{b_{m+1}} + \frac{a_{m+2}}{b_{m+2}} + \frac{a_{m+3}}{b_{m+3}} + \cdots \ .$$

An important role is played in continued fraction theory by the expression h_n defined by

$$h_n = -S_n^{-1}(\infty) = B_n/B_{n-1} \ .$$

It is easily shown, from the difference equations (DN6), that

$$\text{(DN10)} \qquad h_1 = 1 \quad \text{and} \quad h_n = b_n + \frac{a_n}{b_{n-1}} + \frac{a_{n-1}}{b_{n-2}} + \cdots + \frac{a_2}{b_1}, \ n = 2, 3, 4, \ldots \ .$$

In some cases the a_n , b_n are functions of a complex variable z . To emphasize this dependence on z we sometimes write

$$a_n(z) \ , \ b_n(z) \ , \ S_n^{(m)}(z,w) \ , \ A_n(z) \ , \ B_n(z) \ , \ f_n^{(m)}(z) \ , \ f^{(m)}(z) \ , \text{ etc.}$$

If in a continued fraction all $b_n = 1$ or all $a_n = 1$, the notation outlined above shall nevertheless be employed. If two or more continued fractions are considered at the same time

s_n^* , $S_n^{*(m)}$, $f^{*(m)}$,... might be used for $K(a_n^*/b_n^*)$ and

s_n' , $S_n^{(m)'}$, $f^{(m)'}$ ... for $K(a_n'/b_n')$.

SURVEY OF CONTINUED FRACTION METHODS
OF SOLVING MOMENT PROBLEMS
AND RELATED TOPICS

William B. Jones and W. J. Thron

1. Introduction. There is a constellation of interrelated topics:

Orthogonal polynomials;

Gauss quadrature;

Integral representation of continued fractions;

Determination of functions having given power series as asymptotic expansions;

Expansions of functions in series of orthogonal polynomials;

Solutions of certain three-term recurrence relations.

Today most of these topics are usually studied without reference to continued fraction theory. Nevertheless, all of them either arose or received important impetus from the theory of continued fractions. Thus Szegö [20, p. 54] asserts that: "historically the orthogonal polynomials $\{p_n(x)\}$ originated in the theory of continued fractions. This relationship is of great importance and is one of the possible starting points of the treatment of orthogonal polynomials."

Wynn in two statements [25, p. 190,191] attempts to delineate the role of continued fractions even more sharply. His first statement is: "many theories originating from the study of continued fractions have, upon reflection, been found to have little to do with them." He next asserts: "The theory of continued fractions has been preeminently an avenue to new and unexpected results...." The results to be described here provide a further confirmation of Wynn's thesis.

The interrelations in the constellation are intricate but, unfortunately, confusing. This is in part due to the fact that many mathematicians have made contributions using the tools, the language and the outlook of their trade, be it continued fractions, orthogonal polynomials, functional analysis or others. Not surprisingly they were, with few notable exceptions, somewhat narrow in their knowledge of the literature and quite ignorant of the earlier history. We too must plead guilty of such narrowness, although we are attempting to acquire a wider point of view and to learn more about the beginnings of the various topics in the constellation. However, at this writing our knowledge is still much more fragmentary than we would like it to be.

Arising from an investigation of correspondence and convergence properties of general T-fractions, we have been led to formulate strong Stieltjes and Hamburger moment problems (SSMP and SHMP, respectively). We showed that certain sequences of Laurent polynomials (L-polynomials), closely related to the denominators of the approximants of the positive T-fractions, are orthogonal with respect to a distribution function which can be derived from the positive T-fraction. In addition we were able to obtain new results for other topics in the constellation.

To facilitate the description of our results we first make a series of definitions.

A general T-fraction is a continued fraction of the form

$$\frac{F_1 z}{1 + G_1 z +} \frac{F_2 z}{1 + G_2 z +} \frac{F_3 z}{1 + G_3 z +} \cdots , \tag{1.1a}$$

where $F_n \neq 0$ for all n. It can also be written in the following equivalent forms:

$$\frac{F_1}{z^{-1} + G_1 +} \frac{F_2}{1 + G_2 z +} \frac{F_3}{z^{-1} + G_3 +} \frac{F_4}{1 + G_4 z +} \cdots , \tag{1.1b}$$

$$\frac{z}{e_1 + d_1 z +} \frac{z}{e_2 + d_2 z +} \frac{z}{e_3 + d_3 z +} \cdots , \tag{1.1c}$$

where $e_n \neq 0$ for all n. Here

$$e_1 = 1/F_1 , \quad e_{2n-1} = \prod_{k=1}^{n-1} F_{2k} / \prod_{k=1}^{n} F_{2k-1} , \quad n = 2,3,4,\ldots, \tag{1.2a}$$

$$e_{2n} = \prod_{k=1}^{n} F_{2k-1} / \prod_{k=1}^{n} F_{2k} , \quad n = 1,2,3,\ldots, \tag{1.2b}$$

$$d_n = G_n e_n , \quad n = 1,2,3\ldots . \tag{1.2c}$$

If all $F_n > 0$ and $G_n > 0$, then (1.1a) (and all forms equivalent to it) is called a positive T-fraction. If $d_{2n-1} > 0$, $e_{2n} > 0$ and e_n, d_n are real for all $n \geq 1$, then (1.1c) (and all forms equivalent to it) is called a semi-positive T-fraction.

For a function $f(z)$ holomorphic at $z = 0$, let us denote by $\Lambda_0(f)$ its Taylor series expansion at 0. Let

$$\sum_{k=0}^{\infty} \alpha_k z^k$$

be a formal power series, and let $\{R_n(z)\}$ be a sequence of rational functions holomorphic at $z = 0$. Then we say that the sequence $\{R_n(z)\}$ corresponds to the series $\Sigma \alpha_k z^k$ at $z = 0$ if the formal power series $\Lambda_0(R_n) - \Sigma \alpha_k z^k$ has the form

$$\Lambda_0(R_n) - \sum_{k=0}^{\infty} \alpha_k z^k = g_{m_n} z^{m_n} + g_{m_n+1} z^{m_n+1} + \cdots ,$$

where $m_n \to \infty$ as $n \to \infty$. A continued fraction $K(a_n(z)/b_n(z))$ is said to correspond to a series if the sequence of approximants corresponds to the series.

Analogous definitions are used for correspondence at $z = a$ and, in particular, for $a = \infty$.

A general T-fraction (1.1a) (at least if all $G_n \neq 0$) corresponds to formal powers series L_0 at $z = 0$ and L_∞ at $z = \infty$. It is convenient to write these series as

$$(1.3) \qquad L_0 = \sum_{k=1}^{\infty} - c_{-k} z^k \quad \text{and} \quad L_\infty = \sum_{k=0}^{\infty} c_k z^{-k} .$$

By $\Phi^c(a,b)$ we shall mean the family of all real-valued, functions $\psi(t)$ defined on $a < t < b$, which are bounded, monotone non-decreasing with infinitely many points of increase on (a,b), and for which the integrals

$$(1.4) \qquad c_n = \int_a^b (-t)^n \, d\psi(t)$$

exist for all integers $n \geq 0$. (This additional condition is meaningful if $a = -\infty$ and/or $b = +\infty$, possibilities which we do admit.) The family of functions $\psi \in \Phi^c(a,b)$ for which the c_n in (1.4) also exist for all negative integers n we shall denote by $\Phi(a,b)$. The functions $\psi \in \Phi^c(a,b)$ or $\Phi(a,b)$ are called distribution functions and the c_n defined by (1.4) are called moments with respect to the distribution ψ.

The classical Stieltjes moment problem defined by Stieltjes in 1894 consists in finding conditions on the moments $\{c_n\}_0^\infty$ which would insure the existence of a function $\psi \in \Phi^c(0,\infty)$ for which (1.4) holds, for all $n = 0,1,2,\ldots$, with $a = 0$, $b = \infty$. Stieltjes found necessary and sufficient conditions for the existence of such a ψ. He also found necessary and sufficient conditions for the ψ to be unique. In 1920 Hamburger extended the problem to the interval $(-\infty,\infty)$. This is the classical Hamburger moment problem. In the solutions to the two problems Hankel determinants $H_0^{(m)} = 1$,

$$(1.5) \qquad H_k^{(m)} = \begin{vmatrix} c_m & c_{m+1} & \cdots & c_{m+k-1} \\ c_{m+1} & c_{m+1} & \cdots & c_{m+k} \\ \vdots & \vdots & & \vdots \\ c_{m+k-1} & c_{m+k} & & c_{m+2k-2} \end{vmatrix} , \quad k = 1,2,3,\ldots$$

as well as J-fractions

$$(1.6) \qquad \frac{k_1}{\ell_1 + z -} \; \frac{k_2}{\ell_2 + z -} \; \frac{k_3}{\ell_3 + z -} \cdots$$

are of importance.

We return now to the main theme of this article. General T-fractions (1.1a) with all $G_n \neq 0$ correspond to formal power series (1.3) at $z = 0$ and $z = \infty$. Thus associated with a general T-fraction (with $G_n \neq 0$) is a double sequence $\{c_n\}_{-\infty}^{\infty}$ of numbers and it makes sense to ask whether a function $\psi \in \Phi(0,\infty)$

exists for which the c_n with $n = 0, \pm 1, \pm 2, \ldots$ are the moments with respect to ψ. This is the strong Stieltjes moment problem (SSMP). It was posed and solved by Waadeland and the authors in [13] by means of positive T-fractions. They also showed that a positive T-fraction has an integral representation of the form

$$G(z) = \int_0^\infty \frac{zd\psi(t)}{z+t},$$

where $\psi \in \Phi(0,\infty)$, and that the function $G(z)$ has the power series (1.3) (to which the positive T-fraction corresponds) as asymptotic expansions at 0 and ∞, respectively. In a later paper [12] the present authors identified and studied the orthogonal functions associated with the SSMP. They turn out to be Laurent polynomials (L-polynomials)

(1.7) $$a_0x^{-n} + \cdots + a_{2n}x^n .$$

Let the n th denominator of the positive T-fraction in the form (1.1b) be denoted by $V_n(z)$; then the orthogonal L-polynomials are given by $Q_0(z) = 1$,

(1.8) $$Q_{2m-1}(z) = (-1)^m V_{2m-1}(-z),\ Q_{2m}(z) = (-1)^m V_{2m}(-z),\quad n = 1,2,3,\ldots .$$

To pose the strong Hamburger moment problem (SHMP) is the natural next step. It is defined and solved, but without the use of continued fractions, by Njåstad and the authors in [9]. Only a partial solution of the SHMP can be obtained using continued fractions. This was also worked on jointly with Njåstad and is in the process of being written. We present here an outline of the solution in terms of semi-positive T-fractions, various ramifications such as integral representations, and the question when does correspondence imply asymptoticity. In addition, generalized approximants, which play a role in resolving the question of uniqueness of the solution to the SHMP, are also discussed. The generalized approximants arise from a "modification" (in the sense of [23] in these Proceedings) of semi-positive T-fractions.

An overview of the contents of this article is as follows. In Section 2 we give a summary of the historical background of the topics to be discussed here. In Section 3, general T-fractions, their correspondence to power series at $z = 0$ and $z = \infty$, as well as the partial fraction decomposition of semi-positive T-fractions are presented. In that section the generalized approximants are introduced and convergence of semi-positive T-fractions are taken up.

In Section 4 the results of the preceding section are used to obtain integral representations for all convergent subsequences of generalized approximants of semi-positive T-fractions.

Section 5 is devoted to solutions of the SSMP and SHMP. In Section 6 L-polynomials, orthogonality, Favard's theorem on recurrence relations, and related topics are considered. In particular, the identification of sequences of orthogonal L-polynomials with the denominators of positive T-fractions will be described.

Section 7 is concerned with Gaussian quadratures and convergence results that can be obtained from them. In Section 8 the discussion shifts back to semi-positive T-fractions. Here sufficient conditions for limit functions of convergent subsequences of generalized approximants to have the series L_0 and L_∞ as asymptotic expansions will be derived.

2. Summary of early history. Even though Legendre discovered the sequence of polynomials named after him in 1782, and was aware of the orthogonality property of the sequence with even subscripts as early as 1785, it was really Gauss who got the subject started.

In an article in 1812, Gauss studied hypergeometric functions and obtained, among other results, a continued fraction expansion for ratios of hypergeometric functions. In a second paper in 1814 he posed and solved a new quadrature problem (earlier work had been done by Cotes and Newton among others), namely, to find an approximation to an integral

$$\int_{-1}^{+1} f(t)dt$$

of the form

$$\sum_{k=1}^{n} \lambda_k^{(n)} f(\tau_k^{(n)}) , \tag{2.1}$$

where $\lambda_k^{(n)}$ and $\tau_k^{(n)}$ are to be determined in such a way that the approximation is exact for all polynomials $f(t)$ of degree not greater than $2n-1$. The proof makes use of the continued fraction expansion

$$\int_{-1}^{+1} \frac{dt}{z+t} = \log\left(\frac{z+1}{z-1}\right) = \frac{2}{z} - \frac{1/3}{z} - \frac{2^2/3\cdot5}{z} - \frac{3^2/5\cdot7}{z} - \cdots , \tag{2.2}$$

which was known to Gauss from his work in 1812. Let $K_n(z)/L_n(z)$ be the $2n$ th approximant of (2.2). Then the roots of $L_n(z)$ are all real and distinct and

$$\frac{K_n(z)}{L_n(z)} = \sum_{k=1}^{n} \frac{\lambda_k^{(n)}}{z+\tau_k^{(n)}} , \tag{2.3}$$

where the $\lambda_k^{(n)}$ and $\tau_k^{(n)}$ are exactly the constants needed in (2.1). Thus the Gaussian quadrature formula can be proved and the constants involved in it can be obtained from continued fraction considerations. Gauss actually computed some of the $\lambda_k^{(n)}$ and $\tau_k^{(n)}$.

Gauss considered this work important and expected that it would be used extensively in practical problems. As it turned out Gaussian quadrature and its generalizations were found to be of considerable theoretical interest throughout the 19th century. After the advent of computers it again attracted the attention of applied mathematicians.

Jacobi in a series of papers, proved the quadrature formula without using continued fractions. In the second of his papers he pointed out that the $L_n(z)$ are indeed the Legendre polynomials.

The people who extended the Gauss quadrature formula during the nineteenth century, among whom Christoffel, Heine, Tchebycheff and Stieltjes are probably the most notable (but Mehler, Radon, Markoff and Posse should also be mentioned), made essential use of continued fraction considerations.

The pattern developed as follows (using the notation introduced by Stieltjes only toward the end of the period). To obtain an approximation to

$$\int_a^b f(t)\, d\psi(t) ,$$

where $\psi(t) \in \Phi^c(a,b)$, one obtains a J-fraction (1.6) which converges to the integral

$$\int_a^b \frac{d\psi(t)}{z+t} ,$$

for $z \notin [-b, -a]$. Let $K_n(z)/L_n(z)$ be the n th approximant of this J-fraction. The constants $\lambda_k^{(n)}$, $\tau_k^{(n)}$ in the quadrature formula

$$\int_a^b f(t)\, d\psi(t) \approx \sum_{k=1}^{n} \lambda_k^{(n)} f(\tau_k^{(n)})$$

are determined by the partial fraction decomposition of $K_n(z)/L_n(z)$ as in (2.3). In the general case, the $\lambda_k^{(n)}$, $\tau_k^{(n)}$ depend on ψ , but are independent of f . The approximation is exact if $f(z)$ is a polynomial of degree at most $2n-1$.

Using the correspondence between the integral and the J-fraction, one then can prove the following:

(2.4) $$\int_a^b t^k L_m(-t)\, d\psi(t) = 0 , \ k < m , \quad m = 1,2,3,\ldots,$$

(2.5) $$\int_a^b L_n(-t) L_m(-t)\, d\psi(t) = \prod_{\nu=1}^{n} k_\nu \, \delta_{n,m} ,$$

where $\delta_{n,m}$ is the Kronecker δ. One also has

(2.6) $$K_n(z) = \int_a^b \frac{L_n(z)-L_n(-t)}{z+t} d\psi(t) .$$

If one sets $P_n(z) = (-1)^n L_n(-z)$, it follows that $\{P_n(z)\}$ is a sequence of orthogonal polynomials with respect to $\psi(t)$, normalized so that the coefficient of z^n in $P_n(z)$ is 1 . It is for this reason that (to paraphrase Gautschi [5, p. 82]) throughout the 19th century orthogonal polynomials were generally viewed as the denominators $L_n(z)$ of the nth approximant of a J-fraction. It is now easily seen that

(2.7) $$P_n(z) = (z-\ell_n) P_{n-1}(z) - k_n P_{n-2}(z) , \quad n = 2,3,\ldots ,$$

where $k_n > 0$ and $\ell_n \in \mathbb{R}$. The result, that a sequence $\{P_n(z)\}$ satisfying (2.7) is the sequence of orthogonal functions with respect to some ϕ, is usually attributed to Favard who stated it in 1935. There are other claimants to having been the first to have obtained this result. However, as Chihara [2, p. 209] puts it very well: "This multiple discovery is not surprising since the theorem is really implicitly contained in the theory of continued fractions. It seems quite likely that mathematicians who worked with continued fractions were well aware of the theorem, but never bothered to formulate it explicitly. Nevertheless, the explicit formulation was a real contribution since most workers in orthogonal polynomials tend to avoid continued fractions whenever possible."

A very detailed survey of Gauss-Christoffel quadrature formulae was recently given by Gautschi [5].

We now turn to other major topics in our constellation. One of these is the expansion of functions (both real-valued as well as analytic) in series of orthogonal polynomials (or as Blumenthal in 1898 still said, "continued fraction denominators").

Expansions of this kind are important for many reasons. One of these where continued fractions enter "naturally" is an interpolation problem of Tchebycheff of 1858 which involves determination of the finite J-fraction equal to the sum

$$\sum_{\nu=1}^{n} \frac{\lambda_\nu}{z+\tau_\nu} .$$

The interpolation problem can then be answered in terms of a finite sum $\sum c_\nu L_\nu(z)$, where the $L_\nu(z)$ are the "continued fraction denominators". Tchebycheff also considered the limiting situation where one determines the infinite J-fraction which is equal to an integral of the form

$$\int_a^b \frac{\phi(t)dt}{z+t} ,$$

at least in certain special cases. Other mathematicians who worked on this problem during the 19th century were Heine, Pincherle, Darboux and Blumenthal.

R. Murphy in 1833-35 was probably the first to study a moment problem. He referred to it as "the inverse method of definite integrals". In this context he encountered polynomials satisfying the orthogonality condition. Recognizing the importance of the condition, he used the term "reciprocal functions" for what we call today orthogonal functions. He was not as we mistakenly stated in [11, p. 6] the originator of the name "orthogonal". According to Gautschi [5, p. 78], "The name 'orthogonal' for function systems came into use only later, probably first in E. Schmidt's 1905 Göttingen dissertation;..." Murphy was interested, among others, in the following problem. If

$$\int_0^1 t^k f(t)dt = 0 , \quad k = 0,1,2,\ldots,n-1 ,$$

what can be said about $f(t)$? Clearly this is the question, to what extent a finite set of moments determines the function $f(t)$. Tchebycheff starting in 1855 took an interest in moment problems. (Some of our information is taken from the brief historical review in the book, The Problem of Moments by Shohat and Tamarkin.) Among others, he was interested in the question whether from

$$\int_{-\infty}^{\infty} x^n p(x)dx = \int_{-\infty}^{\infty} x^n e^{-x^2} dx , \quad n = 0,1,2,\ldots ,$$

one could conclude that $p(x) = e^{-x^2}$. According to Shohat and Tamarkin, "Tchebycheff's main tool is the theory of continued fractions which he uses with extreme ingenuity." He also obtained the approximation for

$$\int_a^x f(t)dt \quad , \quad a < x < b$$

given the moments

$$\int_a^b t^n f(t)dt \quad .$$

Stieltjes in 1894 was able to pull together the work on moments of his predecessors in a very satisfying manner at least as far as the interval $[0,\infty)$ was concerned. He used and refined the tools which had been introduced by the mathematicians we have mentioned here. Stieltjes found necessary and sufficient conditions for the existence of a solution $\psi \in \Phi^c(0,\infty)$. He also described a way to obtain the solution by first obtaining a continued fraction expansion of

$$G(z) = \int_0^\infty \frac{d\psi(t)}{z+t} \quad ,$$

valid for all z not on the negative real axis. By an inversion process he then arrived at $\psi(t)$. It was not until 1920 that the general moment problem was solved by Hamburger using J-fractions. Further work on the moment problem by M. Riesz, R. Nevanlinna, Carleman, and Hausdorff in the early nineteen twenties does not make use of continued fractions.

Another motivation of Stieltjes' work was the "summing" of the divergent series to which the continued fraction corresponds. The divergent series then becomes an asymptotic expansion of the function (represented by the integral) to which the continued fraction converges. Stieltjes had written his thesis in 1886 on asymptotic series (he called them semi-convergent) the same year in which Poincaré wrote a fundamental paper on the subject.

The history of integral representation of continued fractions is sketched in Section 4.

3. Semi-positive T-fractions. In [13] we gave references for the history of general T-fractions. In that article we also proved the general theorem for correspondence, which, in terms of L_0 and L_∞ of (1.3), and without loss of generality, can be stated as follows.

Theorem 3.1. *Let*

$$L_0 = \sum_{m=1}^{\infty} - c_{-m} z^m \quad \text{and} \quad L_\infty = \sum_{m=0}^{\infty} c_m z^{-m}$$

be given. Then there exists a general T-fraction

$$\underset{n=1}{\overset{\infty}{K}}\left(\frac{F_n z}{1+G_n z}\right) , \quad F_n \neq 0 , \quad G_n \neq 0 , \quad n = 1,2,3,\ldots$$

corresponding to L_0 at 0 and to L_∞ at ∞ if and only if the Hankel determinant (defined in (1.5)) satisfy

$$H_n^{(-n+1)} \neq 0 \quad \text{and} \quad H_n^{(-n)} \neq 0 , \quad n = 0,1,2,\ldots .$$

When these conditions hold, the F_n and G_n are given by

$$\text{(3.1a)} \qquad F_1 = -\frac{H_1^{(-1)}}{H_0^{(1)}} , \quad F_n = -\frac{H_{n-2}^{(-n+3)} H_n^{(-n)}}{H_{n-1}^{(-n+2)} H_{n-1}^{(-n)}} , \quad n = 2,3,4,\ldots ,$$

$$\text{(3.1b)} \qquad G_1 = -\frac{H_1^{(-1)}}{H_1^{(0)}} , \quad G_n = -\frac{H_{n-1}^{(-n+2)} H_n^{(-n)}}{H_n^{(-n+1)} H_{n-1}^{(-n+1)}} , \quad n = 2,3,4,\ldots .$$

In [13] we derived integral representations for positive T-fractions. Here we shall obtain similar representations for semi-positive T-fractions. Semi-positivity can be characterized in three different ways. The conditions defining semi-positivity arose naturally in our investigation of the SHMP. It is an interesting speculation whether, without this connection, one would have thought of these conditions which are so useful in establishing that general T-fractions satisfying them have integral representations. Whether there are other classes of general T-fractions for which there are integral representations (with the integral extending over the real line) is an open question.

In the sequel F_n, G_n, d_n, e_n and c_n will always be assumed to be real numbers. Further since F_n, G_n shall be different from zero, so shall $d_n \neq 0$ and $e_n \neq 0$.

Theorem 3.2. The general T-fraction (1.1a) is semi-positive if and only if

$$\text{(3.2)} \qquad F_{2k-1}F_{2k} > 0 \quad \text{and} \quad F_{2k-1}/G_{2k-1} > 0 , \quad k = 1,2,3,\ldots ,$$

or if and only if, in terms of the coefficients of the corresponding power series,

$$\text{(3.3)} \quad H_{2n}^{(-2n+1)} \neq 0 , \quad H_{2n+1}^{(-2n-1)} \neq 0 , \quad H_{2n+1}^{(-2n)} > 0 , \quad H_{2n}^{(-2n)} > 0 , \quad n = 0,1,2,\ldots .$$

Proof. From (1.2) one deduces for $n \geq 1$,

$$\text{(3.4a)} \qquad d_{2n-1} = e_{2n-1}G_{2n-1} = \frac{\prod_{k=1}^{n-1}(F_{2k-1}F_{2k})}{\left(\prod_{k=1}^{n-1}F_{2k-1}\right)^2}\left(\frac{G_{2n-1}}{F_{2n-1}}\right) ,$$

$$\text{(3.4b)} \qquad e_{2n} = \left(\prod_{k=1}^{n}F_{2k-1}\right)^2 \Big/ \prod_{k=1}^{n}(F_{2k-1}F_{2k}) .$$

From these formulas one concludes, inductively, that

$$\text{(3.5)} \qquad d_{2n-1} > 0 , \quad e_{2n} > 0 , \quad n = 1,2,3,\ldots$$

implies (3.2) . Conversely, (3.2) implies (3.5) . From (3.1) one obtains

(3.6a) $$F_1F_2 = \frac{H_2^{(-2)}}{H_1^{(0)}} \; , \quad F_{2k-1}F_{2k} = \frac{H_{2k-3}^{(-2k+4)}H_{2k}^{(-2k)}}{H_{2k-2}^{(-2k+2)}H_{2k-1}^{(-2k+2)}} \; , \quad k = 2,3,\ldots \; ,$$

(3.6b) $$\frac{F_1}{G_1} = H_1^{(0)} \; , \quad \frac{F_{2k-1}}{G_{2k-1}} = \frac{H_{2k-3}^{(-2k+4)}H_{2k-1}^{(-2k+2)}}{(H_{2k-2}^{(-2k+3)})^2} \; , \quad k = 2,3,\ldots \; .$$

Hence (3.3) implies (3.2) . The equations (3.6) can also be used to show that (3.2) implies (3.3) . ■

To study the value behavior of semi-positive T-fractions it is convenient to consider them in the form (1.1c) . Let $C_n(z)$, $D_n(z)$ be defined by

(3.7a) $$C_0(z) = 0 \; , \quad C_{-1}(z) = 1 \; , \quad D_0(z) = 1 \; , \quad D_{-1}(z) = 0 \; ,$$

(3.7b) $$C_n(z) = (e_n+d_nz)C_{n-1}(z) + zC_{n-2}(z) \; , \quad n = 1,2,3,\ldots \; ,$$

(3.7c) $$D_n(z) = (e_n+d_nz)D_{n-1}(z) + zD_{n-2}(z) \; , \quad n = 1,2,3,\ldots \; ,$$

so that $C_n(z)$ and $D_n(z)$ are the nth numerator and denominator, respectively, of (1.1c) . Then $C_n(z)$ and $D_n(z)$ are polynomials in z of degree n and

(3.8) $$S_n(z,w) = \frac{C_n(z)+wC_{n-1}(z)}{D_n(z)+wD_{n-1}(z)} \quad \text{(see (DN5))} \; .$$

The nth approximant of (1.1c) , in particular, becomes

$$S_n(z,0) = \frac{C_n(z)}{D_n(z)} \; .$$

We now define

$$V_0(z) = [w : -\pi + \arg z < \arg w < \arg z] \quad \text{for} \quad 0 < \arg z < \pi \; ,$$

and

$$V_0(z) = [w : \arg z < \arg w < \pi + \arg z] \quad \text{for} \quad -\pi < \arg z < 0 \; .$$

For all z not on the real axis $V_0(z)$ is thus defined to be an open half-plane. $V_1(z)$ is defined to be the upper half-plane if z is in the upper half-plane and the lower half-plane if z is in the lower half-plane.

In terms of these regions we now have, for any semi-positive T-fraction and for z not real,

(3.9a) $$e_{2n} + d_{2n}z \in V_0(z) \; ,$$

(3.9b) $$\frac{z}{e_{2n}+d_{2n}z+V_0(z)} \subset V_1(z) \; ,$$

(3.9c) $$e_{2n-1} + d_{2n-1}z \in V_1(z) \; ,$$

(3.9d) $$\frac{z}{e_{2n-1}+d_{2n-1}z+V_1(z)} \subset V_0(z) \; .$$

The conclusions are still valid if, on the left in (3.9) , $V_0(z)$ and $V_1(z)$ are replaced by their closures. Since $z\tau$, τ real, are the boundary points of $V_0(z)$ and since the real axis is the boundary of $V_1(z)$, one concludes that

(3.10a) $$S_{2n}(z,\tau z) \in V_0(z) ,$$

(3.10b) $$S_{2n-1}(z,\tau) \in V_0(z) ,$$

for all non-real z and all real τ. In particular,

$$\frac{C_n(z)}{D_n(z)} \in V_0(z) \quad \text{for} \quad z \notin \mathbb{R} .$$

It follows that all zeros of $C_n(z)$ and $D_n(z)$ must be real. The same is true for the zeros of

$$C_n(z,\tau) = \begin{cases} C_{2k}(z) + \tau z C_{2k-1}(z) , & \text{if } n = 2k , \\ C_{2k-1}(z) + \tau C_{2k-2}(z) , & \text{if } n = 2k-1 , \end{cases}$$

$$D_n(z,\tau) = \begin{cases} D_{2k}(z) + \tau z D_{2k-1}(z) , & \text{if } n = 2k , \\ D_{2k-1}(z) + \tau D_{2k-2}(z) , & \text{if } n = 2k-1 , \end{cases}$$

for all real τ.

The quotients

$$\frac{C_n(z,\tau)}{D_n(z,\tau)}$$

we shall call the <u>generalized approximants of</u> the semi-positive T-fraction.

To prove that all zeros of $D_n(z,\tau)$, τ real, are simple, we consider the partial fraction decomposition of the generalized approximants

(3.11) $$\frac{C_n(z,\tau)}{D_n(z,\tau)} = z \sum_{\nu=1}^{r_n} \left[\frac{k_\nu^{(1)}}{z-t_\nu} + \frac{k_\nu^{(2)}}{(z-t_\nu)^2} + \cdots + \frac{k_\nu^{(m_\nu)}}{(z-t_\nu)^{m_\nu}}\right] .$$

Here the t_ν are the (distinct) zeros of $D_n(z,\tau)$ (for fixed τ) and hence they are real numbers. It follows that the $k_\nu^{(\mu)}$ are also real. We now sketch the argument that $m_\nu = 1$ and $k_\nu > 0$ for $t_\nu > 0$. A similar argument can be given for $t_\nu \leq 0$. (Zero can be a zero of $D_n(z,\tau)$ only if $n = 2k - 1$ and $\tau = -e_{2k-1}$.) Set

$$z = t_\nu + \varepsilon e^{i\pi/2m_\nu} , \quad \varepsilon > 0 .$$

Then for ε sufficiently small the term

$$\frac{k_\nu^{(m_\nu)} z}{(z-\tau_\nu)^{m_\nu}}$$

is the dominant term in $C_n(z,\tau)/D_n(z,\tau)$. Hence

$$\frac{C_n(t_\nu+\varepsilon e^{i\pi/2m_\nu},\tau)}{D_n(t_\nu+\varepsilon e^{i\pi/2m_\nu},\tau)} \sim \frac{t_\nu k_\nu^{(m_\nu)}}{\varepsilon^{m_\nu} e^{i\pi/2}} = \frac{-it_\nu k_\nu^{(m_\nu)}}{\varepsilon^{m_\nu}} .$$

Since this value must lie in the region $V_0(t_\nu+\varepsilon e^{i\pi/2m_\nu})$, we conclude that $k^{(m_\nu)}$ must be positive. Repeating the argument for $z = t_\nu + \varepsilon e^{i\pi/m_\nu}$, for $m_\nu > 1$, we are led to the conclusion that $k_\nu^{(m_\nu)} < 0$. Hence $m_\nu > 1$ is impossible and all zeros of $D_n(z,\tau)$ are simple. There are in general n zeros (except when $n = 2k$ and $\tau = -d_{2k}$). The partial fraction decomposition thus becomes

$$\frac{C_n(z,\tau)}{D_n(z,\tau)} = z \sum_{\nu=1}^{n(\tau)} \frac{k_\nu}{z-t_\nu}, \quad t_\nu \text{ real, } k_\nu > 0, \quad \nu = 1,2,\ldots,n .$$

We also note that

$$\sum_{\nu=1}^{n(\tau)} k_\nu = \lim_{z\to\infty} \frac{C_n(z,\tau)}{D_n(z,\tau)} = 1/d_1 > 0 .$$

We have proved the following result.

Theorem 3.3. *For the generalized approximants of a semi-positive T-fraction the following partial fraction expansion is valid*

$$\frac{C_n(z,\tau)}{D_n(z,\tau)} = z \sum_{\nu=1}^{n(\tau)} \frac{k_\nu}{z-t_\nu} .$$

Here $k_\nu > 0$, $\nu = 1,2,\ldots,n(\tau)$, $\sum_{\nu=1}^{n(\tau)} k_\nu = 1/d_1 > 0$, *and* t_ν *is real for* $\nu = 1,2,\ldots,n(\tau)$. $n(\tau) = n$ *unless* $n = 2k$ *and* $\tau = -d_{2k}$.

Since positive T-fractions are semi-positive, all results for semi-positive T-fractions apply to positive T-fractions. However, for positive T-fractions stronger results are valid. Thus for positive T-fractions it can be proved (as was done in [13]) that all t_ν are *negative.*

To study the convergence of (1.1c) it is desirable to transform the general T-fraction to the form $\delta_1 K(1/b_n)$ and use one of the convergence criteria developed for that case. For z in the upper half-plane we can write $\delta_1 = -iz$ and

$$b_{2n} = (e_{2n}+d_{2n}z)iz^{-1} = ie_{2n}z^{-1} + id_{2n} ,$$

$$b_{2n-1} = (e_{2n-1}+d_{2n-1}z)(-i) = -ie_{2n-1} - izd_{2n-1} .$$

Then $\mathrm{Re}(b_n) > 0$ for all $n \geq 1$, so that if one imposes additional conditions so that either

(3.12) $\qquad -\pi/2 + \varepsilon_1 < \arg b_{2n-1} < \pi/2 - \varepsilon_1 , \quad n = 1,2,3,\ldots ,$

or

(3.13) $\qquad -\pi/2 + \varepsilon_2 < \arg b_{2n} < \pi/2 - \varepsilon_2 , \quad n = 1,2,3,\ldots ,$

then by [17, p. 66] (see also [11, p. 90]) either the odd part or the even part of the semi-positive T-fraction (1.1c) converges. Sufficient for (3.12) to hold is $|e_{2n-1}| < K_1 d_{2n-1}$, and (3.13) will be satisfied if $|d_{2n}| < K_2 e_{2n}$. The case in which z is in the lower half-plane can be treated in a similar manner. We have proved the following theorem.

Theorem 3.4. *The odd (even) part of the semi-positive T-fraction (1.1c) converges if there exists a positive constant K_1 (K_2) such that, for all $n \geq 1$*

$$|e_{2n-1}| < K_1 d_{2n-1} \quad (|d_{2n}| < K_2 e_{2n}) .$$

From Theorem 4.32 of [10] one concludes that (1.1c) will converge if $\sum_1^\infty e_{2\nu} = \infty$ or if $\sum_1^\infty d_{2\nu-1} = \infty$, for in that case condition (C) of that theorem will hold provided $z \notin \mathbb{R}$. This establishes the next result.

Theorem 3.5. *Sufficient for the convergence of the semi-positive T-fraction (1.1c) for $z \notin \mathbb{R}$ is either*

$$\sum_{\nu=1}^{\infty} e_{2\nu} = \infty \quad \text{or} \quad \sum_{\nu=1}^{\infty} d_{2\nu-1} = \infty .$$

The continued fraction may converge even if neither of the two series diverges.

As was shown in [13] a much stronger result holds for positive T-fractions. Then the divergence of one of the series $\sum d_\nu$, $\sum e_\nu$ is necessary and sufficient for the convergence of the continued fraction.

4. *Integral representation of semi-positive T-fractions.* That some integrals of the form

$$\int_a^b \frac{f(t)}{z+t} dt \tag{4.1}$$

have continued fraction expansions was certainly known to Gauss. In the middle of the 19th century, Christoffel, Heine, Tchebycheff, Posse and Markoff made further contributions. The emphasis appears to have been on showing that integrals of the form (4.1) with more and more general $f(t)$ had J-fraction expansions. Tchebycheff also considered J-fraction expansions for series

$$\sum \frac{\theta_k}{z+t_k} \tag{4.2}$$

both finite and infinite. By introducing what are now called Stieltjes integrals, Stieltjes in 1894 unified both of these concepts. He also looked at the problem from the other direction, that is, to determine which continued fractions have integral representations. This question is of course closely related to finding solutions of the moment problem.

Stieltjes expressed the nth approximant of the continued fraction as an integral by first showing that it had a partial fraction decomposition

$$\frac{A_n(z)}{B_n(z)} = \sum_{\nu=1}^{m_n} \frac{k_\nu^{(n)}}{z+t_\nu^{(n)}} ,$$

in which $k_\nu^{(n)}$ are all positive, $\sum_1^{m_n} k_\nu^{(n)} = c$ is independent of n , and the $t_\nu^{(n)}$ are also all positive. Assuming the subscripts of the $t_\nu^{(n)}$ are chosen so that

$$t_1^{(n)} < t_2^{(n)} < \cdots < t_{m_n}^{(n)} ,$$

he defined

$$\phi_n(t) = \begin{cases} 0 & , \text{ for } 0 \le t \le t_1^{(n)} \\ \sum_{\nu=1}^{\mu} k_\nu^{(n)} & , \text{ for } t_\mu^{(n)} < t \le t_{\mu+1}^{(n)} , \quad \mu + 1 \le m_n \\ c & , \text{ for } t_{m_n}^{(n)} < t < \infty . \end{cases}$$

Then clearly

$$\frac{A_n(z)}{B_n(z)} = \int_0^\infty \frac{d\phi_n(t)}{z+t} .$$

The question then becomes whether there are subsequences $\{\phi_{n_k}\}$ such that $\phi_{n_k} \to \phi$ and such that

$$\lim_{k\to\infty} \int_0^\infty \frac{d\phi_{n_k}(t)}{z+t} = \int_0^\infty \frac{d\phi(t)}{z+t} .$$

Stieltjes was able to provide the necessary proofs in the case of S-fractions, in part because he was able to analyze their convergence behavior completely.

For J-fractions the situation becomes considerably more complicated. In this case the $t_\nu^{(n)}$ may not be positive, but they are still real. Also relatively little is known about the convergence behavior of J-fractions. Thus one must first of all show the existence of convergent subsequences. This was done by Grommer in 1914 by a selection process. For the main idea of this process, Grommer gives credit to Hilbert, whose student he was. The process allows him to assert the existence of a subsequence $\{\phi_{n_k}\}$ which converges to a monotone, non-decreasing bounded function $\phi(t)$ on $-\infty < t < +\infty$. In a different context, Helly in 1912 had given a simpler approach to such a selection process. Helly had a second result which insured that

$$\lim_{k\to\infty} \int_a^b f(t)d\phi_{n_k}(t) = \int_a^b f(t)d\phi(t)$$

for a large class of functions $f(t)$. However, this theorem is valid only for finite intervals (a,b). Grommer was able to prove that

$$\lim_{k\to\infty} \int_{-\infty}^\infty \frac{d\phi_{n_k}(t)}{z+t} = \int_{-\infty}^\infty \frac{d\phi(t)}{z+t} .$$

The proof was later simplified by Hamburger. In 1966 the present authors [10] obtained an integral representation for ordinary T-fractions

$$K\left(\frac{z}{1+d_n z}\right)$$

with $d_n > 0$, $n > 1$. This was done by using the, by now familiar, pattern of obtaining a partial fraction decomposition of the nth approximant and then following the Stieltjes-Grommer-Helly path to an integral representation. In 1980

Waadeland and we [13] derived integral representations for positive T-fractions by the same method.

The results of Section 3 now allow us to obtain integral representations for convergent subsequences of semi-positive T-fractions. To this end we state a general theorem which brings out the essential ingredients involved in deriving an integral representation. It should be pointed out that there are a number of factors which play no role in the proof of the theorem, even though they may be important in other respects. These are:

(a) The $R_n(z)$ being approximants of a continued fraction;

(b) Increasing degrees for the denominators;

(d) Correspondence to a formal power series at ∞ (and possibly at 0);

(e) Existence of moments.

Theorem 4.1. *Let $\{R_n(z)\}$ be a sequence of rational functions such that each $R_n(z)$ has a partial fraction decomposition of the form*

$$R_n(z) = \sum_{\nu=1}^{k_n} \frac{M_\nu^{(n)}}{z + t_\nu^{(n)}} ,$$

where $M_\nu^{(n)} > 0$ for $\nu = 1,2,\ldots,k_n$, $\sum_1^{k_n} M_2^{(n)} \leq B$. Moreover, the $t_\nu^{(n)}$ for $\nu = 1,2,\ldots,k_n$, are all real and distinct. Then there exists a subsequence $\{n_m\}$ and a monotone non-decreasing, bounded function $\psi(t)$ on $-\infty < t < \infty$ such that

$$\lim_{m\to\infty} R_{n_m}(z) = \int_{-\infty}^{\infty} \frac{d\psi(t)}{z+t} ,$$

uniformly for z in a bounded region with a positive distance from the real axis.

The proof is essentially contained in [17, p. 207-211] and so will not be repeated here. It uses Stieltjes' idea of writing $R_n(z)$ as an integral

$$\int_{-\infty}^{\infty} \frac{d\psi_n(t)}{z+t} ,$$

the Helly-Grommer selection process and the Grommer-Hamburger proof that

$$\lim_{m\to\infty} \int_{-\infty}^{\infty} \frac{d\psi_{n_m}(t)}{z+t} = \int_{-\infty}^{\infty} \frac{d\psi(t)}{z+t} .$$

The integrals

$$\int_{-\infty}^{\infty} \frac{d\psi_n(t)}{z+t}$$

are really not improper and thus there is no question about their existence. That the integral

$$\int_{-\infty}^{\infty} \frac{d\psi(t)}{z+t}$$

exists follows easily from the monotonicity and boundedness of $\psi(t)$ on $-\infty < t < +\infty$.

From Theorem 3.3 and 4.1 we can now conclude:

Theorem 4.2. *Let* $C_n(z,\tau)/D_n(z,\tau)$ *be the generalized* nth *approximant of a semi-positive* T-*fraction. Then for every sequence* $\{\tau_k\}$, $\tau_k \in \mathbb{R}$, *there exists a sequence* $\{n_k\}$ *and a monotone non-decreasing, bounded function* $\psi(t)$ *on* $-\infty < t < +\infty$ *such that*

$$\lim_{k\to\infty} \frac{C_{n_k}(z,\tau_k)}{D_{n_k}(z,\tau_k)} = z\int_{-\infty}^{\infty} \frac{d\psi(t)}{z+t},$$

uniformly with respect to z *in any bounded region* $\mathfrak{D}$ *with a positive distance from the real line. If* $\{C_{n_j}(z,\tau_j)/D_{n_j}(z,\tau_j)\}$ *converges for all* $z \in \mathfrak{D}$, *then it converges to an integral*

$$z\int_{-\infty}^{\infty} \frac{d\psi(t)}{z+t}.$$

5. *Solution of Strong Moment Problems.* By the *strong Stieltjes moment problem* (SSMP) we mean the following: For a given double sequence of real numbers $C = \{c_n\}_{-\infty}^{\infty}$, does there exist a distribution function $\psi(t) \in \Phi(0,\infty)$ such that

$$c_n = \int_0^{\infty} (-t)^n d\psi(t), \quad n = 0,\pm1,\pm2,\ldots \quad ? \tag{5.1}$$

Such a function ψ is called a *solution* to the SSMP for C. Using continued fraction methods similar to those described in Sections 3 and 4, Waadeland and the authors [13] proved the following:

Theorem 5.1. *Let* $C = \{c_n\}_{n=-\infty}^{\infty}$ *be a given double sequence of real numbers and let* $H_k^{(m)}$ *denote the Hankel determinants associated with* C (*see* (1.6)). *Let* L_0 *and* L_∞ *be formal power series defined by* (1.3). *Then the following three statements are equivalent*:

(A) *The strong Stieltjes moment problem for* C *has a solution.*

(B) *The following determinant conditions hold*:

$$H_{2n}^{(-2n)} > 0, \quad H_{2n+1}^{(-2n)} > 0, \quad n = 0,1,2,\ldots, \tag{5.2a}$$

$$H_{2n}^{(-(2n-1))} > 0, \quad H_{2n+1}^{(-(2n+1))} < 0, \quad n = 0,1,2,\ldots. \tag{5.2b}$$

(C) *There exists a positive* T-*fraction* (1.1a) *corresponding to* L_0 *at* 0 *and to* L_∞ *at* ∞.

By a further utilization of integral representations of positive T-fractions, Waadeland and the authors [13] obtained the following result on the uniqueness of the solution to the SSMP.

Theorem 5.2. *Let* $C = \{c_n\}_{-\infty}^{\infty}$ *be a double sequence of real numbers for which there exists at least one solution to the strong Stieltjes moment problem. Then the solution is unique if and only if the positive* T-*fraction* (1.1a) *corresponding to* L_0 *at* 0 *and to* L_∞ *at* ∞ (*see* (1.3)) *is convergent; that is, if and only if*

$$\sum_{n=1}^{\infty} e_n = \infty \quad \text{or} \quad \sum_{n=1}^{\infty} d_n = \infty, \tag{5.3}$$

where the positive coefficients, e_n *and* d_n *are defined by* (1.2).

The <u>strong Hamburger moment problem</u> (SHMP) is defined as follows: For a given double sequence of real numbers $C = \{c_n\}_{-\infty}^{\infty}$, does there exist a distribution function $\psi(t) \in \Phi(-\infty,\infty)$ such that

(5.4) $$c_n = \int_{-\infty}^{\infty} (-t)^n d\psi(t) \quad , \quad n = 0,\pm 1,\pm 2,\ldots \quad ?$$

Such a function ψ is called a solution to the SHMP for C. The use of continued fractions in the study of the general SHMP appears not to be possible. This is due in part to the fact that the related orthogonal L-polynomials (see Section 6) do not always satisfy three-term recurrence relations. Nevertheless, using orthogonal L-polynomials, Gaussian quadrature and two theorems of Helly (referred to in Section 4), Njåstad and the authors [9] have proved the following:

<u>Theorem</u> 5.3. <u>Let $C = \{c_n\}_{-\infty}^{\infty}$ be a given double sequence of real numbers and let $H_k^{(m)}$ denote the associated Hankel determinants (see (1.6)). Then the strong Hamburger moment problem for C has a solution if and only if conditions (5.2a) are satisfied.</u>

Necessary and sufficient conditions for the uniqueness of the solution to the SHMP are not yet known. However, two other methods for proving Theorem 5.3 have recently been called to our attention. One due to Christian Berg [1] employs a Hahn-Banach argument. Another proof, which we believe is due to W.B. Gragg, makes use of the classical moment theorem of Hamburger. A sketch of the latter proof is included here. We consider only the proof of sufficiency of condition (5.2a). By the Hamburger moment theorem, given a sequence $\{c_n\}_{n=-2k}^{\infty}$, there exists a distribution function $\nu_k(t) \in \Phi^c(-\infty,\infty)$ such that

(5.5) $$c_{-2k+m} = \int_{-\infty}^{\infty} (-t)^m d\nu_k(t) \quad , \quad m = 0,1,2,\ldots$$

if and only if

(5.6) $$H_j^{(-2k)} > 0 \quad , \quad j = 1,2,3,\ldots \quad .$$

Using Jacobi's identity

$$H_n^{(k-1)} H_n^{(k+1)} - H_{n-1}^{(k+1)} H_{n+1}^{(k-1)} = [H_n^{(k)}]^2$$

and an induction argument or properties of quadratic forms, one can show that (5.6) is implied by (5.2a). Let $\mu_k(t)$ be defined by

(5.7) $$\mu_k(t) = \int_{-\infty}^{t} s^{2k} d\nu_k(s) \quad , \quad k = 0,1,2,\ldots \quad .$$

Then from [23, p. 12] and (5.5) it follows that

(5.8) $$\int_{-\infty}^{\infty} (-t)^{-2k+m} d\mu_k(t) = \int_{-\infty}^{\infty} (-t)^m d\nu_k(t) = c_{-2k+m} \quad ,$$
$$\text{for } k = 0,1,2,\ldots \text{ and } m = 0,1,2,\ldots \quad .$$

In particular, (5.8) implies

(5.9) $$\int_{-\infty}^{\infty} d\mu_k(t) = c_0 \quad , \quad k = 0,1,2,\ldots \quad .$$

Thus the total variation of each $\mu_k(t)$ is $c_0 > 0$. Thus by Helly's selection principle there exists a subsequence $\{\mu_{k_j}\}$ and a bounded non-decreasing function $\mu(t)$ such that

$$\lim_{k_j\to\infty} \mu_{k_j}(t) = \mu(t) \quad \text{for} \quad -\infty < t < \infty \ .$$

Then by a modification of Helly's second theorem referred to in Section 4,

$$\lim_{k_j\to\infty} \int_{-\infty}^{\infty} (-t)^n d\mu_{k_j}(t) = \int_{-\infty}^{\infty} (-t)^n d\mu(t) \quad , \quad n = 0,\pm 1,\pm 2,\ldots \ .$$

This together with (5.8) implies that

(5.10) $$c_n = \int_{-\infty}^{\infty} (-t)^n d\mu(t) \quad , \quad n = 0,\pm 1,\pm 2,\ldots \ .$$

Using (5.10) and properties of quadrature forms it can now be shown that $\mu(t)$ has infinitely many points of increase. This completes the proof.

6. Orthogonal L-polynomials. A function $R(z)$ of the form

(6.1) $$R(z) = \sum_{j=k}^{m} r_j z^j \quad , \quad r_j \in \mathbb{R} \quad , -\infty < k \le m < +\infty$$

is called a (k,m) Laurent polynomial (or L-polynomial) in the complex variable z. The set $\mathcal{R}$ of all L-polynomials forms a linear space over $\mathbb{R}$ with respect to the usual definitions of vector addition and scalar multiplication. A basis for $\mathcal{R}$ is given by the sequence $1, 1/z, z, 1/z^2, z^2, \ldots$. We let $\mathcal{R}_{2m}$ denote the subspace spanned by

$$z^{-m}, z^{-m+1}, \ldots, 1, z, \ldots, z^m$$

and $\mathcal{R}_{2m-1}$ the subspace spanned by

$$z^{-m}, z^{-m+1}, \ldots, 1, z, \ldots, z^{m-1} \ .$$

Theorem 6.1. If $\phi(t) \in \Phi(-\infty,\infty)$, then

(6.2) $$(R,S) = \int_{-\infty}^{\infty} R(t)S(t)d\phi(t) \quad , \quad R,S \in \mathcal{R}$$

defines an inner product on $\mathcal{R}$.

Proof. Linearity, symmetry and homogeneity follow directly from properties of the Riemann-Stieltjes integral. To prove positivity we note that if $Q \in \mathcal{R}$, then

$$(Q,Q) = \int_{-\infty}^{\infty} [Q(t)]^2 d\phi(t) \ge 0 \ .$$

Moreover, since ϕ has infinitely many points of increase, $(Q,Q) = 0$ if and only if $Q(t) \equiv 0$. ■

It is clear from Theorem 6.1 that, if $\phi(t) \in \Phi(a,b)$, where $-\infty < a < b < +\infty$, then

(6.3) $$(R,S) = \int_a^b R(t)S(t)d\phi(t) \quad , \quad R,S \in \mathcal{R}$$

defines an inner product on $\mathcal{R}$. In this case ϕ can be extended to $(-\infty,\infty)$ by setting $\phi(t) = 0$ if $t \notin (a,b)$. Let $\|R\| = (R,R)^{1/2}$ denote the norm of R.

Theorem 6.2. Let $\psi(t) \in \Phi(a,b)$ with $-\infty < a < b < +\infty$, let $C = \{c_n\}_{-\infty}^{\infty}$ where

(6.4) $$c_n = \int_a^b (-t)^n d\psi(t) \quad , \quad n = 0,\pm 1,\pm 2,\ldots$$

and let $H_k^{(m)}$ denote the Hankel determinants associated with C. Let $\{R_n(z)\}$ denote the sequence of L-polynomials denoted by

(6.5a) $$R_{2n}(z) = \frac{(-1)^n}{H_{2n}^{(-2n)}} \begin{vmatrix} c_{-2n} & \cdots & c_{-1} & (-z)^{-n} \\ \vdots & & \vdots & \vdots \\ c_{-1} & \cdots & c_{2n-2} & (-z)^{n-1} \\ c_0 & \cdots & c_{2n-1} & (-z)^n \end{vmatrix} \quad , \quad n = 0,1,2,\ldots$$

(6.5b) $$R_{2n+1}(z) = \frac{(-1)^n}{H_{2n+1}^{(-2n)}} \begin{vmatrix} c_{-2n-1} & & c_{-1} & (-z)^{-n-1} \\ \vdots & & \vdots & \vdots \\ c_{-1} & \cdots & c_{2n-1} & (-z)^{n-1} \\ c_0 & \cdots & c_{2n} & (-z)^n \end{vmatrix} \quad , \quad n = 0,1,2,\ldots .$$

Then: (A) The L-polynomials $\{R_n(z)\}_0^{\infty}$ are orthogonal with respect to $\psi(t)$ and

(6.6) $$\|R_{2n}\|^2 = \frac{H_{2n+1}^{(-2n)}}{H_{2n}^{(-2n)}} \quad , \quad \|R_{2n+1}\|^2 = \frac{H_{2n+2}^{(-(2n+2))}}{H_{2n+1}^{(-2n)}} \quad , \quad n = 0,1,2,\ldots .$$

(B) If

(6.7) $$R_{2n}(z) = \sum_{j=-n}^{n} r_{2n,j} z^j \quad , \quad R_{2n+1}(z) = \sum_{j=-n-1}^{n} r_{2n+1,j} z^j \quad , \quad j = 0,1,2,\ldots \quad ,$$

then

(6.8a) $$r_{2n,n} = r_{2n+1,-n-1} = 1 \quad , \quad n = 0,1,2,\ldots \quad ,$$

and

(6.8b) $$r_{2n,-n} = \frac{H_{2n}^{(-(2n-1))}}{H_{2n}^{(-2n)}} \quad , \quad r_{2n+1,n} = \frac{H_{2n+1}^{(-(2n+1))}}{H_{2n+1}^{(-2n)}} \quad , \quad n = 0,1,2,\ldots .$$

Proof: Using (6.4) and (6.5) one can show that, for each $n = 0,1,2,\ldots$,

(6.9a) $$(z^k, R_{2n}) = 0 \quad , \quad -n \le k \le n-1 \quad ,$$

(6.9b) $$\|R_{2n}\|^2 = (R_{2n}, R_{2n}) = (z^n, R_{2n}) = H_{2n+1}^{(-2n)}/H_{2n}^{(-2n)} \quad ,$$

(6.9c) $$(z^k, R_{2n+1}) = 0 \quad , \quad -n \le k \le n \quad ,$$

(6.9d) $$\|R_{2n+1}\|^2 = (R_{2n+1}, R_{2n+1}) = (z^{-n-1}, R_{2n+1}) = H_{2n+2}^{(-(2n+2))}/H_{2n+1}^{(-2n)} .$$

By Theorem 5.1(B), the Hankel determinants in (6.6) are all positive. This proves (A). The proof of B is immediate from the Hankel determinants involved. ■

It can be seen from (6.8) that

(6.10) $$r_{2n,-n} \neq 0 \quad \text{and} \quad r_{2n+1,n} \neq 0 \quad , \quad n = 0,1,2,\ldots$$

if and only if

(6.11) $H_{2n}^{(-(2n-1))} \neq 0$ and $H_{2n+1}^{(-(2n+1))} \neq 0$, $n = 0,1,2,\ldots$.

We note that (6.11) is implied by (3.3) . In the following we shall assume that (6.11) holds. In this case it is convenient to normalize the orthogonal L-polynomials of Theorem 6.2 , obtaining orthogonal L-polynomials $Q_n(z)$ as follows:

(6.12) $Q_{2n}(z) = (1/r_{2n,-n})R_{2n}(z)$, $Q_{2n+1}(z) = R_{2n+1}(z)$, $n = 0,1,2,\ldots$.

We shall write the $Q_n(z)$ in the form

(6.13) $Q_{2n}(z) = \sum_{j=-n}^{n} q_{2n,j}z^j$, $Q_{2n+1}(z) = \sum_{j=-n-1}^{n} q_{2n+1,j}z^j$, $n = 0,1,2,\ldots$.

We obtain then

<u>Theorem</u> 6.3. <u>Let</u> $\psi(t) \in \Phi(a,b)$ <u>with</u> $-\infty \leq a < b \leq +\infty$ <u>and let</u> $C = \{c_n\}_{-\infty}^{\infty}$ <u>be defined by</u> (6.4) . <u>Suppose that</u> (6.11) <u>holds and let</u> $\{Q_n(z)\}$ <u>be defined by</u> (6.12) <u>where</u> $\{R_n(z)\}$ <u>is defined by</u> (6.5) . <u>Then:</u>

(A) $\{Q_n(z)\}$ <u>is a sequence of orthogonal</u> L-<u>polynomials with respect to</u> ψ , <u>normalized such that</u>

(6.14a) $q_{2n,-n} = q_{2n+1,-n-1} = 1$, $n = 0,1,2,\ldots$,

(6.14b) $q_{2n,n} = H_{2n}^{(-2n)}/H_{2n}^{(-(2n-1))} \neq 0$, $q_{2n+1,n} = H_{2n+1}^{(-(2n+1))}/H_{2n+1}^{(-2n)} \neq 0$, $n = 0,1,2,\ldots$

(B) <u>The</u> $Q_n(z)$ <u>satisfy the system of three-term recurrence relations</u>

(6.15a) $Q_0(z) = 1$, $Q_1(z) = z^{-1} + (c_{-1}/c_0)$,

(6.15b) $Q_{2n}(z) = (1-G_{2n}z)Q_{2n-1}(z) - F_{2n}Q_{2n-2}(z)$, $n = 1,2,3,\ldots$,

(6.15c) $Q_{2n+1}(z) = (z^{-1}-G_{2n+1})Q_{2n}(z) - F_{2n+1}Q_{2n-1}(z)$, $n = 1,2,3,\ldots$,

<u>where the</u> F_n <u>and</u> G_n <u>are defined by</u> (3.1) .

(C) <u>If</u> $V_n(z)$ <u>denotes the</u> nth <u>denominator of the general</u> T-<u>fraction</u> (1.1b) , <u>then</u>

(6.16) $Q_{2n}(z) = (-1)^n V_{2n}(-z)$, $Q_{2n+1}(z) = (-1)^{n+1} V_{2n+1}(-z)$, $n = 0,1,2,\ldots$.

(D) <u>For each</u> $n \geq 1$, $Q_{2n}(z)$ <u>has exactly</u> 2n <u>zeros; they are real, distinct and lie in the interval</u> (a,b) .

(E) <u>For each</u> $n \geq 0$, $Q_{2n+1}(z)$ <u>has exactly</u> 2n+1 <u>zeros; they are real and distinct and all but at most one of them lie in the open interval</u> (a,b) . <u>If</u> 0 <u>is not in</u> (a,b) , <u>then all zeros of</u> $Q_{2n+1}(z)$ <u>lie in</u> (a,b) .

<u>Proof</u>. Part (A) is an immediate consequence of Theorem 6.2 . (B) can be verified by the orthogonality of the $Q_n(z)$ with a standard argument for orthogonal polynomials. (C) follows by comparing (6.15) with the difference equations satisfied by the denominators $V_n(z)$ of (1.1b) . It is clear from (6.13) and (6.14) that, for each $n \geq 1$, $Q_n(z)$ has exactly n zeros. We shall prove (E) ; the proof of (D) is analogous and hence omitted. We showed

in Section 3 that the roots are real and simple but not necessarily in (a,b). Let n be a given positive integer and let λ denote the number of distinct real zeros of odd order of $Q_{2n+1}(z)$ that lie in the open interval (a,b). Denote these by $t_1,t_2,\ldots,t_\lambda$. First we consider

$$I_{2n-1} = \int_a^b Q_{2n-1}(t) \prod_{j=1}^{\lambda} (1-t/t_j)/t^{n-2} d\psi(t)$$

$$= \int_a^b S_{2n-1}(t) \prod_{j=1}^{\lambda} (1-t/t_j)^2/t^{2n-2} d\psi(t) \quad .$$

Here $S_{2n-1}(t)$ is a polynomial in t that does not change sign on (a,b); hence the integral does not vanish since $\psi(t)$ has infinitely many points of increase on (a,b). On the other hand I_{2n-1} can be written in the form

$$I_{2n-1} = \int_a^b Q_{2n-1}(t)(t^{-n+2}+a_{-n+3}t^{-n+3}+\cdots+a_{\lambda-n+2}t^{\lambda-n+2})d\psi(t) \quad .$$

Thus by orthogonality I_{2n-1} would be zero if $\lambda - n + 2 \leq n - 1$; i.e., $\lambda \leq 2n - 3$. It follows that $\lambda \geq 2n - 2$ and hence that $Q_{2n-1}(t)$ has at least $2n - 2$ distinct zeros in (a,b). Since the complex zeros of $Q_{2n-1}(t)$ occur in conjugate pairs, the one remaining zero must be real. This proves the first part of (E). To prove the second part of (E) we suppose that $0 \notin (a,b)$ and let λ be defined as above. Then consider

$$I'_{2n-1} = \int_a^b Q_{2n-1}(t) \prod_{j=1}^{\lambda} (1-t/t_j)/t^{n-1} d\psi(t)$$

$$= \int_a^b S_{2n-1}(t) \prod_{j=1}^{\lambda} (1-t/t_j)^2/t^{2n-1} d\psi(t) \quad .$$

Again $S_{2n-1}(t)$ is a polynomial in t that does not change sign on (a,b) and, since $0 \notin (a,b)$, t^{2n-1} also does not change sign on (a,b). Therefore $I'_{2n-1} \neq 0$, since $\psi(t)$ has infinitely many points of increase on (a,b). On the other hand, I'_{2n-1} can be written as

$$I'_{2n-1} = \int_a^b Q_{2n-1}(t)(t^{-n+1}+b_{-n+2}t^{-n+2}+\cdots+b_{\lambda-n+1}t^{\lambda-n+1})d\psi(t) \quad .$$

Hence by orthogonality I'_{2n-1} would be zero if $\lambda - n + 1 \leq n - 1$; i.e., $\lambda \leq 2n - 2$. It follows that $\lambda = 2n - 1$, which proves the second part of (E). ■

One can say even more about the orthogonal L-polynomials $Q_n(z)$ of Theorem 6.3 when $\psi(t) \in \Phi(0,\infty)$. This case is considered in the following:

<u>Theorem</u> 6.4. <u>Let</u> $\psi(t) \in \Phi(a,b)$ <u>where</u> $0 < a < b < +\infty$ <u>and let the orthogonal</u> L-<u>polynomials</u> $Q_n(z)$ <u>with respect to</u> $\psi(t)$ <u>be defined as in Theorem</u> 6.3 <u>and</u> (6.13). <u>Then</u>: (A) <u>In addition to the assertions of Theorem</u> 6.3 <u>we also have</u>

(6.17) $$q_{2n,n} > 0 \,, \quad q_{2n+1,n} < 0 \,, \quad n = 0,1,2,\ldots$$

<u>and</u>

(6.18) $$F_n > 0 \ , \quad G_n > 0 \ , \quad n = 1,2,3,\ldots \ .$$

(B) For each $n \geq 1$, $Q_n(z)$ has exactly n zeros $t_j^{(n)}$, $n = 1,2,\ldots,n$. They are distinct, positive, real numbers ordered such that

$$0 \leq a < t_1^{(n)} < t_2^{(n)} < \cdots < t_n^{(n)} < b \ .$$

(C) Let $A_n(z)$ and $B_n(z)$ denote the nth numerator and denominator, respectively, of the positive T-fraction (1.1a). Then $A_n(z)$ and $B_n(z)$ are polynomials in z of degree n. The zeros $\tau_j^{(n)}$ of $B_n(z)$ are given by $\tau_j^{(n)} = -t_j^{(n)}$, $j = 1,2,\ldots,n$. The nth approximant of (1.1a) has a partial fraction decomosition of the form

(6.20) $$\frac{A_n(z)}{B_n(z)} = \sum_{j=1}^{n} \frac{z\pi_j^{(n)}}{z+t_j^{(n)}} \ , \quad n = 1,2,3,\ldots \ ,$$

where

(6.21) $$\sum_{j=1}^{n} \pi_j^{(n)} = F_1/G_1 \quad \text{and} \quad \pi_j^{(n)} > 0 \quad \text{for} \quad 1 \leq j \leq n \ .$$

Proof. (A) and (B) are immediate consequences of Theorems 5.1 and 6.3. (C) is proved in [13, Theorem 3.2]. ■

As shown in Section 3 there is also a partial fraction decomposition in Theorem 6.3. Our next theorem is the converse of Theorem 6.3(B) when $0 \leq a < b \leq +\infty$. It is the analogue of a theorem for classical orthogonal polynomials attributed to Favard [3].

Theorem 6.5. Let $\{Q_n(z)\}_0^\infty$ be any sequence of L-polynomials satisfying a system of three-term recurrence relations of the form (6.15) where $F_n > 0$, $G_n > 0$ for all $n \geq 1$. Then there exists a distribution function $\psi(t) \in \Phi(0,\infty)$ such that $\{Q_n(z)\}_0^\infty$ is the sequence of orthogonal L-polynomials with respect to $\psi(t)$ normalized as in Theorem 6.3.

Proof. The numbers F_n, G_n determine a positive T-fraction (1.1a). If $A_n(z)$ and $B_n(z)$ denote the nth numerator and denominator, respectively of (1.1a), then by the determinant formula [11, p. 20] one obtains

$$\frac{A_{n+1}(z)}{B_{n+1}(z)} - \frac{A_n(z)}{B_n(z)} = \frac{(-1)^n z^{n+1}}{B_n(z)B_{n+1}(z)} \ .$$

From this it is clear that the positive T-fraction (1.1a) corresponds to formal power series of the form (1.3) at 0 and ∞, respectively. Moreover, by Theorem 3.1 the F_n, G_n and c_n are related to each other by (3.1). By Theorem 5.1 the strong Stieltjes moment problem for $C = \{c_n\}_{-\infty}^{\infty}$ has a solution $\psi(t) \in \Phi(0,\infty)$; hence

(6.22) $$c_n = \int_0^\infty (-t)^n d\psi(t) \ , \quad n = 0,\pm1,\pm2,\ldots \ .$$

Therefore (6.3) defines an inner product for $\mathcal{R}$. It can be seen that the $Q_n(z)$ satisfy (6.16), by comparing (6.15) with the difference equations satisfied by the $V_n(z)$. It can also be seen that the $Q_n(z)$ can be written in the form (6.13) and satisfy the normalization (6.14a). It follows then from Theorem 6.3(C) that the $Q_n(z)$ are orthogonal with respect to $\psi(t)$. ■

We conclude this section with the following result which shows that the denominators of a positive T-fraction give rise to a sequence of orthogonal L-polynomials.

Theorem 6.6. *Let $V_n(z)$ denote the nth denominator of a positive T-fraction (1.1b) with $F_n > 0$, $G_n > 0$ for all $n \geq 1$. Let $\{Q_n(z)\}_0^\infty$ be defined by (6.16) and $Q_0(z) \equiv 1$. Then there exists a distribution function $\psi(t) \in \Phi(0,\infty)$ such that $\{Q_n(z)\}_0^\infty$ is a sequence of orthogonal L-polynomials with respect to $\psi(t)$ normalized by (6.13) and (6.14a).*

Proof. The difference equations satisfied by the $V_n(z)$ insure that the $Q_n(z)$ satisfy (6.15). Hence the assertion is an immediate consequence of Theorem 6.5. ■

7. Gaussian Quadrature. We present here a new case in the development of Gaussian quadaratures which also is suggested by continued fractions.

Theorem 7.1. *Let $\psi(t) \in \Phi(a,b)$ with $0 \leq a < b \leq +\infty$. Let n be a positive integer and let $Q_n(z)$ denote the nth orthogonal L-polynomial with respect to $\psi(t)$ normalized as in Theorem 6.3. Let $t_j^{(n)}$, $j = 1,2,\ldots,n$, denote the zeros of $Q_n(z)$. Then:*

(A) *For every $F(z) \in \mathcal{R}_{2n-1}$,*

$$\int_a^b F(t)d\psi(t) = \sum_{j=1}^{n} w_j^{(n)} F(t_j^{(n)}) \tag{7.1a}$$

where the Christoffel numbers $w_j^{(n)}$ are defined by

$$w_j^{(n)} = \frac{1}{Q_n'(t_j^{(n)})} \int_a^b \frac{Q_n(t)}{t-t_j^{(n)}} d\psi(t) \quad , \quad j = 1,2,\ldots,n \ . \tag{7.1b}$$

(B) *Let $\{P_n(z)\}_0^\infty$ denote the sequence of L-polynomials defined by the recurrence relations*

$$P_0(z) = 1 \quad , \quad P_1(z) = -F_1 \ , \tag{7.2a}$$

$$P_{2n}(z) = (1-G_{2n}z)P_{2n-1}(z) - F_{2n}P_{2n-2}(z) \quad , \quad n = 1,2,3,\ldots \ , \tag{7.2b}$$

$$P_{2n+1}(z) = (z^{-1}-G_{2n+1})P_{2n}(z) - F_{2n+1}P_{2n-1}(z) \quad , \quad n = 1,2,3,\ldots \ . \tag{7.2c}$$

Here the F_n and G_n are determined by $\psi(t)$ as in Theorem 6.3(B). Then

$$w_j^{(n)} = \frac{P_n(t_j^{(n)})}{t_j^{(n)} Q_n'(t_j^{(n)})} \quad , \quad j = 1,2,\ldots,n \ , \tag{7.3}$$

$$w_j^{(n)} > 0 \ \text{ for } \ j = 1,2,\ldots,n \quad , \quad \sum_{j=1}^{n} w_j^{(n)} = F_1/G_1 \ , \tag{7.4}$$

and

(7.5) $$\frac{P_n(z)}{Q_n(z)} = \sum_{j=1}^{n} \frac{zw_j^{(n)}}{z-t_j^{(n)}} .$$

Proof. Let $F(z) \in \mathcal{R}_{2n-1}$ be given and let

(7.6) $$L_{n,j}(z) = \frac{Q_n(z)}{(z-t_j^{(n)})Q_n'(t_j^{(n)})} , \quad j = 1,2,\ldots,n .$$

Then it can be seen that

(7.7) $$L_{2m-1,j}(z) \in \mathcal{R}_{2m-1} , \quad L_{2m,j}(z) \in \mathcal{R}_{2m-1}$$

and

(7.8) $$L_{n,k}(t_j^{(n)}) = \delta_{kj} \quad \text{(Kronecker delta)}.$$

Let

(7.9) $$R(z) = \sum_{k=1}^{n} L_{n,k}(z)F(t_k^{(n)}) ,$$

so that $R(z) \in \mathcal{R}_n$ and

(7.10) $$R(t_j^{(n)}) = \sum_{k=1}^{n} L_{n,k}(t_j^{(n)})F(t_k^{(n)}) = F(t_j^{(n)}) , \quad j = 1,2,\ldots,n .$$

Therefore $F(z) - R(z)$ is in $\mathcal{R}_{2n-1}$ and has zeros at $t_1^{(n)}, t_2^{(n)}, \ldots, t_n^{(n)}$. It follows that there exists $S(z) \in \mathcal{R}_{n-1}$ such that

(7.11) $$F(z) - R(z) = Q_n(z)S(z) .$$

To see this note that

$$F(z) - R(z) = z^{-n}P(z) \prod_{j=1}^{n} (z-t_j^{(n)})$$

where $P(z)$ is a polynomial in z of degree at most $n-1$. If n is even (i.e. $n = 2m$), then for some constant c

$$F(z) - R(z) = cQ_n(z)P(z)z^{-n} ;$$

hence $S(z) = cP(z)z^{-n} \in \mathcal{R}_{2m-1} = \mathcal{R}_{n-1}$. A similar argument holds if n is odd. Now using (7.11) we have

(7.12) $$\int_a^b F(t)d\psi(t) = \int_a^b R(t)d\psi(t) + \int_a^b S(t)Q_n(t)d\psi(t) .$$

The second integral on the right in (7.12) vanishes by orthogonality since $S(t) \in \mathcal{R}_{n-1}$. Thus by (7.6) and (7.9) we obtain

$$\int_a^b F(t)d\psi(t) = \sum_{j=1}^{n} F(t_j^{(n)}) \int_a^b L_{n,j}(t)d\psi(t)$$

$$= \sum_{j=1}^{n} w_j^{(n)}F(t_j^{(n)})$$

which proves (A) . To prove (B) we first show that

(7.13) $$P_n(z) = z\int_a^b \frac{Q_n(z)-Q_n(t)}{z-t} d\psi(t) , \quad n = 0,1,2,\ldots .$$

In fact, if we let $\hat{P}_n(z)$ denote the right side of (7.13), then it can be shown by induction (using orthogonality and (6.15)) that the $\hat{P}_n(z)$ satisfy the recurrence relations (7.2). Thus (7.13) follows since the $P_n(z)$ are completely determined by (7.2). From (7.13) we obtain

$$P_n(t_j^{(n)}) = t_j^{(n)} \int_a^b \frac{Q_n(t)}{t-t_j^{(n)}} d\psi(t) \quad , \quad j = 1,2,\ldots,n \quad .$$

This together with (7.1b) gives (7.3). Now let $A_n(z)$ and $B_n(z)$ denote the nth numerator and denominator, respectively, of the corresponding positive T-fraction (1.1a), and let $U_n(z)$, $V_n(z)$ denote the nth numerator and denominator, respectively, of the equivalent positive T-fraction (1.1b). Then it is readily shown that for $m = 0,1,2,\ldots$

(7.14a) $$P_{2m}(z) = (-1)^m U_{2m}(-z) = z^{-m} A_{2m}(-z) \quad ,$$

(7.14b) $$P_{2m+1}(z) = (-1)^{m+1} U_{2m+1}(-z) = z^{-m-1} A_{2m+1}(-z) \quad ,$$

(7.14c) $$Q_{2m}(z) = (-1)^m V_{2m}(-z) = z^{-m} B_{2m}(-z) \quad ,$$

(7.14d) $$Q_{2m+1}(z) = (-1)^{m+1} V_{2m+1}(-z) = z^{-m-1} B_{2m+1}(-z) \quad .$$

Thus by Theorem 6.4(C) we have, for $n = 1,2,3,\ldots$,

(7.15) $$\frac{P_n(z)}{Q_n(z)} = \frac{A_n(-z)}{B_n(-z)} = \sum_{j=1}^{n} \frac{z\pi_j^{(n)}}{z+\tau_j^{(n)}}$$

so that $P_n(z)/Q_n(z)$ has the same form as (7.5). It follows from this that $t_j^{(n)} = -\tau_j^{(n)}$, $j = 1,2,\ldots,n$. It remains to show that

(7.16) $$\pi_j^{(n)} = w_j^{(n)} \quad , \quad j = 1,2,\ldots,n \quad .$$

It follows from (7.15) that $\pi_j^{(n)}$ is the residue of $P_n(z)/zQ_n(z)$ at $z = -\tau_j^{(n)} = t_j^{(n)}$. Therefore

$$\pi_j^{(n)} = \lim_{z \to t_j^{(n)}} \frac{(z-t_j^{(n)})P_n(z)}{zQ_n(z)} = \frac{P_n(t_j^{(n)})}{t_j^{(n)} Q_n'(t_j^{(n)})} = w_j^{(n)} \quad ,$$

so that (7.4) then follows from (6.21). ■

An application of Theorem 7.1 and [20, Theorem 15.2.2] gives the folowing convergence result which is the analogue of a theorem proved by Stieltjes [19].

<u>Theorem</u> 7.2. <u>Let</u> $\psi(t) \in \Phi(a,b)$ <u>with</u> $0 < a < b < +\infty$ <u>and let</u> $t_j^{(n)}$ <u>and</u> $w_j^{(n)}$ <u>be defined as in Theorem</u> 7.1. <u>Then</u>

(7.17) $$\int_a^b F(t)d\psi(t) = \lim_{n\to\infty} \sum_{j=1}^{n} w_j^{(n)} F(t_j^{(n)})$$

<u>holds for every function</u> F(t) <u>for which the Riemann-Stieltjes integral on the left side of</u> (7.17) <u>exists</u>.

The next result of this section is obtained an an application of Theorem 7.2. It is the analogue of a theorem of Markoff [14].

Theorem 7.3. *Let* $\psi(t) \in \Phi(a,b)$ *with* $0 < a < b < +\infty$. *Let* $\{F_n\}$ *and* $\{G_n\}$ *denote the sequences of positive coefficients defined as in Theorem* 6.3 *and* 6.4. *Then:* (A) *For every complex number* $z \notin [-b,-a]$ *we have*

$$(7.18) \qquad z\int_a^b \frac{d\psi(t)}{z+t} = \frac{F_1 z}{1+G_1 z} + \frac{F_2 z}{1+G_2 z} + \frac{F_3 z}{1+G_3 z} + \cdots .$$

(B) *The positive* T-*fraction on the right side of* (7.18) *converges uniformly on every compact subset of* $D = [z : z \notin [-b,-a]]$ *to the holomorphic function defined by the integral in* (7.18) .

Proof. (A): Let $z \in D$ be fixed and set

$$F(t) = \frac{z}{z+t} , \quad a \le t \le b .$$

By Theorem 7.2 and (7.15) we obtain

$$(7.19) \qquad z\int_a^b \frac{d\psi(t)}{z+t} = \lim_{n\to\infty} \sum_{j=1}^{n} \frac{zw_j^{(n)}}{z+t_j^{(n)}} = \lim_{n\to\infty} \frac{A_n(z)}{B_n(z)} ,$$

where $A_n(z)/B_n(z)$ is the nth approximant of the positive T-fraction in (7.18) . This proves (A) . To prove (B) we let K be an arbitrary compact subset of D . Then there exist constants $B(K)$ and $\delta(K)$ such that, for all $z \in K$, we have $|z| \le B(K)$ and $|z+t_j^{(n)}| \ge \delta(K)$ for $j = 1,2,\ldots,n$ and $n = 1,2,3,\ldots$. Thus by (7.15) and (7.4)

$$\left|\frac{A_n(z)}{B_n(z)}\right| \le \sum_{j=1}^{n} \frac{|z|\,w_j^{(n)}}{|z+t_j^{(n)}|} \le \frac{B(K)}{\delta(K)} \frac{F_1}{G_1} \quad \text{for all } z \in K , \quad n \ge 1 .$$

It follows that $\{A_n(z)/B_n(z)\}_1^\infty$ is a normal family of holomorphic functions in D . Therefore by the Stieltjes-Vitali theorem ([21, Theorem 20.15 and Remark 20.2] or [8, Theorem 5.3.1]) the sequence $\{A_n(z)/B_n(z)\}$ converges uniformly on K. ■

The coefficients F_n, G_n in (7.18) can be computed in terms of the moments

$$(7.20) \qquad \mu_k = (-1)^k c_k = \int_a^b t^k d\psi(t)$$

by means of the following algorithm due to McCabe and Murphy [16] (see also [11, Section 7.3]).

FG-Algorithm. Given μ_k , $-p-2 \le k \le p-2$, compute

$$F_1^{(m)} = 0 , \quad G_1^{(m)} = \mu_{m-1}/\mu_{m-2} , \quad -p \le m \le p-1$$

and then, for $n = 1,2,\ldots,p-1$, compute

$$F_{n+1}^{(m)} = F_n^{(m+1)} + G_n^{(m+1)} - G_n^{(m)} , \quad n-p-1 \le m \le p-n-1 ,$$

$$G_{n+1}^{(m)} = \frac{F_{n+1}^{(m)}}{F_{n+1}^{(m-1)}} G_n^{(m-1)} , \quad n-p \le m \le p-n-1 .$$

An application of this algorithm produces a triangular table of the form shown in Table 1 . The entries in the central row give the desired coefficients F_n and G_n as follows:

$$F_1 = \mu_{-1}\ , \quad F_n = F_n^{(0)} \quad , \quad n = 2,3,\ldots,p \tag{7.21a}$$

$$G_n = G_n^{(0)} \quad , \quad n = 1,2,\ldots,p \ . \tag{7.21b}$$

If $0 \leq a < b \leq +\infty$, then the operations in the FG-Algorithm can always be

Table 1. Triangular array of output from FG-Algorithm.

$$\begin{array}{llllllll}
0 = F_1^{(-p)} & G_1^{(-p)} & F_2^{(-p)} & & & & & \\
0 = F_1^{(-p+1)} & G_1^{(-p+1)} & F_2^{(-p+1)} & G_2^{(-p+1)} & F_3^{(-p+1)} & & & \\
\cdot\ \ \cdot & \cdot\ \ \cdot & \cdot\ \ \cdot & \cdot\ \ \cdot & \cdot\ \ \cdot & & & \\
\cdot\ \ \cdot & \cdot\ \ \cdot & \cdot\ \ \cdot & \cdot\ \ \cdot & \cdot\ \ \cdot & & & \\
0 = F_1^{(0)} & G_1^{(0)} & F_2^{(0)} & G_2^{(0)} & \cdots\cdots\cdots & & F_p^{(0)} & G_p^{(0)} \\
\cdot\ \ \cdot & \cdot\ \ \cdot & \cdot\ \ \cdot & \cdot\ \ \cdot & \cdot\ \ \cdot & & & \\
\cdot\ \ \cdot & \cdot\ \ \cdot & \cdot\ \ \cdot & \cdot\ \ \cdot & \cdot\ \ \cdot & & & \\
0 = F_1^{(p-2)} & G_1^{(p-2)} & F_2^{(p-2)} & G_2^{(p-2)} & & & & \\
0 = F_1^{(p-1)} & G_1^{(p-1)} & & & & & &
\end{array}$$

carried out; in other words the denominator $F_{n+1}^{(m-1)} \neq 0$ so that there is no danger of dividing by zero when computing $G_{n+1}^{(m)}$. Therefore Theorem 7.3 together with the FG-Algorithm provide an efficient method for obtaining rational approximations of functions that can be expressed in the form of the Riemann-Stieltjes integral in (7.18) . It should be noted that these rational approximations (the approximants of the positive T-fraction in (7.18)) are two-point Padé approximants. Some appications of this type will be described in a subsequent paper on joint work with O. Njåstad.

We conclude this section with examples of Gaussian quadrature formulas (Theorem 7.1) for the distribution function $\psi(t)$ and interval (a,b) given by

$$d\psi(t) = e^{-t-1/t}dt \quad , \quad (a,b) = (0,\infty) \ . \tag{7.22}$$

Calculation of the moments (7.20) for this distribution is facilitated by the recurrence relations

$$\mu_k = \mu_{-k-2} = k\mu_{k-1} + \mu_{k-2} \quad , \quad k = 0,\pm1,\pm2,,\ldots \tag{7.23}$$

and by the series

$$\mu_{-1} = \sum_{k=0}^{\infty} \frac{2\Psi(k+1)}{(k!)^2} \ , \tag{7.24a}$$

$$\mu_n = \sum_{k=0}^{\infty} (-1)^k \frac{(n-k)!}{k!} + (-1)^{n+1} \sum_{k=0}^{\infty} \frac{\Psi(k+1)+\Psi(n+k+1)}{k!(n+k+1)!} \ , \tag{7.24b}$$

where $\Psi(z) = \Gamma'(z)/\Gamma(z)$ can be computed by

$$-\Psi(1) = \gamma = 0.5772\ 15664\ 90153 \ldots \quad \text{(Euler's constant)}$$

$$\Psi(k+1) = \Psi(k) + 1/k \quad , \quad k = 1,2,3,\ldots$$

(see for example [7, p. 307-341]). The moments μ_{-1} and μ_0 can be computed from (7.24) and all remaining moments can be obtained from (7.23). Some values of μ_k are given in Table 2. The weights $w_j^{(n)}$ and abscissas $t_j^{(n)}$ in the quadrature rule (7.1a) can then be computed by methods described in [4], [6] or [18]. Values of $w_j^{(n)}$ and $t_j^{(n)}$ for $j = 1,2,\ldots,n$ and $n = 1,2,\ldots,10$ are given in Table 3. These quadrature rules have been applied

Table 2. Moments $\mu_k = \int_0^\infty t^k e^{-t-1/t} dt$, $-1 \le k \le 10$, with at least 24 significant decimal digits.

k	μ_k
-1	0.22778 77454 99066 87130 54396
0	0.27973 17636 33044 85456 91973
1	0.50751 95091 32111 72587 46369
2	1.29477 07818 97268 30631 84711
3	4.39183 18548 23916 64483 00502
4	18.86209 82011 92934 88563 86719
5	98.70232 28607 88591 07302 34097
6	611.07603 53659 24481 32377 91299
7	4,376.23457 04222 59960 33947 7319
8	35,620.95259 87440 04164 03959 768
9	324,964.80795 91182 97436 69585 64
10	3,285,269.03218 99269 78530 99816 2

to approximate the following three improper integrals:

Case (a)

$$\int_0^\infty t^{-3/2} e^{-t-1/t} dt = \sqrt{\pi}\, e^{-2} = 0.23987\ 55439 ,$$

Case (b)

$$\int_0^\infty \ln t \left(\frac{t^2-1}{t^2}\right) e^{-t-1/t} dt = \mu_{-1} = 0.22778\ 77455 ,$$

Case (c)

$$\int_0^\infty \frac{\ln t}{t} e^{-t-1/t} = 0 .$$

The approximations of these integrals by the quadrature rule (7.1a) are given in Table 4. Also included are the errors of the approximations and the number of significant digits obtained. The number of significant digits SD(n) for an approximation GQ(n) of an integral I is calculated by the standard formula

$$SD(n) = -\log_{10}\left|\frac{I-GQ(n)}{I}\right| .$$

Since $I = 0$ in Case (c), one cannot compute significant digits.

For comparison we computed approximations of the integrals

$$\int_0^\infty f(t) e^{-t} dt$$

using (Gauss) Laguerre quadrature, where we chose $f(t)$ to be

Case (a) $$f(t) = t^{-3/2} e^{-1/t} ,$$

Case (b) $$f(t) = \ln t \left(\frac{t^2-1}{t^2}\right) e^{-1/t} ,$$

Case (c) $$f(t) = \frac{\ln t}{t} e^{-1/t} .$$

Using n-point Laguerre quadrature approximations with $n = 1,2...,10$, we were able to obtain at most one correct decimal for all three cases. We conclude that

<u>Table 3.</u> Absissa $t_j^{(n)}$ and weights $w_j^{(n)}$ for the quadrature formula

$$\int_0^\infty F(t)e^{-t-1/t}dt \approx \sum_{j=1}^{n} w_j^{(n)} F(t_j^{(n)}) .$$

Numbers in parentheses are exponents of 10 for floating decimal representations.

$t_j^{(n)}$	$w_j^{(n)}$
n = 1	
1.22803 69298	2.79731 76363(-1)
n = 2	
1.19015 08379	1.83490 01084(-1)
3.65830 37862	7.71368 98019(-2)
n = 3	
0.36370 40428	1.91048 54778(-2
1.19015 08379	1.83490 01084(-1)
3.65830 37862	7.71368 98019(-2)
n = 4	
0.25095 39244	3.64236 63312(-3)
0.69811 57495	8.67465 27365(-2)
1.96057 07663	1.65110 86597(-1)
5.04848 00675	2.42320 03963(-2)
n = 5	
0.18750 47193	6.53027 20937(-4)
0.45584 46302	2.96585 73142(-2)
1.16729 49502	1.45122 03973(-1)
2.88169 12907	9.77101 52507(-2)
6.51287 66175	6.58797 10401(-3)
n = 6	
0.14783 36197	1.13132 72631(-4)
0.32252 18957	8.51276 55623(-3)
0.74995 75105	8.11564 55820(-2)
1.77588 33321	1.43440 25334(-1)
3.92341 42339	4.48825 14049(-2)
8.03341 29026	1.62664 21354(-3)
n = 7	
0.12109 26307	1.91710 69928(-5)
0.24232 84362	2.19890 67702(-3)
0.51467 87975	3.47491 37223(-2)
1.15144 90395	1.23314 12683(-1)
2.51266 73178	1.01686 82087(-1)
5.06117 00695	1.73888 84764(-2)
9.59792 84128	3.74716 11203(-4)

$t_j^{(n)}$	$w_j^{(n)}$
n = 8	
0.10203 25541	3.19761 43435(-6)
0.19050 32807	5.29569 68101(-4)
0.37369 77285	1.25641 87975(-2)
0.78269 03126	7.57838 99121(-2)
1.66590 79073	1.27456 88683(-1)
3.36093 44269	5.73535 39217(-2)
6.27628 96489	5.95859 44147(-3)
11.19788 85165	8.18887 76326(-5)
n = 9	
0.08785 32104	5.26861 18479(-7)
0.15502 69936	1.21261 51548(-4)
0.28426 63184	4.05414 49297(-3)
0.55693 55447	3.72184 24659(-2)
1.13958 30771	1.08856 48912(-1)
2.28927 30575	1.00218 04558(-1)
4.30405 62309	2.73854 52461(-2)
7.55474 84601	1.86025 20524(-3)
12.82707 79281	1.71664 57849(-5)
n = 10	
0.07694 33775	8.59510 44467(-8)
0.12959 25714	2.67128 51817(-5)
0.22454 60615	1.20584 51046(-3)
0.41341 96516	1.56467 90887(-2)
0.80564 92987	7.11476 87220(-2)
1.59121 12511	1.15327 14160(-1)
3.01222 54997	6.43152 00696(-2)
5.32754 94863	1.15191 27969(-2)
8.88591 74089	5.39692 31390(-4)
14.48082 84476	3.47903 89245(-6)

the quadrature method described in the present paper is much better suited for the types of integrals chosen in these cases.

Table 4. Approximation of $I = \int_0^\infty F(t)e^{-t-1/t}dt$ by quadrature formula $GQ(n) = \sum_{j=1}^{n} w_j^{(n)} F(t_j^{(n)})$. SD(n) denotes the number of significant digits in GQ)n) given by $SD(n) = -\log_{10}|(I-GQ(n))/I|$.

Case (a): $F(t) = t^{-3/2}$, I = 0.23987 55439

n	GQ(n)	I-GQ(n)	SD(n)
3	0.23944 665	4.29(-4)	2.75
4	0.23971 613	-9.61(-5)	3.40
5	0.23984 978	2.58(-5)	3.97
6	0.23988 337	-7.83(-6)	4.49
7	0.23987 293	2.61(-6)	4.96
8	0.23987 648	-9.40(-7)	5.41
9	0.23987 518	3.59(-7)	5.82
10	0.23987 569	-1.44(-7)	6.22

Case (b): $F(t) = \ln t\left(\frac{t^2-1}{t^2}\right)$, I = 0.22778 77455

n	GQ(n)	I-GQ(n)	SD(n)
3	0.22871 474	-9.27(-4)	2.39
4	0.22764 616	1.42(-4)	3.21
5	0.22781 639	-2.86(-5)	3.90
6	0.22778 079	6.96(-6)	4.52
7	0.22778 967	-1.93(-6)	5.07
8	0.22778 715	5.91(-7)	5.59
9	0.22778 794	-1.97(-7)	6.06
10	0.22778 768	6.90(-8)	6.51

Case (c): $F(t) = \frac{\ln t}{t}$, I = 0.0

n	GQ(n)	I-GQ(n)	SD(n)
3	0.00105 818	-1.05(-3)	not defined
4	-0.00025 178	2.52(-4)	
5	0.00006 996	-7.00(-5)	
6	-0.00002 176	2.18(-5)	
7	0.00000 738	-7.38(-6)	
8	-0.00000 269	2.69(-6)	
9	0.00000 104	-1.04(-6)	
10	-0.00000 042	4.20(-7)	

Acknowledgement. The authors wish to express thanks to Allan Steinhardt and Robert Jones for able assistance with the computations discussed above.

8. Correspondence and asymptotic expansions. We first show that the generalized approximants of a general T-fraction (with $G_n \neq 0$) also correspond to a formal power series L_0 at 0 (and L_∞ at ∞). L_0 and L_∞ are defined in (1.3) . Let

$$L_0^{(n)} = -c_{-1}z - \cdots - c_{-n}z^{-n} .$$

Then

$$(8.1)\quad \frac{C_{2n}(z,\tau)}{D_{2n}(z,\tau)} - L_0^{(2n)}(z) = \frac{[C_{2n}(z)-L_0^{(2n)}D_{2n}(z)]+\tau z[C_{2n-1}(z)-L_0^{(2n)}D_{2n-1}(z)]}{D_{2n}(z)+\tau z D_{2n-1}(z)}$$

$$= \frac{\alpha_{2n+1}z^{2n+1}+\cdots+\beta_{2n+1}z\tau z^{2n}}{D_{2n}(z)+\tau z D_{2n-1}(z)} = \gamma_{2n+1}z^{2n+1} + \cdots ,$$

except for two values of τ for which the power of z is reduced by one. A similar argument holds for odd generalized approximants as well as for the correspondence of generalized approximants at ∞.

Next we can prove:

Theorem 8.1. *For* $\tau \in \mathbb{R}$ *and* z *fixed,* $z \in \mathbb{R}$ *, the* $(2m-1)$th *generalized approximant* $S_{2m-1}(z,\tau)$ *of a semi-positive T-fraction describes the boundary of a circular disk* $T_{2m-1}(z)$ *whose radius* $\rho_{2m-1}(z)$ *is given by*

$$(8.2)\qquad \rho_{2m-1}(z) = \frac{|z|^{2m-1}}{2|\mathrm{Im}(z)|\sum_{\nu=1}^{m}(d_{2\nu-1}|D_{2\nu-2}|^2+e_{2\nu-2}|D_{2\nu-3}|^2)|z|^{2(m-\nu)}} .$$

It is also true that

$$S_{2m}(z,z\tau) \in T_{2m-1}(z)$$

and hence

$$\frac{C_{n_k}(z,\tau_k)}{D_{n_k}(z,\tau_k)} \in T_{2m-1}(z) , \text{ for } n_k \geq 2m-1 , \quad \tau_k \in \mathbb{R} .$$

Proof. $S_{2m-1}(z,\tau)$ is the image of $\mathbb{R}$ under a linear fractional transformation and hence is a circle (or straight line). Using an idea of Thron [22] we have, for the circumference of this circle (in between steps are omitted),

$$2\pi\rho_{2m-1}(z) = \int_{-\infty}^{\infty}|S'_{2m-1}(z,\tau)|d\tau$$

$$= \frac{|D_{2m-1}(z)C_{2m-2}(z)-D_{2m-2}(z)C_{2m-1}(z)|}{|D_{2m-2}(z)|^2|\mathrm{Im}(D_{2m-1}(z)/D_{2m-2}(z))|}\int_{-\infty}^{\infty}\frac{dt}{1+t^2}$$

$$= \frac{2\pi|z^{2m-1}|}{|D_{2m-1}(z)\bar{D}_{2m-2}(z)-\bar{D}_{2m-1}(z)D_{2m-2}(z)|} .$$

A recursion formula applied to the denominator then yields (8.2) . The other assertions of the theorem follow from the formulas (3.9) .

For small $|z|$, $z \neq 0$, we have

$$\rho_{2m-1}(z) \sim \frac{|z|^{2m-1}}{2|\mathrm{Im}(z)|(d_{2m-1}|D_{2m-1}(0)|^2+e_{2m-2}|D_{2m-3}(0)|^2)}$$

$$< \eta_m\frac{|z|^{2m-2}}{|\mathrm{Im}(z)/z|} .$$

Thus if $|Im(z)/z| > \lambda$, then

(8.3) $$\rho_{2m-1}(z) < |z|^{2m-2}\eta_m/\lambda \quad \text{for} \quad |z| < \delta_m \ .$$

Now assume that

(8.4) $$\frac{C_{n_k}(z,\tau_k)}{D_{n_k}(z,\tau_k)} \to F(z) \quad , z \in \mathbb{R} \ .$$

Then $F(z) \in T_{2m-1}(z)$, $m > 1$, and hence

(8.5) $$\left|\frac{C_{n_k}(z,\tau_k)}{D_{n_k}(z,\tau_k)} - F(z)\right| < 2\rho_{2m-1}(z) < \frac{2\eta_m}{\lambda}|z|^{2m-2} \ ,$$

for $n_k > 2m-1$ and $|z| < \delta_m$, $|Im(z)/z| > \lambda$. A formula similar to (8.3) can be derived for $|z| > M_m$, $|Im(z)/z| > \lambda$. One has

(8.6) $$\rho_{2m-1}(z) \sim \frac{|z|^{2m-1}}{2|Im(z)|d_{2m-1}|D_{2m-2}(z)|^2} < \frac{\sigma_m|z|^{-(2m-2)}}{\lambda} \ .$$

We are now ready to prove that $L_0(z)$ is the asymptotic expansion of $F(z)$ at $z = 0$ with respect to either

$$R_\lambda^+ = [z : Im(z) > 0 \ , \ |Im(z)/z| > \lambda]$$

or

$$R_\lambda^- = [z : Im(z) < 0 \ , \ |Im(z)/z| > \lambda] \ .$$

The very simple proof was suggested by the proof in a theorem of B.C. Meyer [15] . We have

$$|F(z)-L_0^{(n)}(z)| = \left|F(z) - \frac{C_{n_k}(z,\tau_k)}{D_{n_k}(z,\tau_k)}\right| + \left|\frac{C_{n_k}(z,\tau_k)}{D_{n_k}(z,\tau_k)} - L_0^{(n)}(z)\right|$$

$$< \eta_{m(n)}|z|^{n+1} + \gamma_n|z|^{n+1}$$

for $n_k \geq 2m(n) - 2 \geq n + 1$, as a consequence of (8.5) and (8.1) . The proof that $F(z)$ has $L_\infty(z)$ as its asymptotic expansion at ∞ follows a similar pattern making use of (8.6) . We thus have

Theorem 8.2. *Let*

$$F(z) = \lim_{k\to\infty} \frac{C_{n_k}(z,\tau_k)}{D_{n_k}(z,\tau_k)} \ , \quad z \notin \mathbb{R} \ , \quad \tau_k \in \mathbb{R} \ , \quad k \geq 1 \ ,$$

where $C_{n_k}(z,\tau_k)/D_{n_k}(z,\tau_k)$ *is a generalized approximant of a semi-positive T-fraction. Then* $F(z)$ *has* $L_0(z)$ $(L_\infty(z))$ *as its asymptotic expansion at* $z = 0$ $(z = \infty)$ *with respect to either* R_λ^+ *or* R_λ^-, $\lambda > 0$.

We conclude this section with the remark that formula (8.2) for $\rho_{2m-1}(z)$ may be useful in determining when the SHMP has a unique solution.

We hope to be able to prove that this is the case if and only if $\rho_n(z) \to 0$.

References

1. Berg, Christian, written communication, (December, 1981).

2. Chihara, T.S., Introduction to Orthogonal Polynomials, (Mathematics and Its Applications Ser.), Gordon, (1978).

3. Favard, J., Sur les polynomes de Tchebicheff, C.R. Acad. Sci. Paris 200 (1935), 2052-2053.

4. Gautschi, Walter, Construction of Gauss-Christoffel quadrature formulas, Math. Comp. 22 (1968), 251-270.

5. Gautschi, Walter, A survey of Gauss-Christoffel quadrature formulae, Christoffel Memorial Volume (P.L. Butzer and F. Feher, eds.), Birkhäuser Verlag (1981), 72-147.

6. Golub, G.H. and Welsch, J.H., Calculation of Gauss quadrature rules, Math. Comp. 23 (1969), 221-230. Loose microfiche suppl. A1-A10.

7. Gradshteyn, I.S. and Ryzhik, I.M., Table of integrals, series and products, Academic Press, New York (1980).

8. Hille, E., Analytic Function Theory, Vol. II, Ginn, New York (1962).

9. Jones, William B., Olav Njåstad, and W.J. Thron, Orthogonal Laurent polynomials and the strong Hamburger moment problem, submitted.

10. Jones, William B. and W.J. Thron, Further properties of T-fractions, Math. Annalen, 166 (1966), 106-118.

11. Jones, William B. and W.J. Thron, Continued Fractions: Analytic Theory and Applications, Encyclopedia of Mathematics and Its Applications, No. 11, Addison-Wesley Publishing Company, Reading, Mass. (1980).

12. Jones, William B. and W.J. Thron, Orthogonal Laurent polynomials and Gaussian quadrature, Quantum Mechanics in Mathematics, Chemistry, and Physics, (K.E. Gustafson and W.P. Reinhardt, eds.) Plenum Press, New York (1981), 449-455.

13. Jones, William B., W.J. Thron and H. Waadeland, A strong Stieltjes moment problem, Trans. Amer. Math. Soc., (1980), 503-528.

14. Markoff, A., Deux démonstrations de la convergence de certaines fractions continues, Acta Math., 19 (1895), pp. 93-104.

15. Meyer, Burnett, On continued fractions corresponding to asymptotic series, submitted.

16. McCabe, J.H. and J.A. Murphy, Continued fractions which correspond to power series expansions at two points, J. Inst. Maths. Applics. 17 (1976), 233-247.

17. Perron, O., Die Lehre von den Kettenbrüchen, Band II, B.G. Teubner, Stuttgart (1957).

18. Sack, R.A. and Donovan, A.F., An algorithm for Gaussian quadrature given modified moments, Numer. Math. 18 (1971/72), 465-478.

19. Stieltjes, T.J., Quelques recherches sur la théorie des quadratures dites méchaniques, Ann. Sci. Éc. norm., Paris (3) 1 (1884), 409-426.

20. Szegö, G., Orthogonal Polynomials, Colloquium Publications vol. 23, Amer. Math. Soc., New York (1959).

21. Thron, W.J., The Theory of Functions of a Complex Variable, John Wiley and Sons, Inc., New York (1953).

22. Thron, W.J., Convergence of infinite exponentials with complex elements, Proc. Amer. Math. Soc. 8 (1957), 1040-1043.

23. Thron, W.J. and Waadeland, Haakon, Survey of modifications of continued fractions, these Lecture Notes.

24. Widder, David Vernon, The Laplace Transform, Princeton Univ. Press, Princeton (1946).

25. Wynn, Peter, The work of E.N. Christoffel on the theory of continued fractions, Christoffel Memorial Volume (P.L. Butzer and F. Fehér, eds.) Birkhäuser Verlag (1981), 190-202.

William B. Jones
Department of Mathematics
Campus Box 426
University of Colorado
Boulder, Colorado 80309

W.J. Thron
Department of Mathematics
Campus Box 426
University of Colorado
Boulder, Colorado 80309

MODIFICATIONS OF CONTINUED FRACTIONS,

A SURVEY.

W. J. Thron

Haakon Waadeland

1. Introduction. By a modification of a continued fraction $b_0 + K(a_n/b_n)$ we shall mean a transformation of its sequence of approximants $\{S_n(0)\}$ into a sequence $\{S_n(w_n)\}$ for some sequence $\{w_n\}$. Without further restrictions, in particular on the sequence $\{w_n\}$ the concept is too broad as the following choice of $\{w_n\}$ will illustrate. Let $\{x_n\}$ be an arbitrary sequence in $\hat{\mathbb{C}}$ and set

$$w_n = S_n^{-1}(x_n) .$$

Then the modification of $\{S_n(0)\}$ is $\{x_n\}$, which is surely undesirable. What will constitute reasonable restrictions will be discussed in Section 2, where we shall mainly be concerned with the question whether it would be advantageous to redefine convergence of continued fractions following a suggestion of Hamel [11]. Supporting evidence from the work of Weyl, Hamburger, Hamel, Schur and Phipps will also be discussed in this section.

In Section 3 we shall briefly survey the topic of converging factors.

If, as is done throughout most of these Proceedings, a continued fraction is thought of as being generated by a sequence of l.f.t., then the idea of a modification at least in the simple form $w_n = w$, for all $n \geq 1$, is a natural one. Even if continued fractions are considered in terms of three term recursions relations one is easily led to a modification as follows: if $b_0 + K(a_n/b_n)$ converges to f then

$$f = b_0 + \frac{a_1}{b_1} + \cdots + \frac{a_n}{b_n + f^{(n)}} ,$$

where $f^{(n)}$ is the n th tail (DN9) of the continued fraction. Thus an approximation $w_n \sim f^{(n)}$ could be expected to yield in

$$b_0 + \frac{a_1}{b_1} + \cdots + \frac{a_n}{b_n + w_n}$$

a value closer to f than the n th approximant. This idea of accelerating convergence was used by Glaisher [8], Wynn [41], Hayden [12], Phipps [21] and recently by Gill [3,6,7], the authors [30,33] and L. Jacobsen [14]. It shall be discussed in more detail in Section 5. It is probably in this context that the interpretation of a modification as a method of summability [17,p.327] is most appropriate.

The limit points of the sequence $\{h_n\}$ (DN10) play an important role in the theory of modifications because

$$S_n(w_n) - S_n(0) = \frac{(f_{n-1} - f_n)w_n}{h_n + w_n} .$$

In Section 4 the behavior of $\{h_n\}$ for limit periodic continued fractions is analyzed. It proves to be convenient to look at this problem in the framework of linear difference equations as worked out by Poincaré [24], Pincherle [22] and Perron [19].

Possibly the most interesting use of modifications is for the purpose of analytic continuation. The initial work was done by Waadeland [34,35,36,37,38,39] and Hovstad [13]. The process was also analyzed by Gill [5]. Recently, the authors [31,32] have made some further contributions to the subject. Here, what we call the "wrong" modification plays a critical role. This will be discussed in Sections 6 and 7.

Section 8 contains some speculations on possible future applications of wrong modifications. The results of Jacobsen [14] involving auxiliary continued fractions look promising in this context.

2. A new look at the definition of convergence. After some initial hesitation (see [17,p.10]) mathematicians soon settled on the definition that a continued fraction is convergent if $\lim_{n\to\infty} A_n/B_n$ exists in $\mathbb{C}$. Since it was then not generally assumed that $a_n \neq 0$ for all n, one had to worry some about A_n/B_n being meaningless for an infinite number of n. This can be taken care of by requiring that $B_n \neq 0$ for all but a finite number of n, or by assuming that this is implied if we say that $\lim A_n/B_n$ exists, or, finally, by assuming as we do here, that $a_n \neq 0$ for all $n \geq 1$. A minor change was made a few years ago and that is if $\lim_{n\to\infty} A_n/B_n = \infty$ then the continued fraction is considered to be convergent. Previously it was called unessentially (unwesentlich) divergent in that case.

In 1918 Hamel [10,11] published two papers involving continued fractions, which did not receive much attention from workers in the field. When one of us (Thron) found out about these articles around 1960 he was mainly interested in the fact that Hamel used the l.f.t. approach to continued fractions. That Hamel suggested a new definition of convergence [11,p.42] somehow did not seem to be important. However with the various applications of modifications that are now known as well as certain other arguments that we shall mention below, it now seems desirable to pay a little closer attention to what Hamel had to say. His proposal was that a continued fraction $b_0 + K(a_n/b_n)$ be called convergent if $\{S_n(w_n)\}$ converged for all sequences $\{w_n\}$ with $w_n \in V(\{a_n\},\{b_n\})$. The dependence of V on $\{a_n\}$ and $\{b_n\}$ was not specified. Surprisingly, he also did not require that all sequences have the same limit, nor that the convergence be uniform. Possibly it was because of this vagueness that his remarks were not considered seriously at that time. More recently Phipps [21] also criticized the clasical definition.

Hamel was led to his recommendation by encountering two situations where it seemed more natural to work with sequences $\{S_n(w_n)\}$ than with the sequences of approximants $\{S_n(0)\}$. He therefore complained that to insist on the convergence of $\{S_n(0)\}$ is really very arbitrary ("ist doch eigentlich sehr willkürlich"). One of the situations concerned the representation of functions bounded in the unit circle. Here he found that it was much simpler to write every such function as the limit of a sequence $\{S_n(w_n)\}$ (the a_n, b_n are now functions of z) than as $\lim_{n\to\infty} S_n(0)$. Schur [25] had met with the same phenomenon somewhat earlier except that he used a "continued fraction like" algorithm rather than continued fractions themselves.

From quite a different direction Hamburger [9] in 1920 was led to introduce the concept of <u>complete</u> (vollständig) <u>convergence</u> for J-fractions as follows: A continued fraction converges completely if $\{S_n(t_n)\}$ converges uniformly to the same value for any choice of $\{t_n\}$, as long as the t_n are all real. He was interested in characterizing uniqueness of the solution of his moment problem in terms of J-fractions. It turned out that considering only the ordinary approximants of the J-fraction was inadequate to "span" all possible solutions of the problem. By adding the so called generalized approximants $S_n(t_n)$, t_n real, it became possible to do so.

The use by Weyl [40] as early as 1910 of iterated sequences of l.f.t. $T_n(z) = t_1 \circ \cdots \circ t_n(z)$, where the limiting behavior for all z in a circular disk played an important role, may have had an influence on Hamel and Hamburger also.

Possibly even more important is that many useful continued fraction expansions (among others for ratios of hypergeometric functions) are obtained from three term linear recursion relations (see [17, Chapter 5])

$$P_n = b_n P_{n+1} + a_n P_{n+2}$$

from which one obtains

$$P_n/P_{n+1} = b_n + a_n/(P_{n+1}/P_{n+2}) ,$$

and recursively

$$P_0/P_1 = S_n(P_{n+1}/P_{n+2}) .$$

It then may turn out, as was illustrated by Phipps [20], that $\lim_{n\to\infty} S_n(0)$ is not equal to P_0/P_1.

Finally, it appears that further light can be thrown on all aspects of the theory of modifications by analyzing the possible behavior of $\{S_n(w)\}$. As our illustration in the beginning of this survey indicated the behavior of $\{S_n(w_n)\}$ is too varied to allow classification unless a substantial restriction is imposed on $\{w_n\}$.

A good deal has been known for some time about the possible behavior of sequences of l.f.t. $\{T_n(z)\}$ (see [23],[1],[18],[2]). From [23] and [1] we can deduce that for sequences $\{S_n(w)\}$, that is those satisfying the condition $S_n(\infty) = S_{n-1}(0)$, only three cases can occur, in which the sequence converges for at least one value $w^{(0)}$. The three cases are:

(a) $\{S_n(w)\}$ converges for all $w \in \hat{\mathbb{C}}$. Only two values c_1 and c_2 are taken on by the limit function. The value c_2 is taken on at only one point w^*.

(b) $\{S_n(w)\}$ converges only for w^* and w^+ with $\lim_{n\to\infty} S_n(w^*) = c_1$, $\lim_{n\to\infty} S_n(w^+) = c_2$, $c_1 \neq c_2$.

(c) $\{S_n(w)\}$ converges to only one value for all $w \in D \neq \emptyset$.

To analyze the cases (a) and (c) further let us assume that

$$\lim_{n\to\infty} S_n(w^{(0)}) = \lim_{n\to\infty} S_n(w^{(1)}) = c .$$

Then we have

$$S_n(w^{(0)}) - S_n(w^{(1)}) = \frac{h_n(f_n - f_{n-1})(w^{(1)} - w^{(0)})}{(h_n + w^{(0)})(h_n + w^{(1)})} = r_n ,$$

where $r_n \to 0$, and

$$\text{(2.1)} \qquad S_n(w) - S_n(w^{(0)}) = \frac{h_n(f_n - f_{n-1})(w^{(0)} - w)}{(h_n + w)(h_n + w^{(0)})}$$

$$= r_n \frac{(w^{(0)} - w)(h_n + w^{(1)})}{(h_n + w)(w^{(1)} - w^{(0)})} .$$

It is then clear that for $w \in \hat{\mathbb{C}}$ $\{S_n(w)\}$ converges to c if $w \notin L = [z: z$ is a limit point of $\{-h_n\}]$. We also note that this is true even if $\infty \in L$. If $w = \infty$ (and $\infty \notin L$) the statement is also true. Thus the set D contains $\hat{\mathbb{C}} \sim L$. That D actually may be bigger than $\hat{\mathbb{C}} \sim L$ (and herein lies the difficulty of characterizing all possible D) can be seen as follows. Let $h_{n_k} \to -w'$, then $\{S_n(w')\}$ will still converge to c if $r_{n_k}/(h_{n_k} + w') \to 0$, which is certainly possible.

We now turn to the question when $S_n(w_n) \to c$ uniformly (that is independent of the choice of $\{w_n\}$). An analysis analogous to the one given above shows that this will be the case if, for all $n \geq 1$, and some $d > 0$

$$w_n \in H^{(d)} = [z: N_n(d) \subset \mathbb{C} \sim L,\ |z| < 1/d \ \text{ if } \ \infty \in L\ ]$$

It is also clear that we have uniform convergence only if $\{w_n\} \subset H^{(d)}$ for some $d > 0$. Since L is a closed set $\hat{\mathbb{C}} \sim L$ is open so that, unless $\hat{\mathbb{C}} = L$, there will be $H^{(d)} \neq \emptyset$ for sufficiently small d. These results extend somewhat a theorem of Hayden [12,p.296]. He only considers the case $w^{(0)} = 0$, $w^{(1)} = \infty$. The point $w^{(0)} = 0$ is a special point in our theory since $S_n(0) \to c$ implies $S_n(\infty) \to c$. The theorem is as follows.

Theorem 2.1. Let the continued fraction $b_0 + K(a_n/b_n)$ be such that for two points $w^{(0)}$ and $w^{(1)}$

$$\lim_{n\to\infty} S_n(w^{(0)}) = \lim_{n\to\infty} S_n(w^{(1)}) = c .$$

Define

$$L = [z : z \text{ is a limit point of } \{-h_n\}]$$

and assume that $L \neq \hat{\mathbb{C}}$. Then for $d > 0$ and sufficiently small

$$H^{(d)} = [z : N_z(d) \cap L = \emptyset , |z| < 1/d \text{ if } \infty \in L]$$

is non empty and $S_n(w_n) \to c$ uniformly, if $w_n \in H^{(d)}$ for all $n \geq 1$ and some $d > 0$.

On the basis of this result it seems reasonable to formulate a definition of strong convergence.

Definition 2.1. A continued fraction $b_0 + K(a_n/b_n)$ is called strongly convergent to c if there exist $w^{(0)}$, $w^{(1)}$ such that

$$\lim_{n\to\infty} S_n(w^{(0)}) = \lim_{n\to\infty} S_n(w^{(1)}) = c ,$$

and if $L \neq \hat{\mathbb{C}}$. The sets $H^{(d)} \subset \hat{\mathbb{C}} \sim L$, $d > 0$ will be called the associated convergence sets of the continued fraction.

We conclude this section with an open question and a discussion of the relation between ordinary convergence and strong convergence.

The open question is: Can L be an arbitrary closed set independent of $w^{(0)}$, $w^{(1)}$ and $\{r_n\} \to 0$? To the best of our knowledge the question is still open even if $w^{(0)} = 0$, $w^{(1)} = \infty$. A related question is can case (b) ever occur for continued fractions?

Under (a) $S_n(0) \to c_2$ is not possible since then $S_n(\infty) \to c_2$. Thus in this case strong convergence and ordinary convergence coincide. Case (b) never occurs with ordinary convergence and by our definition it is not subsumed under strong convergence either. In case (c) ordinary convergence implies that $0, \infty \in D$ but they need not be in $\mathbb{C} \sim L$. In particular if $\hat{\mathbb{C}} = L$ then the continued fraction does not converge strongly but does converge in the classical sense. Finally, if $0, \infty \notin D$ then the continued fraction may converge strongly without converging in the ordinary sense.

3. Converging factors. The expression "converging factor" was first used in continued fraction theory by Wynn [41] in 1959. The term had previously appeared in connection with other infinite processes. To Wynn a converging factor is an approximation to the tail of the continued fraction. The term was used by Gill [3] in 1975 for the attractive fixed point of a limit periodic continued fraction, which makes the fixed point an approximation to the tail.

In 1978 Gill [4] sketched a more general theory of converging (or modifying) factors by dividing the set of sequences $\{\mu_n\}$ for which $\lim_{n\to\infty} S_n(\mu_n)$ exists into equivalence classes according to the rule

$$\{\mu_n\} \sim \{w_n\} \Leftrightarrow \lim_{n\to\infty} S_n(\mu_n) = \lim_{n\to\infty} S_n(w_n) .$$

Any sequence $\{z_n\} \sim \{\mu_n\}$ then is called a sequence of modifying factors of $\{S_n(\mu_n)\}$. Clearly, the number of equivalence classes is equal to the number of possible limits. Since for every $c \in \hat{\mathbb{C}}$ there is a sequence $\mu_n = S_n^{-1}(c)$ so that $\lim_{n\to\infty} S_n(\mu_n) = c$, it follows that there is an infinite number of equivalence classes. Gill then proceeds to restate the theorem of Hayden (generalized in Theorem 2.1 of this article) in terms of converging factors equivalent to $\{0\}$ for convergent continued fractions. He thus arrives at sufficient conditions for $\{z_n\}$ to be a sequence of converging factors equivalent to $\{0\}$. A theorem of Phipps [20] which states:

<u>If</u> $b_0 + K(a_n/b_n)$ <u>converges to</u> c <u>and if</u> $\{w_n\}$ <u>satisfies</u>

$$\lim_{n\to\infty} (1 + w_n/h_n)^{-1}(f_n - f_{n-1}) = d \neq 0$$

<u>then</u> $\lim_{n\to\infty} S_n(w_n) = c + d;$

can be used to decide when $\{w_n\} \sim \{0\}$. That there can be at most two equivalence classes containing constant sequences $\{w\}$ follows from the classification discussed in Section 2.

A natural question that now presents itself is: how do regions for converging factors relate to value regions and how do the latter relate to regions $H^{(d)}$. In general value regions $\{V_n\}$ are defined with respect to element regions $\{E_n\}$ (and only for continued fractions $K(a_n/1)$) by requiring

(3.1) $$s_n(V_n) \subset V_{n-1} , \text{ for } a_n \in E_n , n \geq 1$$

and

(3.2) $$E_n \subset V_{n-1} .$$

Since we are concerned here only with a single continued fraction rather than with a whole class, it is reasonable to take $E_n = [a_n]$ for the element regions and then simply speak of a sequence $\{V_n\}$ of <u>value regions for the continued fraction</u> $K(a_n/1)$ if (3.7) holds. We shall omit condition (3.2) so that our value regions are what sometimes are called pre-value regions.

We begin by investigating the sets $H^{(d)}$. Each $H^{(d)}$ contains only a finite number of $-h_n$. Hence

$$-h_n \notin Cl(H^{(d)}) \quad \text{for } n > k(d) , d > 0 .$$

(Here $Cl(A)$ stands for the topological closure of A.) Thus $\infty \in S_n(H^{(d)})$ is impossible — since it is equivalent to $-h_n = S_n^{-1}(\infty) \in H^{(d)}$ — for $n > k(d)$. Beyond this it can be shown that $S_n(H^{(d)})$ is a bounded set in $\mathbb{C}$. This is useful information but it falls short of $H^{(d)}$ being a value region.

Now assume that $\{V_n\}$, with V_0 bounded, is a sequence of value regions for the continued fraction $K(a_n/1)$. Then $-h_n \notin V_n$ so that, if we set

$$D_k = \bigcap_{n=k}^{\infty} V_n , \tag{3.3}$$

then $-h_n \notin D_k$ for $n \geq k$. It follows that $\text{Int } D_k \cap L = \emptyset$. (Here $\text{Int } A$ is the topological interior of the set A.) We thus have the following result.

Theorem 3.1. *Let $\{V_n\}$, with V_0 bounded, be a sequence of value regions for the continued fraction $K(a_n/1)$. Assume that for some k_0 D_{k_0} (as defined in (3.3)) has the property that $\text{Int } D_{k_0} \neq \emptyset$. Then*

$$\text{Int } D_{k_0} \cap L = \emptyset .$$

For converging factors $\{\mu_n\} \sim \{0\}$ one has the following result.

Theorem 3.2. *Let $\{V_n\}$ be a sequence of value regions for the continued fraction $K(a_n/1)$. If in addition $0 \in V_n$, $n \geq k$ and the sequence $\{S_n(V_n)\}$ of nested sets shrinks to a point then a permissible choice for the n th converging factor μ_n in $\{\mu_n\} \sim \{0\}$ is $\mu_n \in V_n$, $n \geq 1$.*

A concrete instance of this general result is given by Gill [7] in these Proceedings. The choice

$$V_n = [v: |v+1| > \sqrt{|a_n|}]$$

does indeed give a sequence of value regions since, if $v \in V_n$, then

$$\frac{|a_n|}{|1+v|} < \sqrt{|a|} ,$$

so that, since it is assumed that

$$\sqrt{|a_n|} + \sqrt{|a_{n-1}|} < 1 ,$$

we have

$$|v'| = \left|\frac{a_n}{1+v}\right| < \sqrt{|a_n|} < 1 - \sqrt{|a_{n-1}|} .$$

Hence

$$|1 + v'| > 1 - |v'| > \sqrt{|a_{n-1}|} .$$

If all $V_n = V$ we simply speak of a *value region for the continued fraction* $K(a_n/1)$.

Assume that V is an open bounded value region and assume that

$$\lim_{n\to\infty} S_n(w^{(0)}) = \lim_{n\to\infty} S_n(w^{(1)}) = c ,$$

where $w^{(0)} \neq w^{(1)}$. It follows from Theorem 3.1 that $V \cap L = \emptyset$ so that for any $w \in V$ $\{S_n(w)\}$ converges. Since $\{S_n(w)\}$ already converges to c for at least two values and since V contains an infinite number of points it now follows that

$S_n(w) \to c$ for all $w \in V$. Hence $c \in \mathrm{Cl}\, V$. We thus have $S_n(w) \to c$ for all $w \in V$ and possibly for other values also (in particular $w^{(0)}$ and/or $w^{(1)}$ may be outside of V). It is possible that there be one value $w^* \notin V$ such that $S_n(w^*) \to c^* \neq c$. It is also possible that the sequence diverges for all w for which it does not converge to c. The additional possibilities are: (a) that the sequence converges for two values only but that the limits are different, or (b) that the sequence converges for at most one point. Case (a) can happen for a periodic continued fraction with two fixed points of equal modulus, where V can be one of the invariant circular disks. Whether case (b) can occur is not known. We now have proved the following result.

Theorem 3.3. Let the open bounded set V be a value region for the continued fraction $K(a_n/1)$ Then the following possibilities exist:

α) $\{S_n(w)\} \to c \in \mathrm{Cl}\, V$ for all $w \in V$,

or

β) $\{S_n(w)\}$ converges for two points only to different limits,

or

γ) $\{S_n(w)\}$ converges for at most one w.

It may be of interest to recall [28,p.120] that case α can occur even though the sequence of nested regions $\{S_n(V)\}$ does not shrink to a point.

4) Applications of the Poincaré - Pincherle - Perron theorem and the convergence of $\{h_n\}$ in the limit periodic case. In 1885 Poincaré [24] considered $m+1$ term difference equations

$$P_n + c_1^{(n)} P_{n-1} + c_2^{(n)} P_{n-2} + \cdots + c_m^{(n)} P_{n-m} = 0\,, \quad n \geq m \tag{4.1}$$

where
$$\lim_{n\to\infty} c_k^{(n)} = c_k\,, \quad k = 1, \cdots, m\,.$$

He showed that

$$\lim_{n\to\infty} P_n/P_{n-1} = y_j\,, \tag{4.2}$$

where y_j is one of the roots of the characteristic equation

$$y^n + c_1 y^{n-1} + \cdots + c_m = 0\,, \tag{4.3}$$

provided all roots are of different absolute value. Poincaré conjectured that "in general" the limit would be the root with the largest absolute value. Later Pincherle [22] and Perron [19] also investigated the subject and showed among others, that for each y_j (provided $c_m \neq 0$) there is a solution $\{P_n^{(j)}\}$ such that $P_n^{(j)}/P_{n-1}^{(j)} \to y_j$. The situation becomes considerably more complicated if some of the roots are of equal modulus.

The theorem and its various extensions and improvements are of interest to us not only because we want to know

$$\lim_{n\to\infty} h_n = \lim_{n\to\infty} B_n/B_{n-1}$$

for limit periodic continued fractions a result which clearly is a special case of the PPP theorem, but also because some of the tools employed in the proof may be helpful. Thus we note (see also [17, Sec. 5.3]) that the space of solutions of a system of $m+1$ term difference equations is a linear vector space of dimension m over $\mathbb{C}$.

Moreover, in the case where all

$$c_k^{(n)} = c_k \text{ , } k = 1, \cdots, m \text{ ; } n \geq 1 \text{ ,}$$

it is easy to see that all sequences $\{y_k^n\}$ are solutions of the system (4.1) and hence the most general solution is (the * indicates the pure periodic case)

$$P_n^* = \sum_{k=1}^{m} \alpha_k y_k^n \text{ .} \tag{4.4}$$

If the y_k are arranged in order of decreasing absolute value

$$|y_1| > |y_2| > \cdots > |y_m| \text{ ,}$$

then it is clear that (4.2) holds for that value j for which $\alpha_j \neq 0$, $\alpha_1 = \cdots = \alpha_{j-1} = 0$.

For the limit periodic case one would expect formulas similar to (4.4) but with additional error terms.

In our paper [32] these ideas are in the background. It might have helped the understanding if we had mentioned this material in that article. So we have in the periodic case

$$A_n^* = -y_1 y_2 (y_1^{n-1} + \cdots + y_2^{n-1}) = \alpha_1 y_1^n + \alpha_2 y_2^n \text{,}$$

where y_1, y_2 are the solutions of the equation

$$y^2 - by - a = 0 \text{ ,}$$

and $\alpha_1 = \dfrac{y_1 y_2}{y_2 - y_1} = \dfrac{-a}{y_2 - y_1}$, $\alpha_2 = \dfrac{a}{y_2 - y_1}$.

Similarly

$$B_n^* = (y_1^n + \cdots + y_2^n) = \beta_1 y_1^n + \beta_2 y_2^n \text{ ,}$$

where

$$\beta_1 = \frac{y_1}{y_1 - y_2} \text{ , } \beta_2 = \frac{-y_2}{y_1 - y_2} \text{ .}$$

In the limit periodic case we obtained for

$$B_n - y_{j+1} B_{n-1} = y_j^n + \sum_{k=0}^{n-1} y_j^{n-1-k} \eta_{k+1} B_k + \sum_{k=0}^{n-1} y_j^{n-1-k} \delta_{k+1} B_{k-1} \text{ ,}$$

where $\delta_n = a_n - a$, $\eta_n = b_n - b$.

This is an example of our earlier assertion that for the limit periodic case one would expect formulas similar to (4.4). It is also of interest to know which solutions of the difference equation have y_j^n as their major term.

To apply these ideas to limit periodic continued fractions with $\lim a_n = a$, $\lim b_n = b$, we first note that the fixed points of the transformation

$$(4.5) \qquad w \to \frac{a}{b+w}$$

are the solutions of the equation

$$x^2 + bx - a = 0$$

while the equation we get from (4.3) with $c_1 = -b$, $c_2 = -a$, $m = 2$ is

$$y^2 - bx - a = 0 .$$

We therefore have

$$x_1 = -y_1 , \; x_2 = -y_2 .$$

In the periodic case

$$\frac{B_n^*}{B_{n-1}^*} - y_1 = \frac{B_n^* - y_1 B_n^*}{B_{n-1}^*} = \frac{y_2^n}{\beta_1 y_1^{n-1} + \beta_2 y_2^{n-1}}$$

$$= \left(\frac{y_2}{y_1}\right)^n \cdot \frac{y_1 - y_2}{1 - \left(\frac{y_2}{y_1}\right)^n} .$$

If we denote by x_1 the attractive fixed point of (4.5) then $-(b+x_1)$ is the repulsive fixed point provided

$$\left| \frac{x_1}{b+x_1} \right| < 1 .$$

Then $x_1 = -y_2$ and $y_1 = b + x_1$ (this is a change from the notation mentioned above) so that one gets

$$\frac{B_n^*}{B_{n-1}^*} - (b+x_1) = \left(\frac{x_1}{b+x_1} \right)^n \frac{b + 2x_1}{(-1)^n - \left(\frac{x_1}{b+x_1} \right)^n} .$$

Making use of this formula we can now prove the following result, where $b_n = b = 1$

<u>Theorem</u> 4.1. <u>For the limit periodic continued fraction</u> $K(a_n/1)$ <u>with</u> $\lim_{n\to\infty} a_n = a \neq 0$, $|\arg (a+1/4)| < \pi$, $a \neq -1/4$, <u>there exists a</u> $d(a)$ <u>such that for</u>

$$d_n = \max_{m\geq n} \left| a_m - a \right| < d(a) , \; n \geq 1 ,$$

$$\left|h_n - (1+x_1)\right| < K(a,p) \sum_{v=1}^{n} p^{n-v} d_v ,$$

<u>where</u>

$$\left|\frac{x_1}{1+x_1}\right| < p < 1 .$$

In particular it follows that $\lim_{n\to\infty} h_n = 1 + x_1$.

Proof: From (4.6) it follows that

$$h_n^* = 1 + \frac{a}{1} + \cdots + \frac{a}{1}$$

satisfies

$$\left|h_n^* - (1+x_1)\right| < \frac{|1+2x_1|}{1-p} \cdot \left|\frac{x_1}{1+x_1}\right|^n .$$

Now define

$$\xi_n = h_n - (1+x_1) , \; \delta_n = a_n - a ,$$

set

$$q = \left|x_1/(1+x_1)\right|$$

and introduce q' so that $q < q' < 1$. Then one can find an $n' = n(a,q')$ such that

$$\left|\xi_{n'}^*\right| = \left|h_{n'}^* - (1+x_1)\right| < \frac{1}{2}(1-q')\left|1 + x_1\right| .$$

Next, there exists a $d_1(a,q')$ such that for $\left|a_m - a\right| < d_1$ for $m = 2,3,\cdots,n'$

$$\left|\xi_{n'}\right| = \left|h_{n'} - (1+x_1)\right| < (1-q')\left|1 + x_1\right| .$$

For $n > n'$, and assuming $\left|\xi_{n-1}\right| < \left|1 + x_1\right|$, one has (using the fact that $a = x_1(1+x_1)$)

$$\left|\xi_n\right| = \left|1 + x_1 - \left(1 + \frac{a+\delta_n}{1+x_1+\xi_{n-1}}\right)\right|$$

$$= \frac{\left|x_1(1+x_1) + x_1\xi_{n-1} - x_1(1+x_1) - \delta_n\right|}{\left|1+x_1 + \xi_{n-1}\right|}$$

$$\le \frac{|x_1|\,|\xi_{n-1}| + d_n}{|1+x_1| - |\xi_{n-1}|} .$$

We next show that

(4.7)
$$\left|\xi_n\right| < (1-q')\left|1 + x_1\right|$$

will hold for all $n > n'$ provided d_n is appropriately restricted. We have

$$\left|\xi_n\right| \le \frac{|x_1|\,|(1-q')|\,|1+x_1| + d_n}{q'\,|1+x_1|}$$

$$= q\left(\frac{1-q'}{q'}\right)\left|1+x_1\right| + \frac{d_n}{q'|1+x_1|}$$

$$< (1-q')\left|1+x_1\right| ,$$

provided (4.7) holds with n replaced by n-1 , and provided

$$\frac{d_n}{q'|1+x_1|} < (1-q')\,|1+x_1|\,\frac{(q'-q)}{q'} .$$

Hence (4.7) will hold if $d_n < d_2(a,q') = (1-q')(q'-q)|1+x_1|^2$ for $n > n'$. Setting $p = q/q'$ so that $0 < p < 1$ one now can show that

$$|\xi_n| < K(a,p) \sum_{v=1}^{n} p^{n-v} d_v \quad \text{for } n \geq n' .$$

That $\xi_{n'}$ satisfies the inequality can be assured by a proper choice of $K(a,p)$. Now assume that (4.8) holds for $n-1 \geq n'$. Then (we shall also assume that $K = K(a,p) > 1/q'|1+x_1|$)

$$|\xi_n| \leq \frac{|x_1||\xi_{n-1}| + d_n}{|1+x_1| - |\xi_{n-1}|} \leq \frac{|x_1|}{|1+x_1|q'}|\xi_{n-1}| + \frac{d_n}{|1+x_1|q'}$$

$$\leq K p \sum_{v=1}^{n-1} p^{n-v-1} d_v + K d_n = K \sum_{v=1}^{n} p^{n-v} d_v .$$

Finally, we show that $\sum_{v=1}^{n} p^{n-v} d_v \to 0$ provided $d_n \to 0$ and $0 < p < 1$. Let $d > d_n$ for all $n \geq 1$ and let $\varepsilon > 0$ be given. Note that

$$\sum_{v=1}^{n-k} p^{n-v} d_v < p^k \frac{d}{1-p} .$$

Determine k so that

$$\frac{p^k d}{1-p} < \frac{\varepsilon}{2} .$$

Next determine n_0 so that $d_{n-m} < \varepsilon/2k$ for $n > n_0$, $0 \leq m \leq k$. It then follows that

$$\sum_{v=1}^{n} p^{n-v} d_v = \sum_{v=1}^{n-k} p^{n-v} d_v + p^{k-1} d_{n-k+1} + \cdots + p^0 d_n$$

$$\leq \frac{p^k d}{1-p} + d_{n-k+1} + \cdots + d_n < \frac{\varepsilon}{2} + \frac{\varepsilon}{2} = \varepsilon .$$

This completes the proof of the theorem.

For $a = -1/4$ the following theorem can be proved.

Theorem 4.2. *If* $|\delta_n| \leq 1/16n(n-1)$, $n \geq 3$, $|\delta_2| \leq 1/8$ *and* $a_n = -1/4 + \delta_n$ *then* $\eta_n = h_n - 1/2$ *satisfies*

$$\left|\eta_n - \frac{5}{8n}\right| < \frac{3}{8n}, \quad n \geq 2 .$$

Proof: Define $H_n = [z: |z - 5/8n| < 3/8n]$. Next observe that

$$\eta_n = h_n - \frac{1}{2} = 1 + \frac{-1/4+\delta_n}{h_{n-1}} - \frac{1}{2} = \frac{1}{2} + \frac{-1/4+\delta_n}{\eta_{n-1}+1/2}$$

$$= \frac{\eta_{n-1}}{2\eta_{n-1}+1} + \frac{2\delta_n}{2\eta_{n-1}+1} .$$

We then have for $v \in H_{n-1}$

$$\left| \frac{v}{2v+1} - \frac{5}{8n} \right| < \frac{3}{8n} - \frac{1}{4n(2n-1)} .$$

This result is most easily obtained by checking on the images of the real axis of the values $v = 1/4(n-1)$ and $v = 1/(n-1)$ in H_{n-1}. Now

$$\left| \frac{2\delta_n}{2\eta_{n-1}+1} \right| < \frac{2 \cdot 2(n-1)}{16n(n-1)(2n-1)} = \frac{1}{4n(2n-1)} ,$$

since

$$| 2\eta_{n-1} + 1 | > 1 + \frac{2}{4(n-1)} = \frac{2(2n-1)}{4(n-1)} = \frac{2n-1}{2(n-1)} .$$

Hence the theorem follows by induction if it can be shown that $\eta_2 \in H_2$. We have

$$\eta_2 = h_2 - \frac{1}{2} = \frac{1}{2} + a_2 = \frac{1}{4} + \delta_2$$

and therefore

$$\left| \eta_2 - \frac{5}{16} \right| = \left| \frac{4}{16} - \frac{5}{16} + \delta_2 \right| < \frac{1}{16} + \frac{1}{8} = \frac{3}{16} .$$

5. Acceleration of convergence. To determine $\{w_n\}$ so that $\{S_n(w_n)\}$ converges substantially faster to f than $\{S_n(0)\}$ is, at least historically, the main reason for studying modifications of continued fractions.

Glaisher [8] appears to have been the first to have suggested a modification for the purpose of accelerating the convergence of the continued fraction

(5.1) $$1 + \frac{1}{1} + \frac{1.2}{1} + \cdots + \frac{n(n-1)}{1} + \cdots ,$$

which is known to converge to $\pi/2$. He suggested choosing $w_n = n$. The reason for his choice is that $w_n = n$ is an approximate solution of the recurrence relation

(5.2) $$w_n = \frac{a_n}{1+w_{n+1}} ,$$

which the tail of the continued fraction must satisfy.

Wynn [41] in 1959 extended the method by suggesting that one obtain approximate solutions of (5.2) of the form

(5.3) $$w_n = \sum_{j=-k}^{m} d_j n^{-j} .$$

Solutions are not in general unique and in particular if one happens to hit on $w_n \sim -h_n$ one obtains a "curious unwinding procedure" where $S_n(w_n) \to \infty$, since $-h_n = S_n^{-1}(\infty)$. This is the explanation suggested by Hayden [12,p308]. Hayden claims, incorrectly, that the $-h_n$ (his "critical" points) form a solution of (5.2). The actual situation is that $w_n = -1 - h_{n-1}$ is a solution of (5.2) so that if one is dealing with a situation where $a_n \to \infty$ sufficiently fast, one may indeed have $w_n \sim -h_{n-1}$ and thus obtain the unwinding procedure.

Apart from the difficulty of determining which choice of the d_j is likely to lead to a good approximation for the tail, there also has not been a systematic study of the acceleration of convergence one can expect to obtain. Finally, we know of no investigation which addresses the question of how large should one take the m in (5.3). Nevertheless the method seems to work in some cases, in particular some involving $a_n \to 0$ and $a_n \to \infty$.

Phipps [21], from a somewhat different point of view and more recently and quite systematically L. Jacobsen [14] have studied the problem of approximating the tails of continued fractions of a more general nature. One of the cases they treat is that of limit periodic continued fractions of period $k > 1$. An account of Jacobsen's work is contained in these Proceedings. Very briefly, she studies continued fractions $K(a_n/1)$, where $|a_n - a_n'| \to 0$ and where the <u>auxiliary continued fraction</u> $K(a_n'/1)$ as well as all of its tails $f^{(n)'}$ are assumed to converge. It is further required that all $f^{(n)'}$ are explicitly known. The modification then is $\{S_n(f^{(n)'})\}$, which, under reasonable conditions, can be shown to converge faster to f than $\{S_n(0)\}$ does.

The most accessible case has been that of limit periodic continued fractions $K(a_n/1)$ of period one. Of the possible cases of $\lim a_n = a$, the present authors [30] treated $|\arg(a + 1/4)| < \pi$, $a \neq 0, \infty$; and $a = -1/4$. We shall describe their results below. The case $a = 0$ is studied by Gill [6] in these Proceedings. We know of no systematic work on the case $a = \infty$, beyond the special cases investigated by Glaisher and Wynn.

The approach in [30] is to use the attractive fixed point x_1 of

$$w \to \frac{a}{1+w}$$

for w_n in the modification. In the case $a = -1/4$ we use the single fixed point $-1/2$ for w_n. If $a = 0$ the attractive fixed point is zero so that the modification is the same as the original sequence. By choosing the attractive fixed point of $w \to s_n(w)$ Gill [6] was able to obtain some acceleration of convergence. It appears that this approach might also be useful in case $a_n \to \infty$.

We now consider the case $|\arg(a+1/4)| < \pi$ $a \neq 0, \infty, -1/4$. From (2.1) one deduces easily

$$\left|\frac{f - S_n(x_1)}{f - S_n(0)}\right| = \left|\frac{h_n}{h_n + x_1}\right| \left|1 - \frac{x_1}{f^{(n)}}\right| . \tag{5.4}$$

For x_1 to be the attractive fixed point of (5.2) it suffices that x_1 be a solution of

$$x^2 + x - a = 0 \tag{5.5}$$

and satisfy

(5.6) $$\left| \frac{x_1}{1+x_1} \right| < 1 .$$

We note that $-(1+x_1)$ is the other solution of (5.5) and is the repulsive fixed point of (5.2).

From Theorem 4.1 we know that $\lim h_n = 1 + x_1$ and we also know how fast the limit is approached. This allows us to improve somewhat on the bound on $|h_n/(h_n + x_1)|$. For all the results that follow we need to have fairly substantial restrictions on $d_n = \max_{m \geq n} |a_m - a|$. In other words our theorems are true for tails of the limit periodic continued fraction.
In [30] we obtained

(5.7) $$\left| \frac{h_n}{h_n+x_1} \right| \leq 1 + \frac{2|x_1|}{D} ,$$

where

(5.8) $$D = D(a) = |a+x_1| - |x_1| ,$$

which is positive because of (5.6). Using Theorem 4.1 (see also [30,p161]) we can conclude that

$$\lim_{n\to\infty} \frac{h_n}{h_n+x_1} = \frac{1+x_1}{1+2x_1} = \frac{1}{2}\left(1 + \frac{1}{\sqrt{1+4a}} \right) .$$

Since it is assumed that $\mathrm{Re}\,\sqrt{1+4a} > 0$ it then follows that for n sufficiently large

$$F_n = \left| \frac{h_n}{h_n+x_1} \right| > \frac{1}{2} .$$

It is also clear that there is no upper bound on $\lim F_n = F(a)$.

For the factor $\left| 1 - x_1 / f^{(m)} \right|$ the estimate obtained in [30] is

$$\left| 1 - \frac{x_1}{f^{(m)}} \right| \leq \frac{4d_n}{|x_1|D(a)} , \quad m \geq n .$$

One thus arrives at the following result

$$\left| \frac{(F - S_n(x_1)}{f - S_n(0)} \right| < G(a) \frac{d_n}{D(a)} ,$$

provided $d_n < d(a)$, for all $n \geq 1$. The constant $G(a)/D(a)$ can become quite large for a near the critical ray $z = -1/4 - t$, $t \in [0,\infty)$.

The case $a = -1/4$, $x_1 = -1/2$ differs from the general case in a number of respects. Among others it is not known when $K(a_n/1)$ will converge. That

(5.9) $$|a_n - (-1/4)| \leq \frac{1}{4(4n^2-1)} , \quad n \geq 1 ,$$

is sufficient for the convergence of the continued fraction follows from a result of Pringsheim [17,p.94] of 1899. Strangely enough Szasz [26] seems not to have been aware of this since he gave a weaker result around 1920. In another article [33] in these Proceedings we consider the bestness of (5.9) and obtain some partial results.

While in the previous case we were able to show that F_n is bounded for a fixed a, this is not the case here. From Theorem 4.2 (in a sense an improvement of Lemma 3.2 of [30]) it follows that

$$\left| \frac{h_n}{h_n - 1/2} \right| < 2n + 1 .$$

Nevertheless the modification $\{S_n(-1/2)\}$ will converge faster than $\{S_n(0)\}$ at least in three cases which were studied in [30]. The results are as follows.

$$\left| \frac{f - S_n(-1/2)}{f - S_n(0)} \right| \leq \begin{cases} \frac{1-\beta}{1+\beta}\left(1 + \frac{\beta}{n} + \frac{2\beta+1}{2n^2}\right), & \text{for } d_n \leq \frac{1-\beta^2}{4(4n^2-1)} , \\ \frac{4d(n+1)(n+2)}{(n+1)^{\alpha+1} - 2d}, & \text{for } d_n \leq \frac{d}{2n^{\alpha+1}} , \ \alpha > 1 , \ d > 0 , \\ \frac{4(n+1)r^{n+1}(2+r)}{(1-r)(1-4r^{n+1})}, & \text{for } d_n \leq r^n , \ 0 < r < 1 . \end{cases}$$

It is possible to accelerate convergence even more by an iteration of modifications. To accomplish this we [33] derive a continued fraction $b_0 + K(a_n/b_n)$ whose nth approximant $S_n(0)$ is $S_n(w_n)$, where $\{S_n(w_n)\}$ is a modification of the continued fractions $K(a_n/b_n)$. In the article [33], which is a part of these Proceedings, various consequences and applications of this construction are developed for special cases.

There has been little work done so far in the important area of combining truncation error bounds with results on acceleration of convergence. The only case in the literature we know about is given in [29]. Here Stieltjes fractions, that is continued fractions of the form

$$K(b_n z/1) , \ b_n > 0 , \ n \geq 1 ,$$

are studied. The result we are referring to is the following.

<u>Let</u> $K(b_n z/1)$ <u>be a limit periodic Stieltjes fraction with</u> $\lim_{m\to\infty} b_n = n > 0$.

<u>Set</u>

$$g_n = \max_{m \geq n} |b_n - b| .$$

<u>Then for all</u> z , <u>with</u> $|\arg z| < \pi/2$

$$|f(z) - S_n(z,x_1(z))| \leq H(b_1,z) g_n \prod_{m=2}^{n} \left(\frac{1}{\sqrt{k_m} + \sqrt{1+k_m}} \right)^2 .$$

<u>Here</u> $H(b_1,z)$ <u>is a positive quantity depending only on</u> b_1 <u>and</u> z ,

$$x_1(z) = -1/2 + \sqrt{1/4 + bz} ,$$

<u>and</u>

$$k_n = \frac{(\cos(1/2 \arg z))^2}{4b_n |z|} .$$

Since the results for limit periodic continued fractions discussed here are usually only valid if the $d_n = \max_{m \geq n} |a_m - a|$ are sufficiently small, it is true that in most applications these results would apply only to certain tails of the continued fraction under consideration. The question thus arises how can the knowledge we have about the behavior of the tails be utilized to make statements about the speed of convergence of the whole continued fraction, or its modification.

The question is taken up both in [17,Theorem 8.13] and in [30,p166]. Not surprisingly a satisfactory answer depends on knowing the location of h_k for fairly large values of k.

Finally, it would be of interest to know when, if ever, for a given $K(a_n/b_n)$ there is a $v_0 \in \hat{\mathbb{C}} \sim L$, provided $L \neq \hat{\mathbb{C}}$, such that

$$\frac{S_n(v_0) - c}{S_n(v) - c} \to 0$$

for all $v \neq v_0$, $v \in \hat{\mathbb{C}} - L$. It is not even clear that in the limit periodic case $v_0 = x_1$ the attractive fixed point of $w \to \frac{a}{b+w}$.

6. Wrong modification. Background and motivation. Given a periodic continued fraction of period one

(6.1) $$\frac{a}{b} + \frac{a}{b} + \frac{a}{b} + \cdots, \quad a \neq 0 .$$

Let x_1 and x_2 be the fixed points of the linear fractional transformation

(6.2) $$w \to \frac{a}{b+w} ,$$

that is the two roots of the quadratic equation

$$w^2 + bw - a = 0 .$$

It is well known [17, Thm.3.1] that the continued fraction (6.1) converges if

$$|x_1| \neq |x_2| \quad \text{or} \quad x_1 = x_2 ,$$

and diverges if

$$|x_1| = |x_2| \quad \text{and} \quad x_1 \neq x_2 .$$

In case of convergence the value of the continued fraction is equal to the fixed point of smallest absolute value, the attractive fixed point.

Since

$$a = -x_1 x_2 \quad \text{and} \quad b = -(x_1 + x_2)$$

the continued fraction (1) can be written in the following way:

(6.1') $$\frac{-x_1x_2}{-(x_1+x_2)} + \frac{-x_1x_2}{-(x_1+x_2)} + \cdots + \frac{-x_1x_2}{-(x_1+x_2)} + \cdots .$$

From this we immediately see that

(6.3) $$S_n(x_1) = x_1 ,$$

which in case of x_1 being the attractive fixed point, is the value of the continued fraction. Using $S_n(x_1)$ instead of $S_n(0)$ then represents a "short-cut" to the value of the continued fraction. This observation is the basis for the method of using the attractive fixed point in accelerating the convergence of a limit periodic continued fraction, as already described earlier in the present article.

But also if x_1 is the repulsive fixed point, or even in the case $|x_1| = |x_2|$, $x_1 \neq x_2$, (6.3) holds. A natural question to ask is if the property that $\{S_n(x_1)\}$ converges is carried over to limit periodic continued fractions, also when x_1 is not the attractive fixed point. As we shall see later the answer is "yes, under certain strong conditions on the rate at which $a_n \to a$ and $b_n \to b$." (Keep in mind that in case of the attractive fixed point no such condition was needed for $K(a_n/1)$, $a_n \to a$. In the parabolic case $a = -1/4$, however, additional conditions were needed.)

Following the established tradition we shall call $\{S_n(x_1)\}$ the right modification if x_1 is the attractive fixed point of the limiting linear fractional transformation $w \to \frac{a}{b+w}$ and the wrong modification if x_1 is the repulsive fixed point. Also in the parabolic case we shall use the word right modification, whereas in the elliptic case we shall use neither.

In the periodic case $\lim_{n\to\infty} S_n(x_1)$ is not the value of the continued fraction when x_1 is repulsive, and there is no reason to believe that it is the value in the limit periodic case except in very special cases. But what is then the reason for computing such a value? Is the value useful for any purpose having to do with the continued fraction? In order to answer these questions we shall go back (in time) to a special case where a certain convergence problem was handled in a way where fixed points were not mentioned, which, however, turned out to be a "fixed point method" of the type we shall discuss here.

In the paper [34] from 1964 are studied T-fractions

(6.4) $$1 + d_0 z + \underset{n=1}{\overset{\infty}{K}} \left(\frac{z}{1+d_n z} \right) ,$$

introduced by Thron [27] in 1948. It is proved that if a function F is holomorphic in a sufficiently large disk $|z| < R$, $R > 1$, $F(0) = 1$, and

(6.5) $$|F(z) - 1| < K \text{ in } |z| < R$$

for a sufficiently small $K > 0$, then the corresponding T-fraction (6.4) is limit periodic with $d_n \to -1$ as $n \to \infty$. Furthermore, the T-fraction converges to $F(z)$ uniformly on any compact subset of the open unit disk $|z| < 1$. In the first version of the theorem one had to take $R > 2$ and $K < \frac{R}{2} - 1$, but some improvements are made in the paper. On the other hand it is proved that no $R < 1$ works. Hovstad [13] proved in 1975 by a refinement of the method, that any $R > 1$

works when matched with a sufficiently small K. The convergence statement of the theorem, that is that the domain, where the T-fraction converges to $F(z)$, is the open unit disk, can not be improved, as the example $F(z) = 1$ shows. This function satisfies (6.5) with any K in any R-disk, whereas its T-fraction expansion

$$1 - z + \frac{z}{1-z} + \frac{z}{1-z} + \cdots \tag{6.6}$$

converges to $1\ (=F(z))$ in $|z| < 1$ but to $-z(\neq F(z))$ in $|z| > 1$ (in both cases uniformly on compact subsets). The unsatisfactory situation that rather strong conditions in a disk $|z| < R$, $R > 1$, imply convergence properties only in the the unit disk was taken up in the paper [35] from 1966. The basic idea of this paper can be described as follows: Computing the value of (6.4) is to compute $\lim_{n\to\infty} S_n(0)$, where

$$S_n(0) = 1+d_0z + \frac{z}{1+d_1z} + \frac{z}{1+d_2z} + \cdots + \frac{z}{1+d_nz}. \tag{6.7}$$

This means to "cut off the tail".

$$\frac{z}{1+d_{n+1}z} + \frac{z}{1+d_{n+2}z} + \cdots, \tag{6.8}$$

or to "replace the tail by 0" (as one does in computing partial sums of a series). But rather than replacing the tail by 0 it should be replaced by what the tail "looks more and more like", namely

$$\frac{z}{1-z} + \frac{z}{1-z} + \frac{z}{1-z} + \cdots, \tag{6.9}$$

or better: by the value (6.9) converges to in $|z| < 1$, namely z. This leads to the following <u>modified</u> approximant:

$$S_n(z) = 1+d_0z + \frac{z}{1+d_1z} + \frac{z}{1+d_2z} + \cdots + \frac{z}{1+(d_n+1)z} \tag{6.10}$$

(Observe that in the particular case of (6.9) with $1 - z$ in front, all $S_n(z)$ are equal to 1, and hence $S_n(z) \to 1$ <u>in the whole plane</u> in this case.)

It turns out that this modification leads to an increase in the region of convergence to the "right" function F. The first result is that

$$\lim_{n\to\infty} S_n(z) = F(z)$$

uniformly on any compact subset of $|z| < \frac{R}{2}$ (under the same conditions as in the paper [12]). The paper contains an improvement to $|z| < \frac{2}{3}R$, and in 1975 Hovstad [13] extended it to any $R' < R$, again matched with the proper strength of the boundedness condition. See also [37].

In [35] the emphasis was on the extension of the convergence region, and nothing was stated explicitely about acceleration, but it is an immediate consequence of the formulas (2.3), (2.4) and the Lemmas 1 and 2 in [35] that

$$\frac{F(z) - S_n(z)}{F(z) - S_n(0)} \to 0 \quad \text{as} \quad n \to \infty \quad \text{in} \quad |z| < 1.$$

So actually there are two benefits of the transition from $S_n(0)$ to $S_n(z)$: acceleration of convergence and extension of region of convergence.

Fixed points are not mentioned in [35], but as we shall see, the method used is a fixed point method. The limiting linear fractional transformation is (since $d_n \to -1$)

$$w \to \frac{z}{1-z+w} . \tag{6.11}$$

This has the fixed points z and -1 . In $|z| < 1$, z is the attractive and -1 the repulsive fixed point, in $|z| > 1$, -1 is attractive and z repulsive. (For $z = -1$ the transformation is parabolic, for $|z| = 1$, $z \neq -1$ it is elliptic.) Hence the modifications are as follows:

In $|z| < 1$: $S_n(z)$ is the right modification; $S_n(-1)$ is the wrong modification

In $|z| > 1$: $S_n(z)$ is the wrong modification; $S_n(-1)$ is the right modification

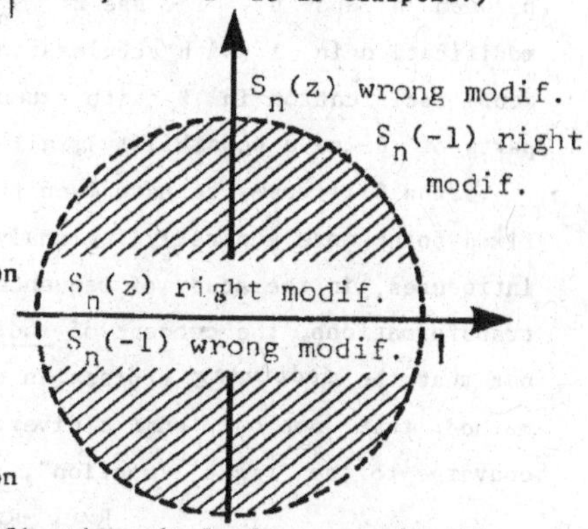

We observe: When $S_n(z)$ is the right modification, it leads to acceleration of convergence, when it is the wrong modification (or a modification that is neither right nor wrong), it leads to analytic continuation of the function represented by the continued fraction (6.4). The use of $S_n(-1)$ for the two purposes is not mentioned in [35], but is briefly touched upon in the paper [36]. There it is proved that if a function G , holomorphic in a domain $|z| > \rho$ $(\rho < \frac{1}{2})$ except for a pole of order 1 at infinity, and subject to a sufficiently strong boundedness condition

$$\left| \frac{G(z)}{z} + 1 \right| < K \quad \text{for} \quad |z| > \rho ,$$

then it has a corresponding T-fraction (6.4) (correspondence at ∞) with $d_n \to -1$, that converges to $G(z)$ uniformly on compact subsets of the upper Riemann hemisphere $|z| > 1$. The use of $S_n(-1)$ in this case will accelerate the convergence to $G(z)$ in $|z| > 1$ and extend the region of convergence to $G(z)$ to $|z| > \rho'$ for some $\rho' \in (\rho , 1)$.

The above discussion of convergence of T-fractions provides an answer to the questions about the purpose of the wrong modification. At least in the special case discussed it increased the region of convergence and led to analytic continuation of the function represented by the continued fraction. In the next section we shall see some newer and more general results on the use of the wrong modification. Observe that (at least so far) the use of the wrong modification heavily depends upon the existence of a variable in the continued fraction. This is not the case for the use of the right modification.

7. Analytic continuation and singularities by use of the wrong modification. In general, if we have a limit periodic continued fraction where the elements are complex functions of a complex variable, the fixed points of the limiting linear fractional transformation, $x_1(z)$ and $x_2(z)$ are functions of that variable. We may then have a situation related to the one described in the preceding section: In one part A of the plane $x_1(z)$ is attractive and $x_2(z)$ repulsive, in another part B it is the opposite. In "nice cases" $|x_1(z)| = |x_2(z)|$ on a separating curve between A and B. The use of $S_n(x_1(z))$ hence represents the right modification in A (with acceleration of convergence as a possible benefit) and the wrong modification in B (with extension of the region of convergence from A to parts of B as a possibility), all under certain more or less strong conditions.

John Gill seems to have been the first one to suggest use of the repulsive fixed point more generally for analytic continuation. In the paper [4] he introduces, in the study of sequences of compositions of linear fractional transformations, the concept of modifying factor (modifying sequence) and points out that the modifying process in some instances may be regarded as a summability method, that "not only sums a divergent continued fraction, but causes it to converge to the 'right' function", and illustrates this with the example

$$\frac{Rz}{R+z} + \frac{-Rz}{R+z} + \frac{-Rz}{R+z} + \cdots, \ R > 0 ,$$

with z as a modifying factor. (This is essentially the same example as the one mentioned in the remark after (6.10).) In the paper [5] he continues along the same line and proves among other things Theorem 1 below for compositions

$$T_n(w) = t_1 \circ t_2 \circ \cdots \circ t_n(w)$$

of non-parabolic linear fractional transformations t_n with finite fixed points α_n, β_n, so that

$$\frac{t_n(w) - \alpha_n}{t_n(w) - \beta_n} = K_n \frac{w - \alpha_n}{w - \beta_n} ,$$

and such that $\alpha_n \to \alpha$, $\beta_n \to \beta$, $K_n \to K$, where α and β are finite and distinct:

THEOREM 7.1. *Suppose* $|K| < 1$, *or* $|K| = 1$, $K \neq 1$ *and* $\prod_{n=1}^{\infty} |K_n| = 0$. *If there exists an* $h_0 > 0$ *such that*

(i) $$\sum_{n=1}^{\infty} |\alpha_n - \alpha_{n-1}| < \infty ,$$

(ii) $$\sum_{n=1}^{\infty} [|\beta_n - \beta_{n-1}| \cdot \prod_{h_0+1}^{n} |K_{j-1}|^{-1}] < \infty ,$$

(iii) $$\lim_{n\to\infty} [\left|\mu_n - \beta_n\right| \cdot \prod_{h_0+1}^{n} \left|K_{j-1}\right|^{-1}] = 0 ,$$

are all satisfied, then $\lim_{n\to\infty} T_n(\mu_n)$ exists.

The relevance of this to our present discussion may be indicated as follows. If we, for simplicity's sake, restrict ourselves to the case $|K| < 1$: α and β are the fixed points of the limit transformation $t = \lim t_n$, α the attractive, β the repulsive one. The conditions (i) and (ii) have to do with how fast one wants α_n to go to α and β_n to β . (Observe the stronger conditions on the repulsive fixed points.) The condition (iii) has to do with the modifying sequence $\{\mu_n\}$ itself. It obviously tends to β , and $T_n(\mu_n)$ may be regarded as a "generalized wrong modification". If the t_n's are linear fractional transformations of the form $a_n/(b_n+w)$, that is, transformations that generate continued fractions, and we take (if possible) $\mu_n = \beta$, we have the wrong modification in the earlier described sense. Based upon this and an example related to the one mentioned earlier from [5,p78] he "suspects that, in the case of certain continued fractions, the use of the repulsive fixed point as a modifying factor might analytically extend a function from Ω into Γ ." (Ω and Γ being the same as A and B in the beginning of our present section.) Gill points out, that in the T-fraction-case described in Section 6 of the present article convergence of $\{S_n(z)\}$ to a continuous function in the annulus $1 < |z| < \frac{R}{2}$ is a simple consequence of Theorem 7.1. One difficulty with Theorem 7.1 is that it, in many cases, cannot "help us across the border" between A and B . In smooth cases the limit transformation is elliptic on the border, and, unless $\prod_{n=1}^{\infty} |K_n| = 0$, this case is not covered in the theorem. A simple illustration is the continued fraction

$$\frac{z}{1-z} + \frac{z}{1-z} + \cdots .$$

For this continued fraction it is easily seen, that with $\mu_n = z$ for all n the conditions of the theorem are satisfied in $|z| > 1$. Hence $\{S_n(z)\}$ converges there. On the other hand the theorem fails to work on $|z| = 1$.

In two articles [31] andd [32] from 1980 and 1981 we studied the wrong modification for limit periodic continued fractions.

(7.1) $$\mathop{K}_{n=1}^{\infty} \left(\frac{a_n}{b_n}\right) , \text{ with } a_n \to a , b_n \to b .$$

The two papers have a different approach. In the paper [31] we do not aim directly at the convergence of $\{S_n(x)\}$, x being a fixed point of the limit transformation, but only at boundedness. This is obtained by putting strong conditions on the rate at which $a_n \to a$ and $b_n \to b$. The following lemma on boundedness is one of the two crucial points in the proof of the results in [31].

LEMMA 7.2. *Set*

$$\delta_n = a_n - a \, , \quad d_n < \max_{v \geq n} |\delta_v| \, ,$$

$$\eta_n = b_n - b \, , \quad e_n > \max_{v \geq n} |\eta_v| \, ,$$

and let x *be one of the solutions of the equation*

$$x^2 + bx - a = 0 \, .$$

Finally, set

$$P = \left| \frac{x}{b+x} \right| < \infty \, .$$

If there exists a Q , $0 < Q < 1$, $QP < 1$, *so that*

$$d_n + (|x| + |b + x|) e_n = k_n Q^n \, , \quad n \geq 1 \, ,$$

where $\sum_{n=1}^{\infty} k_n$ *converges, then there is an* m_0 , *independent of* N , *such that*

$$|S_N^{(m)}(x) - x| \leq \frac{Q^{m+1}}{|b+x|} \cdot \sum_{n=1}^{\infty} k_n \, , \quad m \geq m_0 \, .$$

Observe that this result is not restricted to the repulsive fixed point, but covers the other cases as well. This is important in the application for analytic continuation, since one needs to operate in a connected part of the plane having nonempty intersections with both A and B. (*We even need more than that.*)

The other crucial point is *correspondence*. In the paper the investigation is limited to two types of continued fraction expansions, general T-fractions and C-fractions. In both cases it is proved that the sequence of *modified* approximants corresponds to the formal power series to which the continued fraction in question corresponds.

From the boundedness and the correspondence the convergence follows by using a recent result of Jones and Thron [16] to the effect that if a sequence $\{R_n(z)\}$ of functions holomorphic for $z \in D$ is uniformly bounded on compact subsets of the region D, if further $0 \in D$ and if finally the sequence corresponds to a formal power series L at $z = 0$, then $\{R_n(z)\}$ converges, uniformly on compact subsets of D, to a holomorphic function $f(z)$ with $L = L_0(f)$.

Before stating the main result of [31] for general T-fractions some motivating remarks indicating *why* we are studying such general T-fractions may be of use. We first recall that under certain conditions there exists for a pair (L, L^*) of formal series

$$L = \sum_{k=1}^{\infty} c_k z^k \, , \quad L^* = \sum_{k=-\infty}^{0} c_k z^{-k} \, ,$$

a general T-fraction

$$\mathop{K}_{n=1}^{\infty}\left(\frac{F_n z}{1+G_n z}\right),$$

corresponding to L at 0 and L^* at infinity. If L and L^* happen to be expansions of <u>functions</u> f and g, we shall write $L = L_0(f)$, $L^* = L_\infty(g)$.

In two recent papers [38,39] Waadeland studied the behavior of general T-fractions

$$\mathop{K}_{n=1}^{\infty}\left(\frac{F_n z}{1+G_n z}\right), \quad F_n \neq 0, \; G_n \neq 0, \; n \geq 1, \tag{7.2}$$

and obtained results analogous to the one for ordinary T-fractions. In the first paper he proved that if $L = L_0(f)$, $L^* = L_\infty(g)$, where $f(z)$ and $g(z)$ satisfy certain boundedness conditions at $z = 0$ and $z = \infty$, respectively, then there exists a general T-fraction which corresponds to L at $z = 0$ and to L^* at $z = \infty$. The general T-fraction satisfies

$$\lim_{n\to\infty} F_n = \lim_{n\to\infty} (-G_n) = F \neq 0,$$

and converges to $f(z)$ for $|z| < 1/|F|$ and to $g(z)$ for $|z| > 1/F$.

The second result is the following. Given $1 < R' < R$ there exists a $K(R,R') > 0$ such that if (here we have normalized F to be 1)

$$|F_n - 1| < K/R^n, \quad |G_n + 1| < K/R^n, \quad n \geq 1, \tag{7.3}$$

then the general T-fraction (7.2) whose elements satisfy (7.3) corresponds at $z = 0$ to an $L = L_0(f)$, where $f(z)$ is holomorphic for $|z| < R'$ and corresponds at $z = \infty$ to an $L^* = L_\infty(g)$, where $g(z)$ is holomorphic for $|z| > 1/R'$.

The main result in [31] for general T-fractions is:

<u>THEOREM</u> 7.3. <u>Let the general T-fraction</u>

$$\mathop{K}_{n=1}^{\infty}\left(\frac{F_n z}{1+G_n z}\right), \quad F_n \neq 0, \; G_n \neq 0, \; n \geq 1$$

<u>satisfy</u>

$$|F_n - 1| < K/R^n, \quad |G_n + 1| < K/R^n, \quad n \geq 1,$$

<u>for some</u> $R > 1$ <u>and</u> $K > 0$. <u>Let</u> L <u>and</u> L^* <u>be the formal power series to which the</u> T-<u>fraction corresponds at</u> 0 <u>and</u> ∞, <u>respectively. Then</u>

$$L = L_0(f),$$

<u>where</u> $f(z)$ <u>is meromorphic for all</u> $|z| < R$, <u>and</u>

$$f(z) = \lim_{n\to\infty} S_n(z,z), \quad |z| < R.$$

<u>Similarly,</u>

$$L^* = L_\infty(g) ,$$

<u>where</u> $g(z)$ <u>is meromorphic for</u> $|z| > 1/R$, <u>and</u>

$$g(z) = \lim_{m\to\infty} S_n(z, -1) , \quad |z| > 1/R .$$

For limit periodic regular C-fractions $K(\alpha_n z/1)$, $\alpha_n \to \alpha \neq 0$ we may without loss of generality assume $\alpha = 1$. The two modifications are in this case $S_n(x(z))$, where

$$x(z) = -\frac{1}{2} \pm \sqrt{\frac{1}{4} + z} ,$$

with + for right and - for wrong modification. Here $+\sqrt{v}$ shall mean the root in the right half plane (including the positive imaginary axis but excluding the negative imaginary axis). It is proved that under the conditions

$$(7.5) \qquad |\alpha_n - 1| = \gamma_n Q^n , \; 0 < Q < 1 , \; \sum_{n=1}^{\infty} \gamma_n = D ,$$

"D sufficiently small" we can find a sector with vertex at $z = -1/4$ and with a non-empty intersection with the sets $\{z : \text{Im } z > 0 , \text{Re } z < -1/4\}$ and $\{z : \text{Im } z < 0 , \text{Re } z < -1/4\}$ such that in the sector the modification $\{S_n(x(z))\}$ with + in (7.4) for $\text{Im } z \geq 0$ and - for $\text{Im } z < 0$ converges uniformly on compact subsets of the sector to a holomorphic function. In the upper part of the sector this is the right modification and hence the value of the continued fraction, in the lower half plane it is the analytic continuation of that function. In a similar way we can obtain an analytic continuation "from below" across the ray $x \leq -1/4$ of the negative real axis. The condition "D sufficiently small" can be replaced by $D < \infty$ at the cost of replacing the continued fraction by a tail of the continued fraction, in which case the value of the continued fraction itself is meromorphic. This result in turn leads to one of the rare results on the location of singular points of an analytic function defined in terms of continued fractions: Under the above conditions the meromorphic function to which the continued fraction converges in $\mathbb{C} \sim T$ $(T = \{z : \text{real and } z \leq -1/4\})$ has a branch point of order 1 at $z = -1/4$.

That limit periodicity alone is not sufficient for this conclusion to be valid is illustrated by

$$\log(1+z) = \frac{z}{1 + \mathop{K}\limits_{n=1}^{\infty} (\beta_n z/1)} ,$$

where

$$\beta_{2n+1} = \frac{1}{4} + \frac{1}{8n + 4} , \; n \geqq 0 ,$$

$$\beta_{2n} = \frac{1}{4} - \frac{1}{8n + 4} , \; n \geqq 1 ,$$

which has a logarithmic branch point at $z = -1$.

In the paper [32] the approach is different. The key result there is that if x and $-(b+x)$ are the fixed points of the limit tranformation

$$s(w) = \frac{a}{b+w}$$

and satisfy a condition

$$r < \left| \frac{x}{b+w} \right| < \frac{1}{r}$$

for some positive r, then the conditions

$$|a - a_n| \leq \gamma_n r^n , \; |b - b_n| \leq \gamma_n r^n ,$$
$$\sum_{n=1}^{\infty} \gamma_n \leq D , \; D \text{ sufficiently small,}$$

imply convergence of the sequence $\{S_n(x)\}$. Furthermore an estimate of $\left|\lim_{n\to\infty} S_n(x) - x\right|$ is given. This in turn implies essentially the same results on general T-fractions and regular C-fractions as in the paper [31].

In computing the wrong modification one often runs into trouble because of instability. Under the above conditions only very mild additional conditions are needed to make $\{S_n(w)\}$ tend to the value of the continued fraction for all $w \neq x$ the repulsive fixed point. In trying to compute $S_n(x)$, where x is the repulsive fixed point, a slight roundoff error in x may cause convergence of the process to the value of the continued fraction instead of the "wrong value". In the paper [33] the Examples 5.2 and 5.3 illustrate this.

8. Possible further development. The right modification, as used for convergence acceleration, has been extended to more general cases by L. Jacobsen in [14]. Here a certain sequence $\{f^{(n)'}\}$ of tails takes the place of the attractive fixed point. The following questions arise quite naturally: Is it possible to generalize the wrong modification in a similar way? And will it serve any meaningful purpose? So far nothing has been done towards answering these questions. Before we can indicate any direction in which to go, we must make up our mind about what a wrong modification should mean in a more general setting. One point of view would be to regard $\{S_n(\mu_n)\}$ as a modification for any "useful" sequence $\{\mu_n\}$ (the term useful being left undefined). This is essentially what Gill does in the limit-periodic case in studying rather generally $\{S_n(\mu_n)\}$ for sequences $\{\mu_n\}$ converging to one of the fixed points α or β. Here we shall, however, adopt another point of view, namely the one upon which the generalized right modification in [14] is based, to generate the sequence $\{\mu_n\}$ by using an auxiliary continued fraction. If in the description we stick to the continued fractions $K(a_n/1)$, an auxiliary continued fraction would be $K(a'_n/1)$, where $a_n - a'_n \to 0$ as $n \to \infty$, and where we know "very much" about the auxiliary continued fraction. In [14] μ_n is chosen to be $f^{(n)'}$, the nth tail of the auxiliary continued fraction. This gives the technical advantage of having at our

disposal the recursion formula $f^{(n)'} = a'_{n+1}/(1+f^{(n+1)'})$. It seems to be a good idea to require $\mu_n = g^{(n)'}$, where the g's satisfy the same recursion formula. This is in accordance with the modifications for limit periodic continued fractions, in which case the auxiliary continued fraction is the periodic one with $a'_n = a$ for all n, $a = \lim_{n\to\infty} a_n$. In this case $g^{(n)'}$ is the attractive or repulsive fixed point of the transformation $w \to a/(1+w)$, and in both cases the recursion formula holds. But there is an infinity of sequences for which the recursion formula holds. Which of them should be chosen to make a meaningful wrong modification? Here we can only give a partial answer by means of examples. Out of several we shall restrict ourselves to two.

1. The continued fraction $K(a_n/1)$ is "near" a k-periodic continued fraction $K(a'_n/1)$ in the sense that $a_n - a'_n \to 0$ as $n \to \infty$. Take the latter to be the auxiliary continued fraction. (We assume that it converges, and also that the quadratic equation that determines its value has distinct roots.) The right modification is given by taking μ_n equal to the nth tail, the wrong one by taking μ_n equal to the nth "wrong tail". The right and wrong tails are the two roots of the quadratic equation determining the value of the nth tail. Which one is the right and which one is the wrong one is told in [17, Thm. 3.1]. For $k = 2$ it is easy to see, that the continued fraction can be written in the form

$$\frac{-xy}{1} + \frac{-(x+1)(y+1)}{1} + \frac{-xy}{1} + \frac{-(x+1)(y+1)}{1} + \cdots$$

Its value, if it converges, is x or y. Assume it is x. Then for the right modification we have to use for $\{\mu_n\}$ the sequence

$$x\ ,\ -(1+y)\ ,\ x\ ,\ -(1+y)\ ,\cdots$$

and for the wrong modification

$$y\ ,\ -(1+x)\ ,\ y\ ,\ -(1+x)\ ,\cdots$$

in both cases starting with μ_0. (In [5] useful relations between right and wrong tails of k-periodic continued fractions are given.) It is likely that under certain conditions on the rate at which $a_n - a'_n \to 0$ the wrong modification will converge. It is also likely that for regular C-fractions of this type ("near" a k-periodic regular C-fraction) similar results on analytic continuation and singularities as in the limit periodic case can be proved.

2. For ratios of certain hypergeometric functions we have limit-periodic C-fraction expansions. The C-fraction expansion $K(\beta_n z/1)$ of $\log(1+z)$ may serve as an example. The tails are also ratios of hypergeometric functions and the "wrong tails" are their analytic continuations. This is a simple consequence of th recursion formulas (6.1.7a) and (6.1.7b) in [17]. For a C-fraction $K(\alpha_n z/1)$ suc that $\alpha_n - \beta_n \to 0$ sufficiently fast (geometric at a certain rate may suffice?) we use $K(\beta_n z/1)$ as an auxiliary continued fraction and the sequence of "first wrong tails", the sequence of "second wrong tails" and so forth as the sequence $\{\mu\}$. It seems likely that we can obtain results of the following type: A C-fraction

"sufficiently close to" the C-fraction of $\log(1+z)$ represents a function with a logarithmic singularity at $z = -1$.

References

1. J. D. De Pree and W. J. Thron, On sequences of Moebius transformations, Math. Zeitschr. 80 (1962), 184-193.
2. John Gill, Infinite compositions of Möbius transformations, Trans. Amer. Math. Soc. 176(1973) 479-487.
3. John Gill, The use of attractive fixed points in accelerating the convergence of limit periodic continued fractions, Proc. Amer. Math. Soc. 47 (1975) 119-126.
4. John Gill, Modifying factors for sequences of linear fractional transformations, Kgl. Norske Vid. Selsk. Skr. (Trondheim) (1978) No. 3, 1-7.
5. John Gill, Enhancing the convergence region of a sequence of bilinear transformations, Math. Scand. 43 (1978), 74-80.
6. John Gill, Convergence acceleration for continued fractions $K(a_n/1)$ with $\lim a_n = 0$, these Lecture Notes.
7. John Gill, Truncation error analysis for continued fractions $K(a_n/1)$ where $\sqrt{|a_n|} + \sqrt{|a_{n-1}|} < 1$, these Lecture Notes.
8. J. W. L. Glaisher, On the transformation of continued products into continued fractions, Proc. London Math. Soc. 5(1873/4), 85.
9. H. Hamburger, Über eine Erweiterung des Stieltjesschen Momentenproblems I, Math. Ann. 81(1920) 235-319.
10. G. Hamel, Eine charakteristische Eigenschaft beschränkter analytischer Funktionen, Math. Ann. 78 (1918), 257-269.
11. G. Hamel, Über einen limitärperiodischen Kettenbruch, Arch. d. Math. u. Phys. 27 (1918) 37-43.
12. T. L. Hayden, Continued fraction approximation to functions, Numer. Math. 7 (1965) 292-309.
13. Rolf M. Hovstad, Solution of a convergence problem in the theory of T-fractions, Proc. Amer. Math. Soc. 48 (1975) 337-343.
14. Lisa Jacobsen, A method for convergence acceleration of continued fractions $K(a_n/1)$, these Lecture Notes.
15. Lisa Jacobsen, Some periodic sequences of circular convergence regions, these Lecture Notes.
16. William B. Jones and W. J. Thron, Sequences of meromorphic functions corresponding to a formal Laurent series, SIAM J. Math. Anal. 10 (1979), 1-17.
17. William B. Jones and W. J. Thron, Continued Fractions: Analytic theory and applications, Encyclopedia of Mathematics and its Applications vol 11, Addison Wesley, Reading, Mass. 1980.
18. M. Mandell and Arne Magnus, On convergence of sequences of linear fractional transformations, Math. Zeitschr. 115 (1970), 11-17.
19. Oskar Perron, Über einen Satz des Herrn Poincaré, J. reine angew. Math. 136 (1909), 17-37.
20. T. E. Phipps, Jr., A continued fraction representation of eigenvalues, SIAM Rev. 13 (1971), 390-395.

21. T. E. Phipps, Jr., A new approach to evaluation of infinite processes, NOLTR 71-36, Naval Ordnance Laboratory, White Oak, Silver Spring, Maryland 1971.

22. S. Pincherle, Sur la génération de systemes récurrents au moyen d'une equation linéaire differentielle, Acta Math. 16 (1893), 341-363.

23. George Piranian and W. J. Thron, Convergence properties of sequences of linear fractional transformations, Michigan Math. J. 4 (1957) 129-135.

24. Henri Poincaré, Sur les equations linéaires aux différentielles ordinaires et aux différences finies, Amer. J. Math. 7 (1885), 203-258.

25. I. Schur, Über Potenzreihen die im Innern des Einheitskreises beschränkt sind, J. reine angew. Math. 147 (1917) 205-232.

26. O. Szasz, Collected mathematical papers, Cincinnati, 1955.

27. W. J. Thron, Some properties of continued fractions $1 + d_0z + K(z/(1+d_nz))$, Bull. Amer. Math. Soc. 54 (1948), 206-28.

28. W. J. Thron, Convergence of sequences of linear fractional transformations and of continued fractions, J. Indian Math. Soc. 27 (1963), 103-127.

29. W. J. Thron, A priori truncation error estimates for Stieltjes fractions, Christoffel Memorial volume, Birkhäuser, Basel (1981).

30. W. J. Thron and Haakon Waadeland, Accelerating convergence of limit periodic continued fractions $K(a_n/1)$, Numer. Math. 34 (1980) 155-170.

31. W. J. Thron and Haakon Waadeland, Analytic continuation of functions defined by means of continued fractions, Math. Scand. 47 (1980) 72-90.

32. W. J. Thron and Haakon Waadeland, Convergence questions for limit periodic continued fractions, Rocky Mountain J. Math. 11 (1981), 641-657.

33. W. J. Thron and Haakon Waadeland, On a certain transformation of continued fractions, these Lecture Notes.

34. Haakon Waadeland, On T-fractions of functions, holomorphic and bounded in a circular disk, Norske Vid. Selsk. Skr. (Trondheim)(1964), No. 8, 1-19.

35. Haakon Waadeland, A convergence property of certain T-fraction expansions, Norske Vid. Selsk. Skr. (Trondheim) (1966), No. 9, 1-22.

36. Haakon Waadeland, On T-fractions of certain functions with a first order pole at the point of infinity, Norske Vid. Selsk. Forh. 40 (1967), No. 1.

37. Haakon Waadeland, T-fractions from a different point of view, Rocky Mountain J. Math. 4 (1974) 391-393.

38. Haakon Waadeland, General T-fractions corresponding to functions satisfying certain boundedness conditions, J. Approximation Theory 26 (1979) 317-328.

39. Haakon Waadeland, Limit periodic general T-fractions and holomorphic functions, J. Approximation Theory 27 (1979) 329-345.

40. H. Weyl, Über gewöhnliche Differentialgleichungen mit Singularitäten und die zugehörigen Entwicklungen willkürlicher Funktionen, Math. Ann. 68 (1910) 220-269.

41. P. Wynn, Converging factors for continued fractions, Numer. Math. 1 (1959) 272-320.

W.J. Thron
Department of Mathematics
Campus Box 426
University of Colorado
Boulder, Colorado 80309
U.S.A.

Haakon Waadeland
Institutt for Matematikk og Statistilsk
Universitetet i Trondheim
7055 Dragvoll
Norway

CONVERGENCE ACCELERATION FOR CONTINUED FRACTIONS $K(a_n/1)$ WITH $\lim a_n = 0$

John Gill

Converging factors, $\{\mu_n\}$, of the continued fraction

(1) $$\frac{a_1}{1}\,+\,\frac{a_2}{1}\,+\cdots+\,\frac{a_n}{1}\,+\cdots,$$

where each a_n is a non-zero complex number, are complex numbers such that

$$\lim_{n\to\infty} S_n(\mu_n) = \lim_{n\to\infty} S_n(0) = f ,$$

provided the latter limit exists.

Various investigations have revealed that in the case in which (1) is limit-periodic (i.e., $\lim a_n = a$) and $a \neq 0$, convergence can be accelerated by the judicious application of converging factors. See, for instance, [1], [2], [3], and [4].

Thron and Waadeland [3] developed

Theorem 1. *Let* $a_n \to a \neq 0$, $|\arg(a + \frac{1}{4})| < \pi$. *Assume that for all* $n \geq 1$,

$$|a_n - a| \leq \min \{\tfrac{1}{2}(|a + \tfrac{1}{4}| + \tfrac{1}{4} - |a|), |a|/2\} .$$

Set $d_n = \max_{m\geq n}|a_m - a|$, $a = \alpha(\alpha+1)$, *where* $|\alpha| < |\alpha+1|$. *Then*

$$\left|\frac{f - S_n(\alpha)}{f - S_n(0)}\right| \leq 2\, d_n \frac{|a| + |\frac{1}{2} + a + \sqrt{\frac{1}{4} + a}|}{|a|(\frac{1}{4} + |\frac{1}{4} + a| - |a|)} , \quad \mathrm{Re}(\sqrt{\tfrac{1}{4} + a}) > 0 .$$

Thus, under certain conditions, $\mu_n \equiv \alpha$ guarantees an acceleration of convergence of (1) when $a \neq 0$.

The following theorem [2] provides a fundamental sufficiency condition with regard to the selection of converging factors of limit-periodic fractions of the form (1).

Theorem 2. *Let* $\lim a_n = a = \alpha(\alpha + 1)$, *where* $|\alpha| < |\alpha + 1|$. *If* $\lim |\mu_n - (\alpha + 1)| > 0$, *then* $\mu_1, \mu_2, \cdots$ *are converging factors of* (1).

If $a = 0$, then $\alpha = 0$ and no benefit ensues if $\mu_n \equiv \alpha$. However, there are circumstances in which the choice $\mu_n = \alpha_{n+1}$, $n \geq 1$, is of value. Here, $a_n = \alpha_n(\alpha_n + 1)$, $|\alpha_n| < |\alpha_n + 1|$. Theorem 2 guarantees $\lim S_n(\alpha_{n+1}) = \lim S_n(0) = f$. If $\alpha_n \to \alpha = 0$ somewhat reluctantly (e.g., $|\alpha_n - \alpha_{n+1}| < |\alpha_n|$), then f lies in a small disc, D_ε, centered at α_1, [2]. In addition, $S_1(\alpha_2)$ may lie in D_ε, closer to f than α_1, whereas $S_1(0)$ may lie outside D_ε. In the pursuit of this geometrical idea through the chain of S_n-compositions, one is led to the following theorem.

<u>Theorem</u> 3. <u>If</u> (i) $\max_{m \geq n} \left|\alpha_m - \alpha_{m+1}\right| \leq \varepsilon_n \left|\alpha_{n+1}\right|$, $n = 1,2,\cdots$, <u>where</u> $0 \leq \varepsilon_n \leq 1$, <u>and</u> (ii) $0 < \left|\alpha_m\right| < \sigma_n \leq \frac{1}{5}$, $m \geq n$, $n \geq 1$, <u>are satisfied, then</u>

$$\left|S_n(\alpha_{n+1}) - f\right| < \frac{\sigma_n \varepsilon_n}{(1-5\sigma_n)^2} \cdot \left|S_n(0) - f\right| ,$$

<u>where</u> $\lim_{n\to\infty} \sigma_n = 0$.

Proof: With h_n as defined in (DN 10) the following equation is easily obtained (see [3]).

$$\left|\frac{f - S_n(\alpha_{n+1})}{f - S_n(0)}\right| = \left|\frac{f^{(n)} - \alpha_{n+1}}{f^{(n)}}\right| \cdot \left|\frac{h_n}{h_n + \alpha_{n+1}}\right| . \tag{2}$$

Let us first consider the expression $\left|f^{(n)} - \alpha_{n+1}\right|$ in (2). Set

$$\rho_{m-1} = f^{(m-1)} - \alpha_m,\ d_m = \left|\left|\alpha_{m+1} + 1\right| - \left|\alpha_m\right|\right| ,$$

$$g_m = \left|\alpha_m(\alpha_m - \alpha_{m+1})\right| ,\ D_n = \min_{m \geq n} d_m ,\ \text{and}\ G_n = \max_{m \geq n} g_m$$

$$m \geq 1 ,\ n \geq 1 .$$

Then

$$\left|\rho_{m-1}\right| = \left|\frac{\alpha_m(\alpha_m+1)}{1 + f^{(m)}} - \alpha_m\right| \leq \frac{\left|\alpha_m\right|(\left|\alpha_m - \alpha_{m+1}\right| + \left|\rho_m\right|)}{\left|1 + \alpha_{m+1}\right| - \left|\rho_m\right|} . \tag{3}$$

As in [3], we wish to find $R_n > 0$ such that $\left|\rho_m\right| \leq R_n$ for $m \geq n$. Assuming $\left|\rho_m\right| \leq R_n$ in (3), we have

$$\left|\rho_{m-1}\right| \leq \frac{\left|\alpha_m\right|(\left|\alpha_m - \alpha_{m+1}\right| + R_n)}{\left|1 + \alpha_{m+1}\right| - R_n} .$$

The expression on the right is $\leq R_n$ provided $g_m \leq d_m R_n - R_n^2$. Since $g_m \leq G_n$ and $D_n \leq d_m$ for $m \geq n$, $g_m \leq d_m R_n - R_n^2$ if $G_n \leq D_n R_n - R_n^2$. This last last inequality is satisfied if $R_n = G_n D_n/(D_n^2 - 2G_n)$, as can be routinely verified by showing that $\left|\alpha_n\right| < \frac{1}{5}$ implies $4G_n < D_n^2$.

Now, $\lim_{n\to\infty} f^{(n)} = 0$ and $\lim_{n\to\infty} \alpha_{n+1} = 0$ imply $\lim_{m\to\infty} \rho_m = 0$. Hence, there exists $k > 0$ for fixed m and n ($m \geq n$) such that $\left|\rho_{m+k}\right| \leq R_n$. Then $\left|\rho_{m-1}\right| \leq R_n$; i.e., $\left|f^{(m-1)} - \alpha_m\right| \leq R_n$, $m \geq n$.

Turning now to the first factor in the right side of (2) , one obtains

$$\left|\frac{f^{(n)} - \alpha_{n+1}}{f^{(n)}}\right| = \frac{1}{\left|\dfrac{\alpha_{n+1}}{f^{(n)} - \alpha_{n+1}} + 1\right|} \leq \frac{1}{\left|\dfrac{\alpha_{n+1}}{f^{(n)} - \alpha_{n+1}}\right| - 1} .$$

Clearly

$$\left|\frac{f^{(n)} - \alpha_{n+1}}{\alpha_{n+1}}\right| \leq \frac{R_n}{|\alpha_{n+1}|}$$

$$\leq \frac{\max_{m\geq n}|\alpha_m - \alpha_{m+1}|}{|\alpha_{n+1}|} \cdot \max_{m\geq n}|\alpha_m| \cdot \frac{D_n}{D_n^2 - 2G_n}$$

$$< \varepsilon_n \sigma_n \frac{1}{1 - 4\sigma_n},$$

since $1 - 2\sigma_n \leq D_n \leq 1$ and $G_n \leq 2\sigma_n^2$.

Therefore,

(4) $$\left|\frac{f^{(n)} - \alpha_{n+1}}{f^{(n)}}\right| < \frac{\varepsilon_n \sigma_n}{1 - 5\sigma_n}.$$

Inverting the second factor in (2), one has

(5) $$\left|\frac{h_n + \alpha_{n+1}}{h_n}\right| \geq 1 - \frac{|\alpha_{n+1}|}{|h_n|} \geq 1 - 2|\alpha_{n+1}| > 1 - 2\sigma_n,$$

since $|h_n| \geq \frac{1}{2}$ in the Worpitzky circle, $a_n \in \{z: |z| < \frac{1}{4}\}$, ([5], page 60), and condition (ii) implies $|a_n| < \frac{1}{4}$. Combining (4) and (5) gives the conclusion of Theorem 3.

Example. Let $|\alpha_1| < 10^{-3}$ and $\alpha_n = (.52)^{n-1}\alpha_1$ for $n \geq 2$. Then $\varepsilon_n < 9.3 \times 10^{-1}$ and $\sigma_n = (.52)^{n-1} \times 10^{-3}$ for $n \geq 2$. Theorem 3 gives, e.g.,

$$|S_2(\alpha_3) - f| < 4.9 \times 10^{-4}|S_2(0) - f|,$$

and

$$|S_{10}(\alpha_{11}) - f| < 2.6 \times 10^{-6}|S_{10}(0) - f|.$$

In general, if the α_n's are quite small, then an improvement on the order of magnitude of $|\alpha_1|$ occurs in the first calculation.

References

1. J. Gill, Modifying Factors for Sequences of Linear Fractional Transformations, K. Norske Vidensk. Selsk. Skr. (1978) No. 3, 1-7.

2. J. Gill, The Use of Attractive Fixed Points in Accelerating the Convergence of Limit Periodic Continued Fractions, Proc. Amer. Math. Soc. 47 (1975), 119-126.

3. W. J. Thron and Haakon Waadeland, Accelerating Convergence of Limit Periodic Continued Fractions $K(a_n/1)$, Numer. Math. 34 (1980), 155-170.

4. Haakon Waadeland, A Convergence Property of Certain T-Fraction Expansions, K. Norske. Vidensk. Selsk. Skr. (1966). No. 9, 1-22.

5. H. S. Wall, Analytic Theory of Continued Fractions, D. Van Norstrand, New York (1948).

John Gill
Department of Mathematics
University of Southern Colorado
Pueblo, Colorado 81001.

TRUNCATION ERROR ANALYSIS FOR CONTINUED FRACTIONS $K(a_n/1)$, WHERE $\sqrt{|a_n|} + \sqrt{|a_{n-1}|} < 1$

John Gill

The value of the convergent continued fraction

(1) $$\frac{a_1}{1+}\frac{a_2}{1+}\cdots+\frac{a_n}{1+}\cdots$$

where the a_n's are complex numbers, may be written

$$\lim_{n\to 0} s_1 \circ s_2 \circ \cdots \circ s_n(0) = f .$$

Each linear fractional transformation, $s_n(z) = a_n/(1+z)$, $n \geq 1$, has an <u>isometric circle</u>:

$$I_n = \{z: |z+1| \leq \sqrt{|a_n|}\},$$

([1], pages 23-27). These circles can be used to develop simple truncation error estimates for the approximants of (1), under conditions appearing in a classical theorem of Pringsheim ([2], page 258).

<u>Converging factors</u> of (1) are complex numbers $\mu_1, \mu_2, \ldots$, such that $\lim S_n(\mu_n) = f$. The geometrical analysis leading to error estimates of (1), in addition, reveals a large class of converging factors, some of which improve convergence of (1).

<u>Theorem</u> 1. <u>If</u>

(i) $$\sqrt{|a_n|} + \sqrt{|a_{n-1}|} \leq 1 , \ n \geq 2 ,$$

<u>and</u>

(ii) $$\prod_{j=2}^{n} \frac{\sqrt{|a_{j-1}|}}{1-\sqrt{|a_j|}} \to 0$$

<u>are satisfied then:</u>

(A) (1) <u>converges</u>.

(B) <u>Complex numbers</u> μ_n <u>satisfying</u> $|\mu_n + 1| > \sqrt{|a_n|}$ <u>are converging factors of</u> (1).

(C)

$$|S_n(\mu_n) - f| < 2\sqrt{|a_n|} \prod_{j=2}^{n} \frac{|a_{j-1}|}{(1-\sqrt{|a_j|})^2} , \ n \geq 2 .$$

<u>Proof:</u> I_n is the isometric circle of $s_n(z)$ ([1], Pages 23-27). Its radius is $\sqrt{|a_n|}$. Geometrically, the action of s_n upon z consists of (a) an inversion in I_n, (b) a reflection in the line $\{z:\mathrm{Re}(z) = -1/2\}$, and, usually, (c) a rotation about $z = 0$.

If $\mu_n \notin I_n$ then $s_n(\mu_n)$ lies in the inverse circle $I'_n = \{z: |z| \leq \sqrt{|a_n|}\}$, because of (a). The condition (i) guarantees that $s_n(\mu_n) \notin I_{n-1}$ and, more generally, that $I'_n \cap I_{n-1} = \phi$.

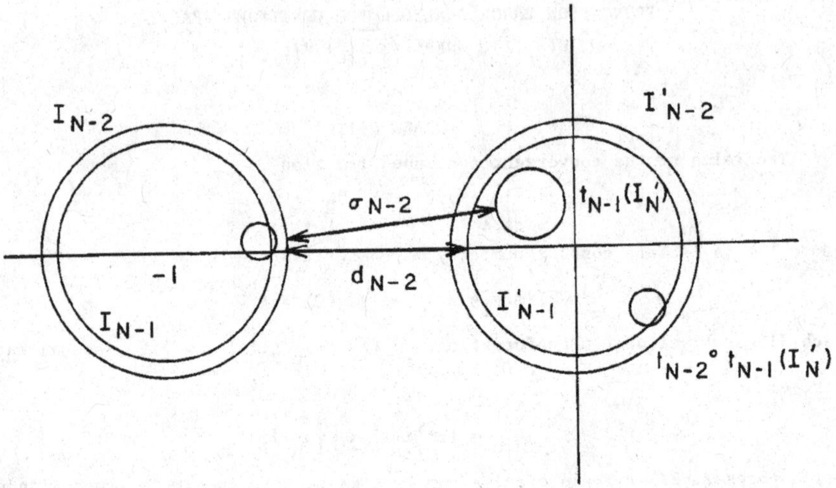

Now, both $s_n(0)$ and $s_n(\mu_n)$ lie in I'_n. Hence, $s_{n-1}\circ s_n(0)$ and $s_{n-1}\circ s_n(\mu_n)$ lie in I'_{n-1}. The image of the circle I'_n under s_{n-1} is a circle $s_{n-1}(I'_n)\subset I'_{n-1}$, (see Fig.). Similarly, $s_{n-2}\circ s_{n-1}(I'_n)$ is a circle contained in I'_{n-2}. From (a), we have

$$\mathrm{Rad}[s_{n-2}\circ s_{n-1}(I'_n)]$$

$$= \mathrm{Rad}[s_{n-1}(I'_n)]\cdot\frac{(\sqrt{|a_{n-2}|})^2}{(\sqrt{|a_{n-2}|}+\sigma_{n-2})(\sqrt{|a_{n-2}|}+\sigma_{n-2}+2\mathrm{Rad}[s_{n-1}(I'_n)])}$$

$$< \mathrm{Rad}[s_{n-1}(I'_n)]\cdot\frac{|a_{n-2}|}{(\sqrt{|a_{n-2}|}+\sigma_{n-2})^2}$$

$$\leq \mathrm{Rad}[s_{n-1}(I'_n)]\cdot\frac{|a_{n-2}|}{(\sqrt{|a_{n-2}|}+d_{n-2})^2}$$

$$= \mathrm{Rad}[s_{n-1}(I'_n)]\cdot\frac{|a_{n-2}|}{(1-\sqrt{|a_{n-1}|})^2}.$$

Repeating this process gives

$$\mathrm{Rad}[S_{n-1}(I'_1)] < \mathrm{Rad}[s_{n-1}(I'_1)]\prod_{j=2}^{n-1}\frac{|a_{j-1}|}{(1-\sqrt{|a_j|})^2}.$$

Since $$\mathrm{Rad}[s_{n-1}(I'_n)] < \sqrt{|a_n|}\cdot\frac{|a_{n-1}|}{(1-\sqrt{|a_n|})^2}\text{, we have}$$

$$\mathrm{Rad}[S_{n-1}(I'_n)] < \sqrt{|a_n|} \cdot \prod_{j=2}^{n} \frac{|a_{j-1}|}{(1-\sqrt{|a_n|})^2} .$$

$$S_n(\mu_n) \in S_{n-1}(I'_n) , \; S_n(0) \in S_{n-1}(I'_n) , \; S_n(I'_{n+1}) \subset S_{n-1}(I'_n) ,$$

and $\mathrm{Rad}[S_{n-1}(I'_n)] \to 0$ as $n \to \infty$ imply

$$\lim_{n\to\infty} S_n(\mu_n) = \lim_{n\to\infty} S_n(0) = f \in S_{m-1}(I'_m) , \; m > 1 .$$

This concludes the proof of Theorem 1.

Consider a_n's of the form $a_n = \alpha_n(\alpha_{n+1})$, where $|\alpha_n| < |\alpha_n + 1|$ for $n \geq 1$.

In addition, we shall refer to a complex number, $a \neq 0$, in the vicinity of the a_n's, and shall set $a = \mu(\mu + 1)$, where $|\mu| < |\mu + 1|$.

The following minor theorem (whose proof is omitted) shows that some of the converging factors displayed in Theorem 1 actually improve convergence.

Theorem 2. If (i) there exists a fixed $\eta > 0$ such that $|\alpha_n| + \eta < \frac{1}{5}$ for $n \geq 1$, (ii) there exists a complex number $a \neq 0$ such that $|\mu| < \frac{1}{5}$, (iii) $\sup_{m \geq n} |a_m - a| < \varepsilon_n < \frac{3|\mu|}{10(1 + |\mu|^{-1})}$, $n \geq 1$, then μ is a converging factor of (1) and

$$|S_n(\mu) - f| < \frac{25}{3} \varepsilon_n |\mu|^{-2} \cdot |S_n(0) - f| \quad \text{for } n \geq 1 .$$

Example. Consider (1) with $|a_n - \frac{1}{6}| < 10^{-k}$, $k > 2$; then $\mu = \frac{\sqrt{15} - 3}{6}$, and Theorem 2 gives

$$|S_n(\mu) - f| < 4 \times 10^{2-k} \cdot |S_n(0) - f| , \; n \geq 1 .$$

If $\alpha_n \to \mu$, then Theorem 2 implies actual acceleration of the convergence process. In general, significant improvement in convergence occurs if $\varepsilon_n \ll |\mu|^2$.

References

1. L. Ford, Automorphic Functions, 2nd Ed., Chelsea, New York (1951).
2. O. Perron, Die Lehre von den Kettenbrüchen, 3. Aufl. 2. Band, B.G. Teubner, Stuttgart (1957).

John Gill
Department of Mathematics
University of Southern Colorado
Pueblo, Colorado 81001

A METHOD FOR

CONVERGENCE ACCELERATION OF

CONTINUED FRACTIONS $K(a_n/1)$

Lisa Jacobsen

1. Introduction. A continued fraction

(1.1) $$K\left(\frac{a_n}{1}\right) = \frac{a_1}{1+}\,\frac{a_2}{1+}\,\frac{a_3}{1+}\cdots$$

is said to converge, if its sequence of approximants $\{f_n\}$ converges, possibly to ∞. Here

(1.2a) $$f_n = \frac{a_1}{1+}\,\frac{a_2}{1+}\cdots+\frac{a_n}{1} = S_n(0) \quad \text{, where}$$

(1.2b) $$S_n(w) = \frac{a_1}{1+}\,\frac{a_2}{1+}\cdots+\frac{a_n}{1+w} \quad ; \quad n = 1,2,3,\ldots$$

The value of $K(a_n/1)$ is then $f = K(a_n/1) = \lim_{n\to\infty} f_n$. In some special cases this value is easy to determine, as for instance in the periodic case, or for some continued fraction expansions of known functions. But in most cases one has to compute the value numerically; i.e. to compute $S_n(0)$ for a sufficiently large n. (n can be determined by truncation error estimates.)

Since the work required to compute f_n increases with n, it will in certain cases be of importance to find a sequence $\{f_n^*\}$ that converges substantially faster to f than $\{f_n\}$; that is $|f-f_n^*|/|f-f_n| \to 0$.

This is not a new idea. In 1873, Glaisher [4] found that for a certain continued fraction $K(a_n/1)$ where $a_n \to \infty$, the sequence $\{S_n(x_n)\}$ converged much faster to the value f of $K(a_n/1)$ than $\{S_n(0)\}$, for an appropriate choice of $\{x_n\}$. In 1959, Wynn [15] extended this idea to other convergent continued fractions $K(a_n/1)$. In 1965 Hayden [5] presented methods of finding appropriate $\{x_n\}$ in some of these cases. Beginning in 1973, Gill [1, 2, 3] studied, among other things, the convergence of limit periodic continued fractions $K(a_n/1)$, $a_n \to a$. He observed that $\{S_n(x_1)\}$, where $x_1 = K(a/1)$, in some cases converged faster to $f = K(a_n/1)$ than $\{S_n(0)\}$. In 1980, Thron and Waadeland [14] proved that

$$\lim_{n\to\infty}\left|\frac{f-S_n(x_1)}{f-S_n(0)}\right| = 0$$

for any continued fraction $K(a_n/1)$ where $a_n \to a \in \mathbb{C} - (-\infty,-\frac{1}{4}]$, and for some $K(a_n/1)$ where $a_n \to -\frac{1}{4}$. They also quantified the improvement by deriving upper bounds for $|f-S_n(x_1)|/|f-S_n(0)|$.

By (1.1) and (1.2b) we see that $f = S_n(f^{(n)})$ where

$$f^{(n)} = \frac{a_{n+1}}{1+}\,\frac{a_{n+2}}{1+}\cdots = \mathop{K}_{m=n+1}^{\infty}\left(\frac{a_m}{1}\right)$$

is the nth tail of the continued fraction $K(a_n/1)$. Hence, it seems feasible to try $f_n^* = S_n(f^{(n)'})$ where $f^{(n)'}$ for $n = 0,1,2,...$ are the known tails of an auxiliary continued fraction $K(a_n'/1)$ such that the elements a_n and a_n' satisfy: $(a_n - a_n') \to 0$. In this article we prove (Theorem 2.4) that the sequence $\{S_n(f^{(n)'})\}$ under certain mild conditions, converges substantially faster to f than $\{f_n\}$, that is

$$\lim_{n\to\infty}\left|\frac{f-S_n(f^{(n)'})}{f-S_n(0)}\right| = 0 \tag{1.3}$$

Therefore $S_n(f^{(n)'})$ will in most cases approximate f with the desired accuracy for a smaller value of n than $f_n = S_n(0)$.

<u>Example</u> 1.1. The continued fraction of period five given by

$$K\left(\frac{a_n'}{1}\right) = \frac{8}{1} + \frac{12}{1} + \frac{8}{1} + \frac{6}{1} + \frac{11}{1} + \cdots$$

has the following tail values:

$$f^{(5n)'} = (\sqrt{11257} - 35)/33 \qquad f^{(5n+1)'} = (\sqrt{11257} - 3)/38$$

$$f^{(5n+2)'} = (3\sqrt{11257} - 65)/74 \qquad f^{(5n+3)'} = (3\sqrt{11257} - 99)/164$$

$$f^{(5n+4)'} = (\sqrt{11257} + 2)/31$$

$$n = 0,1,2,3,... \quad .$$

This continued fraction may be used as a tool for improving the convergence of the continued fraction $K(a_n/1) = K\left(\frac{a_n'+0.3^n}{1}\right)$. Table 1.1 shows the improvement obtained by using the modified approximants $S_n(f^{(n)'})$.

Table 1.1.

n	$S_n(0)$	$S_n(f^{(n)'})$
1	8.3	2.23
2	0.6	2.223
5	2.8	2.2265
10	2.13	2.22649360
20	2.2236	2.22649361320389
24	2.225	2.22649361320389408
50	2.2264935	
75	2.22649361321	
100	2.226493613203892	
117	2.22649361320389408	

We find that $S_n(f^{(n)'})$ gives the value of f with 8 significant digits when $n = 10$, and with 18 significant digits when $n = 24$, whereas the same accuracy in $S_n(0)$ requires $n = 50$ and $n = 117$, respectively. (This computation was done on UNIVAC 1108, RUNIT, The University of Trondheim.)

2. Sufficient conditions. In this section we try to answer the question: under what conditions will (1.3) be satisfied? When A_n, B_n denote the nth numerator and denominator of the continued fraction $K(a_n/1)$ respectively; and $S_n(w)$ is as defined by (DN 1) we get

$$|f - S_n(w_n)| = |S_n(f^{(n)}) - S_n(w_n)| = \frac{|A_{n-1}B_n - A_nB_{n-1}|\,|f^{(n)} - w_n|}{|B_n + B_{n-1}f^{(n)}|\,|B_n + B_{n-1}w_n|}$$

for any sequence $\{w_n\}$ from $\mathbb{C}$. Therefore

(2.1) $$\left|\frac{f - S_n(f^{(n)'})}{f - S_n(0)}\right| = \left|\frac{B_n}{B_n + B_{n-1}f^{(n)'}}\right|\left|\frac{f^{(n)} - f^{(n)'}}{f^{(n)}}\right| = \left|\frac{h_n}{h_n + f^{(n)'}}\right|\left|\frac{f^{(n)} - f^{(n)'}}{f^{(n)}}\right|$$

where, as in (DN 10)

$$h_n = \frac{B_n}{B_{n-1}} = 1 + \frac{a_n}{1 +}\frac{a_{n-1}}{1} + \cdots + \frac{a_2}{1} \quad ; \quad n = 1,2,3,\ldots$$

$(\overset{b}{\underset{n=a}{K}} (a_n/1) = 0$ when $b < a)$

Hence, we see at once that requiring

(2.3a) $$\frac{|h_n|}{|h_n + f^{(n)'}|\,|f^{(n)}|} \le M \quad ; \quad n = 1,2,3\ldots \text{ for some } M < \infty \text{ and}$$

(2.3b) $$\lim_{n\to\infty} |f^{(n)} - f^{(n)'}| = 0$$

is sufficient to satisfy (1.3). Using this idea, we get Theorem 2.4. Before stating this theorem, we make the following definitions:

Definition 2.1. We say that $\{E_n\}_{n=1}^{\infty}$; $\emptyset \ne E_n - \{0\} \subset \mathbb{C}$, is a uniform sequence of convergence regions (u.s.c.), if there exists a sequence $\{\lambda_n\}_{n=1}^{\infty}$ of positive numbers converging to 0, such that $a_n \in E$ for all $n \ge 1$ insures that $K(a_n/1)$ converges and that

(2.4) $$\left| \overset{\infty}{\underset{n=1}{K}} (a_n/1) - \overset{N}{\underset{n=1}{K}} (a_n/1)\right| \le \lambda_N \quad ; \quad N = 1,2,3\ldots$$

If we also have, for any such continued fraction, that

(2.5) $$\left| \overset{\infty}{\underset{n=p+1}{K}} (a_n/1) - \overset{p+N}{\underset{n=p+1}{K}} (a_n/1)\right| \le \lambda_N \quad ; \quad N = 1,2,3,\ldots$$

for all $p \in \mathbb{N}$, we say that $\{E_n\}_{n=1}^{\infty}$ is a totally uniform sequence of convergence regions (t.u.s.c.).

Hence, $\{E_n\}$ is a t.u.s.c. if $\{E_n^{(p)}\}_{n=1}^{\infty}$ are u.s.c.'s for all $p \in \{0,1,2,\ldots\}$ in such a way that the same sequence $\{\lambda_n\}$ can be used for all p, where $E_n^{(p)} = E_{n+p}$; $n = 1,2,3,\ldots,$ $p = 0,1,2,\ldots$.

Definition 2.2. $\{V_n\}_{n=0}^{\infty}$; $\emptyset \ne V_n \subset \hat{\mathbb{C}}$, is a sequence of value regions corresponding to a u.s.c. $\{E_n\}$ if $E_n \subset V_{n-1}$ and $E_n/(1+V_n) \subset V_{n-1}$ for $n = 1,2,3,\ldots$. If besides $V_n \subset V_n'$ for $n = 0,1,2,\ldots$ for any sequence $\{V_n'\}$ of value regions corresponding to $\{E_n\}$, we say that $\{V_n\}$ is the best sequence of value regions corresponding to $\{E_n\}$.

Definition 2.3. $\{E_n\}_{n=1}^{\infty}$ is a CA-sequence if the following three conditions are satisfied:

(i) $\{E_n\}$ is a t.u.s.c.

(2.6) (ii) $0 \notin C\ell(\bigcup_{n=1}^{\infty}(C\ell(V_n) + W_n))$ where $\{V_n\}$ is the best sequence of value regions corresponding to $\{E_n\}$, and $W_n = \{h_n \in \hat{\mathbb{C}};\ a_k \in E_k \text{ for } 2 \leq k \leq n\}$;

(iii) $\bigcup_{n=1}^{\infty} W_n$ is bounded

($C\ell(A)$ denotes the closure of the set A .)

In particular, when $E_n = E_1$; $n = 1,2,3,\ldots$ we say that E_1 is a simple CA-region, and when $E_{2n-1} = E_1$, $E_{2n} = E_2$; $n = 1,2,3,\ldots$, we say that $\langle E_1, E_2 \rangle$ is a set of twin CA-regions. (CA is an abbreviation of convergence acceleration.) (The definition of u.s.c., value regions and best value regions are in accordance with [10, p. 64] , where it also is stressed that the term region is used loosely to mean any subset of $\mathbb{C}$ and $\hat{\mathbb{C}}$ respectively.)

The theorem stating sufficient conditions for convergence acceleration of continued fractions $K(a_n/1)$ by using an auxiliary continued fraction, then is the following:

Theorem 2.4. Let $K(a_n/1)$ and $K(a_n'/1)$ be two convergent continued fractions such that $\lim_{n\to\infty}(a_n - a_n') = 0$. Furthermore let f be the value of $K(a_n/1)$ and $\{f^{(n)'}\}_{n=0}^{\infty}$ be the values of the tails of $K(a_n'/1)$. If $\{a_n\}$ is bounded and has no limit point at 0 , and there exists a CA-sequence $\{E_n\}$ such that $a_n, a_n' \in E_n$; $n = 1,2,3,\ldots$, then

$$\lim_{n\to\infty}\left|\frac{f-S_n(f^{(n)'})}{f-S_n(0)}\right| = 0 \tag{2.7}$$

In the proof of this theorem we make use of the following theorem:

Theorem 2.5. Let $\{E_n\}$ be a t.u.s.c. such that $\bigcup_{n=1}^{\infty} E_n$ is bounded and $E_n - \mathcal{D}_\varepsilon \neq \emptyset$ for all $n > 1$ and some $\varepsilon > 0$, where $\mathcal{D}_\varepsilon = \{z;\ |z| \leq \varepsilon\}$. Furthermore let $K(a_n/1)$ and $K(a_n'/1)$ be continued fractions such that $a_n, a_n' \in E_n$; $n = 1,2,3,\ldots$ and $\lim_{n\to\infty}(a_n - a_n') = 0$. Then

$$\lim_{n\to\infty}(f^{(n)} - f^{(n)'}) = 0 , \tag{2.8}$$

where $f^{(n)}$ and $f^{(n)'}$ are the values of the nth tails of $K(a_n/1)$ and $K(a_n'/1)$ respectively.

Proof. We first prove by induction that

$$\lim_{n\to\infty}(f_m^{(n)} - f_m^{(n)'}) = 0 \text{ for any } m \in \mathbb{N} . \tag{2.9}$$

($f_m^{(n)} = \mathop{K}_{k=n+1}^{n+m}\left(\frac{a_k}{1}\right)$.). For $m = 1$ we have $f_1^{(n)} - f_1^{(n)'} = a_{n+1} - a_{n+1}' \to 0$. For $m > 1$ we get

$$(2.10)\qquad \left|f_m^{(n)} - f_m^{(n)'}\right| \le \frac{\left|a_{n+1}-a'_{n+1}\right| + \left|f_m^{(n)'}\right|\left|f_{m-1}^{(n+1)}-f_{m-1}^{(n+1)'}\right|}{\left|1 + f_{m-1}^{(n+1)}\right|}$$

Here we have $\left|f_m^{(n)'}\right| \le K$ for all $m,n \ge 0$ and some $K < \infty$ because for any continued fraction $K(a_n/1)$ where $a_n \in E_n$ for all $n \in \mathbb{N}$,

$$\left|f^{(n)} - f_1^{(n)}\right| = \left|f^{(n)} - a_{n+1}\right| \le \lambda_1 \Rightarrow \left|f^{(n)}\right| \le L + M \quad\text{and}$$

$$\left|f^{(n)} - f_m^{(n)}\right| \le \lambda_m \Rightarrow \left|f_m^{(n)}\right| \le \left|f^{(n)}\right| + L \le 2L + M \quad\text{where}$$

$$(2.11)\qquad L = \sup\{\lambda_n;\ n \ge 1\}\ ,\quad M = \sup\{|z|;\ z \in \bigcup_{n=1}^{\infty} E_n\}$$

and $\{\lambda_n\}$ is a sequence of positive numbers corresponding to the t.u.s.c. $\{E_n\}$. This means that when $\{V_n\}$ is the best sequence of value regions corresponding to $\{E_n\}$, then $\bigcup_{n=0}^{\infty} V_n$ is bounded. In addition, since $E_{n+1}/(1+V_{n+1}) \subseteq V_n$ and $E_n - \mathcal{D}_\varepsilon \ne \emptyset$ for all $n \ge 1$, there exists a $\delta > 0$ such that $(1+V_n) \cap \mathcal{D}_\delta = \emptyset$ for all $n \ge 0$. Therefore, since $f_{m-1}^{(n+1)} \in V_{n+1}$, we have $\left|1 + f_{m-1}^{(n+1)}\right| > \delta$ for all $m,n \ge 1$. So, by (2.10) we get the following implication: $\left|f_{m-1}^{(n+1)} - f_{m-1}^{(n+1)'}\right| \to 0$ implies that $\left|f_m^{(n)} - f_m^{(n)'}\right| \to 0$ when $n \to \infty$. Induction on m will then give (2.9).

We now prove (2.8). Let $\varepsilon > 0$ be arbitrarily chosen. Since $\lambda_n \to 0$, there exists an $m_0 \in \mathbb{N}$ such that $\lambda_m \le \varepsilon/3$ for every $m \ge m_0$. Choose a fixed $m \ge m_0$. By (2.9) we then know that there exists an $n_0 \in \mathbb{N}$ such that $\left|f_m^{(n)} - f_m^{(n)'}\right| \le \varepsilon/3$ for every $n \ge n_0$. Since

$$\left|f^{(n)}-f^{(n)'}\right| \le \left|f^{(n)}-f_m^{(n)}\right| + \left|f_m^{(n)}-f_m^{(n)'}\right| + \left|f^{(n)'}-f_m^{(n)'}\right| \le \varepsilon\ ;\quad n \ge n_0\ ,$$

we then have $\lim_{n\to\infty}(f^{(n)} - f^{(n)'}) = 0$. ∎

Now it is easy to prove Theorem 2.4.

Proof of Theorem 2.4. Consider equation (2.1). Since $h_n \in W_n$ and $h_n + f^{(n)'} \in W_n + C\ell(V_n)$, the first factor on the right side is bounded. (This follows from the definition of CA-sequences). In addition, $\left|f^{(n)}\right| \ge \varepsilon/(1+L+M)$ where $\varepsilon = \inf\{|a_n|;\ n \in \mathbb{N}\} > 0$ because $\{a_n\}$ has no limit point at 0, $a_n \ne 0$, and M and L, defined by (2.11), are real numbers. Besides, we must have $E_n - \mathcal{D}_{\varepsilon/2} \ne \emptyset$ since $a_n \in E_n - \mathcal{D}_{\varepsilon/2}$, and we can choose $\{E_n\}$ such that $\bigcup_{n=1}^{\infty} E_n$ is bounded since $\{a_n\}$ is bounded. Therefore, by using Theorem 2.5, (2.7) follows. ∎

Remarks.

1) It is sufficient to require $a_n, a'_n \in E_n$ for $n \ge 2$ in the theorem since h_n is independent of a_1. But if that is satisfied, then $\{E_n^*\}$, where

$E_1^* = E_1 \cup \{a_1, a_1'\}$ and $E_n^* = E_n$ for $n \geq 2$, will also be a CA-sequence.

2) The conditions of Theorem 2.4 arose from the conditions (2.3) . Clearly these are not necessary conditions.

3) The applicability of Theorem 2.4 is dependent upon easy ways to describe CA-sequences. The simplest way is the following: If $\{E_n^*\}$ is a CA-sequence and $E_n \subset E_n^*$ for all $n \in \mathbb{N}$, then $\{E_n\}$ is a CA-sequence. Therefore it is of interest to have examples of CA-sequences.

Example 2.1. The sequence $\{E_n\}$ given by $E_n = P_{\alpha,n} \cap \mathcal{D}_M$; $n = 1,2,3,\ldots$ where

(2.12) $$P_{\alpha,n} = \{z \in \mathbb{C} ; |z| - \mathrm{Re}(ze^{-i2\alpha}) \leq 2g_n(1-g_{n+1})\cos^2\alpha\} \quad ; \quad n = 1,2,3,\ldots$$

and

(2.13) $$\mathcal{D}_M = \{z \in \mathbb{C} ; |z| \leq M\}$$

was proved by Thron [12] to be a u.s.c. provided that $|\alpha| < \frac{\pi}{2}$, $M < \infty$, $0 < \kappa < g_n < 1-\kappa$; $n = 1,2,3,\ldots$, and that the series

(2.14) $$\sum_{k=1}^{\infty} \prod_{n=1}^{k} \left(\frac{1}{g_{n+1}} - 1\right)$$

diverges. He also proved [12] that $\{V_n\}$ given by

(2.15) $$V_n = \{z \in \mathbb{C} ; \mathrm{Re}(ze^{-i\alpha}) \geq -g_{n+1}\cos\alpha\} \quad ; \quad n = 0,1,2,\ldots$$

is a corresponding sequence of value regions. By choosing $g_n = \frac{1}{2}(1-\delta)$; $n = 1,2,3,\ldots$, where $0 < \delta < 1$, we get a simple, uniform convergence region

(2.16) $$E_\alpha = \{z \in \mathbb{C} ; |z| - \mathrm{Re}(ze^{-i2\alpha}) \leq \tfrac{1}{2}(1-\delta^2)\cos^2\alpha\} \cap \mathcal{D}_M$$

with a corresponding value region

(2.17) $$V_\alpha = \{z \in \mathbb{C} ; \mathrm{Re}(ze^{-i\alpha}) \geq -\tfrac{1}{2}(1-\delta)\cos\alpha\} .$$

Since the best value region V_α^* corresponding to E_α is contained in $V_\alpha \cap \mathcal{D}_{M^*}$ where $M^* = 2M/(1+\delta)\cos\alpha$, and $W_n \subset 1+V_\alpha^*$ for all $n \geq 1$, (W_n defined by (2.6)), and

$$\inf\{|1+x+y| ; x,y \in V_\alpha\} = \delta > 0 ,$$

we have that E_α is a simple CA-region.

Example 1.1 continued. All the elements a_n and a_n' ; $n = 1,2,3,\ldots$ are contained in the simple CA-region E_α defined by (2.16) with $\alpha = 0$, $M = 12$, and δ arbitrarily chosen from $(0,1)$.

Example 2.2.

(2.18) $$E_1 = \{z \in \mathbb{C} ; |z| \leq \rho^2 - \varepsilon^2\}$$

(2.19) $$E_2 = \{z \in \mathbb{C} ; |z| \geq (1+\rho+\varepsilon)^2\} \cap \mathcal{D}_M$$

where $0 < \rho < \rho + \varepsilon \leq 1$ and $\mathcal{D}_M$ is defined by (2.13) , is a set of twin

CA-regions with a corresponding set of value regions

$$(2.20)\qquad V_0 = \{w \; ; \; |w| \leq \rho-\varepsilon\} \; , \quad V_1 = \{w \; ; \; |w+1| \geq \rho+\varepsilon\} \; .$$

The reasons for this are as follows:

1) $\langle E_1,E_2\rangle$ and $\langle V_0,V_1\rangle$ are corresponding element and value regions because:

$$\frac{E_1}{1+V_1} = \{w \; ; \; |w| \leq \frac{\rho^2-\varepsilon^2}{\rho+\varepsilon}\} = V_0$$

and if $z = re^{i\phi} \in E_2$ and $w = Re^{i\phi} \in V_0$, then

$$\left|1 + \frac{z}{1+w}\right| \geq \frac{r-1-R}{R+1} \geq \frac{(1+\rho+\varepsilon)^2-(1+\rho-\varepsilon)}{1+\rho-\varepsilon} > \rho + \varepsilon \; .$$

That means: $\frac{E_2}{1+V_0} \subset V_1$. Furthermore, $E_1 \subset V_0$ since $\rho^2 - \varepsilon^2 = (\rho+\varepsilon)(\rho-\varepsilon) \leq \rho - \varepsilon$, and $E_2 \subset V_1$ since $(1+\rho+\varepsilon)^2 > 1 + \rho + \varepsilon$.

2) $\langle E_1,E_2\rangle$ is a t.u.s.c. because

$$E_1 \subset E_1^* = \{z \; ; \; z = v^2 \text{ and } |v| \leq \rho\}$$

and

$$E_2 \subset E_2^* = \{z \; ; \; z = v^2 \text{ and } |v \pm i| \geq \rho\}$$

where $\langle E_1^*,E_2^*\rangle$ was proved by Thron (see [10, Theorem 4.46]) to be a set of uniform twin convergence regions. Furthermore, $\langle E_2,E_1\rangle$ also is a set of uniform twin convergence regions. This follows from the fact that for any continued fraction $K(a_n/1)$ where $a_n \in E_{n+1(\mathrm{mod}2)}$ for all n, any $x \in E_1$, $x/(1+K(a_n/1))$ will will be a continued fraction such that

$$|f - f_n| = \frac{|x| \cdot \left| \overset{\infty}{\underset{m=1}{K}} \left(\frac{a_m}{1}\right) - \overset{n-1}{\underset{m=1}{K}} \left(\frac{a_m}{1}\right)\right|}{\left|1 + \overset{\infty}{\underset{m=1}{K}} \left(\frac{a_m}{1}\right)\right| \left|1 + \overset{n-1}{\underset{m=1}{K}} \left(\frac{a_m}{1}\right)\right|} \leq \lambda_n \underset{n\to\infty}{\to} 0$$

Thereby

$$\left| \overset{\infty}{\underset{m=1}{K}} \left(\frac{a_m}{1}\right) - \overset{n-1}{\underset{m=1}{K}} \left(\frac{a_m}{1}\right)\right| \leq \frac{(1+K)^2\lambda_n}{|x|} \underset{n\to\infty}{\to} 0$$

where

$$K = \sup\{|w| \; ; \; w = f_m^{(1)} \; , \; a_\nu \in E_{\nu(\mathrm{mod}2)} \; , \; \nu \; , \; m = 1,2,3,\ldots\} < \infty$$

because when $\langle U_1,U_2\rangle$ are the best value regions corresponding to $\langle E_1,E_2\rangle$, then $U_1 = E_2 \cup \frac{E_2}{1+E_1} \cup \frac{E_2}{1 + \frac{E_1}{1+U_1}}$ is bounded since E_2 is bounded, $-1 \notin E$, and $(-E_1) \cap \mathcal{C}\ell(1+U_1) = \emptyset$.

3) $W_{2n} \subset 1 + V_1$, $W_{2n+1} \subset 1 + V_0$; $n = 1,2,3,\ldots$ (W_n defined by (2.6)), and

$$\inf\{|z+w+1| \; ; \; z \in V_0 \; , \; w \in V_1\} = 2\varepsilon > 0 \; .$$

For further examples, see [6].

3. Truncation error bounds. The next question we want to answer is: how much better is the convergence of the modified approximants $\{S_n(f^{(n)'})\}$ compared to that of the approximants $\{S_n(0)\}$? Upper bounds for the ratio (2.1) will serve as a measure for the improvement.

Theorem 3.1. Let $K(a_n/1)$ be a convergent continued fraction whose value f one wants to approximate, and $K(a_n'/1)$ a convergent continued fraction where all the values $\{f^{(n)'}\}_{n=0}^{\infty}$ of the tails are known. If

(i) There exists a t.u.s.c. $\{E_n\}$ such that $\bigcup_{n=1}^{\infty} E_n$ is bounded, $E_n - \mathcal{S}_\varepsilon \neq \emptyset$ for all n and some $\varepsilon > 0$ where $\mathcal{S}_\varepsilon = \{z \in \mathbb{C} ; |z| \leq \varepsilon\}$ and $a_n, a_n' \in E_n$ for all $n \in \mathbb{N}$,

(ii) there exist constants $D, \mu > 0$ and a bounded sequence $\{t_n\}_{n=0}^{\infty}$ of numbers $\geq \mu$ such that

$$t_{n+1}\left|1 + f^{(n+1)'}\right| - t_n\left|f^{(n)'}\right| \geq D \quad ; \quad n = 0,1,2,\ldots \text{ and}$$

(iii) $\lim_{n\to\infty}(a_n - a_n') = 0$ and

$$|a_n - a_n'| \leq \min\{D^2/4T_0T_1 , |a_n'|/2\} \quad ; \quad n = 1,2,3,\ldots$$

where

$$T_n = \sup\{t_m ; m \geq n\} \quad ; \quad n = 0,1,2,\ldots$$

then

$$(3.1) \qquad \left|\frac{f-S_n(f^{(n)'})}{f-S_n(0)}\right| \leq \left(1 + \frac{|f^{(n)'}|}{\delta_n(f^{(n)'})}\right)\left(2 + 4\,\frac{|f^{(n)'}|T_{n+1}T_{n+2}}{D_{n+1}t_{n+1}}\right)\frac{d_{n+1}}{|a_{n+1}'|} \quad ;$$

for $n = 1,2,3,\ldots$, where

$$d_n = \sup\{|a_m - a_m'| ; m \geq n\} \quad ; \quad n = 1,2,3,\ldots$$

$$D_n = \inf\{t_{m+1}\left|1 + f^{(m+1)'}\right| - t_m\left|f^{(m)'}\right| \quad ; \quad m \geq n\} \quad ; \quad n = 0,1,2,\ldots$$

$$\delta_n(f^{(n)'}) = \inf\{\left|z + f^{(n)'}\right| ; z \in W_n\} \quad ; \quad n = 1,2,\ldots$$

and $W_n = \{h_n \in C ; a_m \in E_m , m = 2,\ldots,n\}$; $n = 1,2,\ldots$.

Before presenting the proof, we shall make some remarks:

1) We do not know, within the conditions of the theorem, that (3.1) converges to 0 - not even the left hand side. However, the right hand side of (3.1) may converge to 0 even if the conditions of Theorem 2.4 are not satisfied. In that case $\{d_n\}$ must converge to 0 sufficiently fast to compensate for possible limit points at 0 for $\delta_n(f^{(n)'})$ or $\{a_n'\}$.

2) Which convergent continued fractions $K(a_n'/1)$ satisfy condition (ii) in the theorem? We see at once that $\left|1 + f^{(n+1)'}\right| \geq \delta$; $n = 0,1,2,\ldots$ for some $\delta > 0$ is necessary. Furthermore, it can be shown that $K(a_n'/1)$, k-periodic, always satisfies (ii) unless $S_k'(x)$ has coinciding fixed points (see [7]) . In

particular, when $K(a_n'/1)$ is 1-periodic $(a_n' = a' \notin (-\infty,-\frac{1}{4}]$ for all $n \in \mathbb{N})$, we can choose $t_n = 1$ for all $n \geq 0$. (Thereby the result coincides with that of Thron and Waadeland [14].)

3) The actual choice of the sequence $\{t_n\}$, when it exists, should be made such that $T_n T_{n+1}/D_n t_n$ is kept as low as conveniently possible. That means for instance in the k-periodic case, that

(3.2) $$t_{kn+p} = t_p = \sum_{j=p}^{k+p-1} \left(\prod_{i=p+1}^{j} \left|1 + f^{(i)'}\right| \cdot \prod_{i=j+1}^{k+p-1} \left|f^{(i)'}\right| \right) ;$$

for $p = 0,\ldots,k-1$ and $n = 0,1,2,\ldots$, is the best possible choice for $\{t_n\}$, because then

(3.3) $$D_n = D = \prod_{m=0}^{k-1} \left|1 + f^{(m)'}\right| - \prod_{m=0}^{k-1} \left|f^{(m)'}\right| \quad ; \quad n \geq 0 .$$

4) The computation of $\delta_n(f^{(n)'})$ depends on the choice of $\{E_n\}$. If $\{E_n\}$ is a CA-sequence, we know that $0 \notin C\ell\left(\bigcup_{n=1}^{\infty} (C\ell(V_n) + W_n)\right)$, where $\{V_n\}$ is the best sequence of value regions corresponding to $\{E_n\}$. Since $f^{(n)'} \in C\ell(V_n)$; $n = 1,2,3,\ldots$, that means that there exists a $\delta > 0$ such that

$$\delta \leq \inf\{|x+y| \; ; \; x \in V_k \, , \; y \in W_k \, , \; k = 1,2,3\ldots\} \leq \delta_n(f^{(n)'})$$

for all $n \geq 1$. By replacing $\delta_n(f^{(n)'})$ by δ in (4.1), we have a value which is valid for any continued fraction $K(a_n'/1)$ where $a_n' \in E_n$; $n = 1,2,\ldots$ An alternative lower bound for $\delta_n(f^{(n)'})$ is easy to get if we know a sequence $\{W_n^*\}$ of regions (or subsets of $\hat{\mathbb{C}}$) such that $W_n \subset W_n^*$; $n = 1,2,3,\ldots$, namely the value we get by replacing W_n with W_n^* in the expression for $\delta_n(f^{(n)'})$. (Such a sequence $\{W_n^*\}$ is for instance known in the Examples 2.1 and 2.2)

5) Ordinarily T_{n+1} and D_{n+1} in (3.1) may be replaced by the constants T_0 and D respectively without significant loss of accuracy.

Proof of Theorem 3.1. Since

(3.4) $$\left|\frac{h_n}{h_n+f^{(n)'}}\right| = \left|1 - \frac{f^{(n)'}}{h_n+f^{(n)'}}\right| \leq 1 + \frac{|f^{(n)'}|}{\delta_n(f^{(n)'})} \quad ; \quad n = 1,2,3,\ldots$$

we only have to establish upper bounds for the second factor on the right hand side of (2.1). Now

$$f^{(m)} - f^{(m)'} = \frac{(a_{m+1}-a_{m+1}') - (f^{(m+1)}-f^{(m+1)'})f^{(m)'}}{(1+f^{(m+1)'}) + (f^{(m+1)}-f^{(m+1)'})} ,$$

so

(3.5) $$\left|f^{(m)} - f^{(m)'}\right| \leq \frac{\left|a_{m+1}-a_{m+1}'\right| + \left|f^{(m+1)}-f^{(m+1)'}\right|\left|f^{(m)'}\right|}{\left|1 + f^{(m+1)'}\right| - \left|f^{(m+1)} - f^{(m+1)'}\right|}$$

when $\left|1 + f^{(m+1)'}\right| > \left|f^{(m+1)} - f^{(m+1)'}\right|$. If we let

(3.6) $$R_n = \frac{D_n d_{n+1}}{\frac{D_n^2}{T_n T_{n+1}} - 2d_{n+1}} \quad ; \; n = 0,1,2,\ldots$$

we have $\frac{1}{2}\left|1 + f^{(m+1)'}\right| \geq R_n/t_{m+1}$ for $m \geq n$. And by using (3.5) we get: If $\left|f^{(m+1)} - f^{(m+1)'}\right| \leq \frac{R_n}{t_{m+1}}$ when $m \geq n$ then:

(3.7) $$\left|f^{(m)} - f^{(m)'}\right| \leq \frac{d_{n+1} + \frac{R_n}{t_{m+1}}\left|f^{(m)'}\right|}{\left|1 + f^{(m+1)'}\right| - \frac{R_n}{t_{m+1}}} \leq \frac{R_n}{t_m}$$

(because this inequality is satisfied when $R_n^2 - R_n D_n + d_{n+1}T_n T_{n+1} \leq 0$ which is true when R_n is given by (3.6), because then

$$R_n^2 - R_n D_n + d_{n+1}T_n T_{n+1} = R_n\left(R_n - D_n + \frac{d_{n+1}T_n T_{n+1}}{R_n}\right) =$$

$$= \frac{R_n d_{n+1}}{D_n \cdot \left(\frac{D_n^2}{T_n T_{n+1}} - 2d_{n+1}\right)}\left[-D_n^2 + 4d_{n+1}T_n T_{n+1}\right] \leq 0$$

by condition (iii) of the theorem.) Since $f^{(m)} - f^{(m)'} \to 0$ (by Theorem 2.5) , we can find to any fixed $n \in \mathbb{N}$ a $m \geq n$ such that $\left|f^{(m+1)} - f^{(m+1)'}\right| \leq R_n/t_{m+1}$ Repeated use of the implication then gives

$$\left|f^{(n)} - f^{(n)'}\right| \leq \frac{R_n}{t_n} = \frac{D_n d_{n+1}}{t_n\left[\frac{D_n^2}{T_n T_{n+1}} - 2d_{n+1}\right]} \quad ; \; n = 0,1,2,\ldots \; .$$

Therefore we get

$$\left|\frac{f^{(n)} - f^{(n)'}}{f^{(n)}}\right| = \left|\frac{(a_{n+1} - a'_{n+1}) - f^{(n)'}(f^{(n+1)} - f^{(n+1)'})}{a'_{n+1} + (a_{n+1} - a'_{n+1})}\right|$$

$$\leq \frac{d_{n+1} + \left|f^{(n)'}\right|\frac{R_{n+1}}{t_{n+1}}}{\frac{1}{2}\left|a'_{n+1}\right|} = 2\,\frac{d_{n+1}}{\left|a'_{n+1}\right|} + 2\,\frac{\left|f^{(n)'}\right|}{\left|a'_{n+1}\right| t_{n+1}} \cdot \frac{D_{n+1}d_{n+2}}{\frac{D_{n+1}^2}{T_{n+1}T_{n+2}} - 2d_{n+2}}$$

$$\leq \left(2 + 4\,\frac{\left|f^{(n)'}\right| T_{n+1}T_{n+2}}{D_{n+1}t_{n+1}}\right)\frac{d_{n+1}}{\left|a'_{n+1}\right|} \; . \qquad \blacksquare$$

As an example of the use of this theorem, we may look again at the continued fractions of Example 1.1:

__Example__ 1.1 continued: Choosing $\{t_n\}$ as described in (3.2) , we get

$$t_{5n} \approx 458 \; , \quad t_{5n+1} \approx 403 \; , \quad t_{5n+1} \approx 349 \; ,$$

$$t_{5n+3} \approx 703 \; , \quad t_{5n+4} \approx 310 \; ,$$

$$T_n = t_3 \approx 703 \quad , \quad D_n = D \approx 450 \quad \text{for all} \quad n \geq 0 \quad .$$

In view of the comments on Example 1.1 in Section 2, we may replace $\delta_n(f^{(n)'})$ in (3.1) by 1 without violating the inequality. Therefore we get, by Theorem 3.1 :

$$(3.8) \qquad \left|\frac{f-S_n(f^{(n)'})}{f-S_n(0)}\right| \leq (1 + |f^{(n)'}|)\left(2 + 4\,\frac{t_3^2|f^{(n)'}|}{Dt_{n+1}}\right)\frac{0.3^{n+1}}{|a'_{n+1}|} = Q_n \cdot 0.3^{n+1}$$

for $n = 1,2,3,\ldots$, where

$$Q_{5n+1} = 11.2 \quad , \quad Q_{5n+2} = 12.9 \quad , \quad Q_{5n+3} = 8.16 \quad ,$$

$$Q_{5n+4} = 13.7 \quad , \quad Q_{5n+5} = 10.0 \quad ; \quad n = 0,1,2,\ldots$$

By Table 1.1 we see that the actual values of the ratio, for different values of n, are as given in Table 3.1. There is no need for being so careful in choosing $\{t_n\}$. We see easily that for instance the following values will work:

$$t^*_{5n+1} = 1.4 \quad , \quad t^*_{5n+2} = t^*_{5n+4} = 2 \quad , \quad t^*_{5n+3} = t^*_{5n+5} = 1 \quad ;$$

for $n = 0,1,2,\ldots$. We then get $D^* \approx 0.6$, and thereby the upper bounds given in the following table:

Table 3.1

n	1	2	5	10	20
$\left\|\frac{f-S_n(f^{(n)'})}{f-S_n(0)}\right\|$	0.0006	0.002	0.00001	10^{-7}	$1.4\cdot 10^{-12}$
$Q_n \cdot 0.3^{n+1}$	1.01	0.35	0.007	$1.8\cdot 10^{-5}$	$1.0\cdot 10^{-10}$
Upper bounds with $\{t^*_n\}$	2.07	0.71	0.013	$3.23\cdot 10^{-5}$	$1.90\cdot 10^{-10}$

The result of Theorem 3.1 can also be used to estimate truncation error bounds for the sequence of modified approximants $\{S_n(f^{(n)'})\}$ in those cases where truncation error bounds for the approximants $\{S_n(0)\}$ are known.

<u>Example</u> 2.1 continued: For continued fractions $K(a_n/1)$ where $a_n \in E_n$ for all $n \in \mathbb{N}$, and $\{E_n\}$ is given by

$$E_n = \{z \; ; \; |z| - \operatorname{Re}(ze^{-i(\psi_n+\psi_{n-1})}) \leq 2kp_{n-1}(\cos\psi_n - p_n)\} \cap \mathcal{D}_M$$

where $M < \infty$, $0 < k < 1$, $p_n > 0$, $\left|p_n e^{i\psi_n} - \frac{1}{2}\right| \leq \frac{1}{2} - \varepsilon$, for all $n \geq 0$, and $0 < \varepsilon < \frac{1}{2}$, the following truncation error bounds are valid [10, Theorem 8.4, p. 305]

$$|f - f_n| \leq \frac{|a_1|(\cos\psi_1 - p_1)}{\left(1 + \frac{\varepsilon^2(1-k)}{M}\right)^{n-1}} \quad ; \quad n = 2,3,4,\ldots$$

By choosing $\psi_n = \alpha$, $p_n = \frac{1}{2}\cos\alpha$, $k = 1-\delta^2$ where $|\alpha| < \frac{\pi}{2}$, $0 < \delta < 1$, we get $E_n = E_\alpha$ for all n , where E_α is defined by (2.16) . Besides, $k = 1 - \delta^2 \in (0,1)$, $p_n > 0$ and

$$\left|p_n e^{i\psi_n} - \frac{1}{2}\right| = \left|\frac{1}{2}\cos\alpha e^{i\alpha} - \frac{1}{2}\right| = [(\frac{1}{2}\cos^2\alpha - \frac{1}{2})^2 + (\frac{1}{2}\cos\alpha\sin\alpha)^2]^{\frac{1}{2}}$$

$$= \frac{1}{2}|\sin\alpha| \leq \frac{1}{2} - \varepsilon \quad \text{for} \quad \varepsilon = \frac{1}{2}[1 - |\sin\alpha|]$$

where $0 < \varepsilon < \frac{1}{2}$. Therefore, truncation error bounds for continued fractions $K(a_n/1)$ where $a_n \in E_\alpha$ for all $n \geq 1$, are given by

$$|f - f_n| \leq \frac{|a_1|\cos\alpha}{2\left(1 + \frac{\delta^2(1-|\sin\alpha|)^2}{4M}\right)^{n-1}} \quad ; \quad n = 2,3,4,\ldots \quad . \tag{3.9}$$

If we know an auxiliary continued fraction $K(a_n'/1)$ such that $a_n - a_n' \to 0$ and $a_n' \in E_\alpha$ for all n , then we get the following truncation error bounds for the modified approximants $S_n(f^{(n)'})$:

$$|f - S_n(f^{(n)'})| \leq \left(1 + \frac{|f^{(n)'}|}{\delta}\right)\left(2 + 4\frac{|f^{(n)'}|T_{n+1}T_{n+2}}{D_{n+1}t_{n+1}}\right)\frac{|a_1|\cos\alpha}{2\left(1+\frac{\delta^2(1-|\sin\alpha|)^2}{4M}\right)^{n-1}}\frac{d_{n+1}}{|a'_{n+1}|}$$

$$\text{for} \quad n = 2,3,4,\ldots$$

with the same notation as in Theorem 3.1 .

Example 1.1 continued: For the continued fraction in Example 1.1 we get

$$|f - S_n(f^{(n)'})| \leq Q_n \cdot 0.3^{n+1} \cdot \frac{8 \cdot \cos 0}{2\left(1 + \frac{(1-0)^2}{4 \cdot 12}\right)^{n-1}} = 0.36Q_n \cdot \left(\frac{72}{245}\right)^{n-1}$$

$$= 0.36Q_n \cdot 0.29^{n-1}$$

where Q_n is as defined earlier.

References

1. John Gill, Infinite compositions of Möbius tranformations. Trans. Amer. Math. Soc. 176 (1973) 479-487.

2. John Gill, The use of attractive fixed points in accelerating the convergence of limit-periodic continued fractions, Proc. Amer. Math. Soc. 47 (1975), 119-126.

3. John Gill, Modifying factors for sequences of linear fractional tranformations, Kong. Norske Vid. Selsk. Skrifter (1978), No. 3 1-7.

4. J.W.L. Glaisher, On the tranformation of continued products into continued fractions, Proc. London Math Soc. 5 (1873/4).

5. T.L. Hayden, Continued fraction approximation to functions, Numer. Math. 7 (1965), 292-309.

6. Lisa Jacobsen, Convergence acceleration for continued fractions $K(a_n/1)$, To appear in Trans. Amer. Math. Soc.

7. Lisa Jacobsen, Some periodic sequences of circular convergence regions, These Lecture Notes.

8. Lisa Jacobsen and Haakon Waadeland, Some useful formulas involving tails of continued fractions, These Lecture Notes.

9. William B. Jones and W.J. Thron, Twin convergence regions for continued fractions $K(a_n/1)$. Trans. Amer. Math. Soc. 150 (1970), 93-119.

10. William B. Jones and W.J. Thron, Continued fractions: Analytic Theory and Applications; Encyclopedia of Mathematics and Its Applications, No. 11; Addison Wesley, Reading, Massachusetts 1980.

11. A. Pringsheim, Vorlesungen über Zahlenlehre, Bd 1/2, Leipzig 1916.

12. W.J. Thron, On parabolic convergence regions for continued fractions, Math. Zeitschr, 69 (1958) 173-182.

13. W.J. Thron, A survey of recent convergence results for continued fractions, Rocky Mountain J. Math. 4 (1974), 273-281.

14. W.J. Thron and Haakon Waadeland, Accelerating convergence of limit periodic continued fractions $K(a_n/1)$, Numer. Math. 34 (1980), 155-170.

15. P. Wynn, Converging factors for continued fractions, Numer. Math. 1 (1959) 272-320.

Lisa Jacobsen

Institutt for Matematikk

Universitetet i Trondheim

7034 Trondheim - NTH

Norway

SOME PERIODIC SEQUENCES OF CIRCULAR CONVERGENCE REGIONS

Lisa Jacobsen

1. Introduction. When dealing with continued fractions $K(a_n/1)$ the following concepts are very useful:

Definition 1.1. *$\{E_n\}_{n=1}^{\infty}$ is a sequence of element regions for continued fractions $K(a_n/1)$, and $\{V_n\}_{n=0}^{\infty}$ is a corresponding sequence of pre-value regions if*

(i) $\emptyset \neq E_n - \{0\} \subseteq \mathbb{C}$; $n = 1,2,3,\ldots$

(ii) $\emptyset \neq V_n \subseteq \hat{\mathbb{C}}$; $n = 0,1,2,$

(iii) $E_n/(1+V_n) \subseteq V_{n-1}$; $n = 1,2,3,\ldots$

If besides $E_n \subseteq V_{n-1}$; $n = 1,2,3,\ldots$, we say that $\{V_n\}$ is a sequence of value regions corresponding to $\{E_n\}$. Furthermore, if $K(a_n/1)$ converges whenever $a_n \in E_n$; $n = 1,2,3,\ldots$, we say that $\{E_n\}$ is a sequence of convergence regions for continued fractions $K(a_n/1)$.

(This definition is given in accordance with [4, p. 64] (for pre-value regions see [5]) where it is also pointed out that the term region is used loosely to mean any subset of the complex plane, extended if necessary.) In particular the terms convergence regions and value regions have proved to be advantageous. Convergence regions for their useful way of describing convergence criteria for continued fractions, and value regions for the following property: When $a_n \in E_n$ for all $n \in \mathbb{N}$, then $f_m^{(n)} \in V_n$ for all $n \geq 0$ and $m \geq 1$.

In this article we shall prove that certain k-periodic sequences of element regions

$$E_{kn+p} = E_p = \{z \in \mathbb{C} \; ; \; |z-\Gamma_p| \leq \rho_p\} \; ; \tag{1.1}$$

$p = 1,\ldots,k$ and $n = 0,1,2,\ldots$ are sequences of convergence regions. In addition we are going to prove that the tails $\{f^{(n)}\}$ of any continued fraction $K(a_n/1)$ with $a_n \in E_n$ for all $n \in \mathbb{N}$, are contained in certain corresponding pre-value regions $\{V_n\}$; $f^{(n)} \in V_n$; $n = 0,1,2,\ldots$. The notations in this article are in accordance with [1] .

2. Main results. When we consider the special case of (1.1) where $\rho_p = 0$; $p = 1,\ldots,k$, the "regions" $\{E_n\}$ will all be one-point sets $E_{kn+p} = E_p = \{\Gamma_p\}$; $p = 1,\ldots,k$, and $n = 0,1,2,\ldots$. Hence $a_n \in E_n$ for all $n \in \mathbb{N}$ implies that $K(a_n/1)$ is the k-periodic continued fraction $K(\Gamma_n/1)$ where $\Gamma_{kn+p} = \Gamma_p$ for $p = 1,\ldots,k$ and $n = 0,1,2,\ldots$. For such continued fractions we have the following results:

Theorem 2.1. *Let $K(a_n/1)$ be a convergent, k-periodic continued fraction*

($a_{kn+p} = a_p$; $p = 1,\ldots,k$ and $n \in \mathbb{N}$.) Furthermore, let x, y denote the two fixed points of $S_k^{(n)}(w)$ such that $x_n = f^{(n)}$; $n = 0,1,2,\ldots$. Finally let

(2.1) $$D = \prod_{n=0}^{k-1} \left|1 + f^{(n)}\right| - \prod_{n=0}^{k-1} \left|f^{(n)}\right| ,$$

(2.2) $$\mathcal{D} = \prod_{n=0}^{k-1} (1 + f^{(n)}) + (-1)^{k-1} \prod_{n=0}^{k-1} f^{(n)} .$$

Then we have:

(A) $D \geq 0$

where the equality sign holds if and only if $x_0 = y_0$.

(B) $y_n = -\frac{B_{k-1}^{(n+1)}}{B_{k-1}^{(n)}}(1 + x_{n+1})$, $1 + y_n = -\frac{B_{k-1}^{(n-1)}}{B_{k-1}^{(n)}} x_{n-1}$ for $n = 0,1,2,\ldots$.

(C) $x_n - y_n = \mathcal{D}/B_{k-1}^{(n)}$; $n = 0,1,2,\ldots$.

(D) the sequence $\{t_n\}_{n=0}^{\infty}$ given by

(2.3) $$t_{kn+p} = t_p = \sum_{j=p}^{k+p-1} \left(\prod_{i=p+1}^{j} \left|1 + f^{(i)}\right| \cdot \prod_{i=j+1}^{k+p-1} \left|f^{(i)}\right| \right)$$

for $p = 0,\ldots,k-1$ and all $n \in \mathbb{N}$, is such that

(2.4) $$t_{n+1}\left|1 + f^{(n+1)}\right| - t_n\left|f^{(n)}\right| = D \quad ; \quad n = 0,1,2,\ldots .$$

(E) the sequence $\{v_n\}_{n=0}^{\infty}$ given by

(2.5) $$v_{kn+p} = v_p = \sum_{j=p}^{k+p-1} \left(\prod_{i=p+1}^{j} \left|1 + y_i\right| \cdot \prod_{i=j+1}^{k+p-1} \left|y_i\right| \right) ;$$

for $p = 0,\ldots,k-1$ and all $n \in \mathbb{N}$, is such that

(2.6) $$v_n\left|y_n\right| - v_{n+1}\left|1 + y_{n+1}\right| = D \quad ; \quad n = 0,1,2,\ldots .$$

(F) $\left|B_{k-1}^{(n)}\right| \leq t_n$, $\left|B_{k-1}^{(n)}\right| \leq v_n$, $|\mathcal{D}| \geq D$; $n = 0,1,2,\ldots$.

(G) For any integer $p \geq 0$, we have

$$S_{kn}^{(p)}(W_p) \to \{x_p\} \text{ when } n \to \infty ,$$

when W_p is any region such that $y_p \notin W_p$.

Proof: (A) By [4, p. 47] $K(a_n/1)$ converges if and only if one of the two conditions

(i) $x_0 = y_0$

(ii) $\left|B_k + B_{k-1}y_0\right|/\left|B_k + B_{k-1}x_0\right| < 1$ and $S_m(0) \neq y_0$ for $m = 1,2,\ldots,k$

is satisfied. Since (see [3])

(2.7) $$x_0 + y_0 = (A_{k-1} - B_k)/B_{k-1}$$

(2.8) $$B_m + B_{m-1}f^{(m)} = \prod_{n=1}^{m} (1 + f^{(n)}) \quad ; \quad m = 1,2,3,\ldots$$

(2.9) $$A_m - B_m f^{(0)} = (-1)^{m-1} \prod_{n=0}^{m} f^{(n)} \quad ; \quad m = 0,1,2,\ldots$$

implies that

(2.10) $$\left|\frac{B_k + B_{k-1}y_0}{B_k + B_{k-1}x_0}\right| = \frac{\prod_{n=0}^{k-1} \left|f^{(n)}\right|}{\prod_{n=1}^{k} \left|1+f^{(n)}\right|} ,$$

statement (A) follows.

(B) follows by straight forward computation by using the following formulas [3] :

(2.11) $$A_m^{(n)} = a_{n+1} B_{m-1}^{(n+1)} \quad ; \quad m,n = 0,1,2,\ldots$$

(2.12) $$B_m^{(n)} = B_{m-1}^{(n+1)} + a_{n+2} B_{m-2}^{(n+2)} \quad ; \quad n = 0,1,2,\ldots \quad , \quad m = 1,2,3,\ldots$$

(2.13) $$B_m^{(n)} = B_{m-1}^{(n)} + a_{n+m} B_{m-2}^{(n)} \quad ; \quad n = 0,1,2,\ldots \quad , \quad m = 1,2,3,\ldots \quad .$$

(C) By (B) we have that

$$x_n - y_n = x_n + \frac{B_{k-1}^{(n+1)}}{B_{k-1}^{(n)}}(1 + x_{n+1}) \quad .$$

Since we have (by [3]) that

(2.14) $$B_{k-1}^{(n+1)}(1 + x_{n+1}) - B_{n-1}^{(n)} x_n = \mathcal{D} \quad ,$$

this gives us statement (C) .

(D) follows by straight forward computation.

(E) Straight forward computation gives

(2.15) $$v_n\left|y_n\right| - v_{n+1}\left|y_{n+1} + 1\right| = \prod_{m=1}^{k} \left|y_m\right| - \prod_{m=1}^{k} \left|1 + y_m\right| \quad .$$

Then (2.6) follows by use of (B) .

(F) By [3] we know that

(2.16) $$B_{k-1}^{(n)} = \sum_{j=n}^{k+n-1} (-1)^{n+k+j-1} \prod_{i=n+1}^{j} (1 + x_i) \cdot \prod_{i=j+1}^{k+n-1} x_i$$

$$= \sum_{j=n}^{k+n-1} (-1)^{n+k+j-1} \prod_{i=n+1}^{j} (1 + y_i) \prod_{i=j+1}^{k+n-1} y_i$$

which gives that $\left|B_{k-1}^{(n)}\right| \leq t_n$, $\left|B_{k-1}^{(n)}\right| \leq v_n$. $|\mathcal{D}| \geq D$ follows directly from the expressions (2.1) and (2.2) .

(G) Let $w \in W_p$ be arbitrarily chosen. Then $\left|w - y_p\right| \geq \delta > 0$ and

$$S_{kn}^{(p)}(w) = \frac{A_{kn}^{(p)} + A_{kn-1}^{(p)} w}{B_{kn}^{(p)} + B_{kn-1}^{(p)} w} = \frac{f_{kn}^{(p)} h_{kn}^{(p)} + f_{kn-1}^{(p)} w}{h_{kn}^{(p)} + w}$$

By Galois' theorem [4, Theorem 3.4, p. 56] we know that $\lim_{n\to\infty} h_{kn}^{(p)} = -y_p$.

Therefore, since $\lim_{n\to\infty} f_n^{(p)} = x_p$, this proves the theorem.

As mentioned in this proof, it is always easy to see whether a k-periodic continued fraction converges or not. Suppose now that we start with a k-periodic continued fraction $K(\Gamma_n/1)$. Then we pick an element a_n from a certain neighborhood of Γ_n for all $n \in \mathbb{N}$. What can now be said about the convergence of $K(a_n/1)$? Will it converge if we pick a_n close enough to Γ_n ? The following theorem gives one answer to this question. It takes care of the case where the neighborhoods are disks centered at Γ_n. In other words, we increase the radii of $E_1,\dots,E_k$.

Theorem 2.2. (A) The k-periodic sequence $\{E_n\}$ given by

(2.17) $$E_{kn+p} = E_p = \{z \in \mathbb{C} ; |z - \Gamma_p| \le \rho_p\} \quad ; \quad p = 1,\dots,k \quad ; \quad n = 0,1,2,\dots$$

is a sequence of convergence regions when $\Gamma_1,\dots,\Gamma_k$ are complex numbers such that

(i) $\Gamma_1 \cdots \Gamma_k \neq 0$,

(ii) the k-periodic continued fraction $K(\Gamma_n/1)$ converges;

$\Gamma_{mk+p} = \Gamma_p \quad ; \quad p = 1,\dots,k \quad ; \quad m = 1,2,3,\dots$

(iii) the fixed points x_n and y_n of the linear fractional transformation

(2.18) $$S_k^{(n)*}(w) = \frac{\Gamma_{n+1}}{1} + \cdots + \frac{\Gamma_{n+k}}{1+w}$$

are distinct; x_n is the nth tail of $K(\Gamma_n/1)$, $n = 0,1,2,\dots$ and

(2.19) $$\rho_{kn+p} = \rho_p = \frac{D^2-\mu^2}{K_p} \quad ; \quad p = 1,\dots,k \quad ; \quad n = 0,1,2,\dots ,$$

(2.20) $$K_p = 4\max\{t_p t_{p-1}, v_p v_{p-1}\} \quad ; \quad p = 1,\dots,k ,$$

(2.21) $$t_{kn+p} = t_p = \sum_{j=p}^{k+p-1} \Big(\prod_{i=p+1}^{j} |1+x_i| \prod_{i=j+1}^{k+p-1} |x_i| \Big) \quad ; \quad p = 1,\dots,k \quad ; \quad n = 0,1,2,\dots ,$$

(2.22) $$v_{kn+p} = v_p = \sum_{j=p}^{k+p-1} \Big(\prod_{i=p+1}^{j} |1+y_i| \prod_{i=j+1}^{k+p-1} |y_i| \Big) \quad ; \quad p = 1,\dots,k \quad ; \quad n = 0,1,2,\dots ,$$

(2.23) $$D = \prod_{n=1}^{k} |1 + x_n| - \prod_{n=1}^{k} |x_n| ,$$

(2.24) $$0 < \mu \le D .$$

(B) The k-periodic sequence $\{V_n\}$ given by

(2.25) $$V_{kn+p} = V_p = \{w \in \mathbb{C} ; |w-x_p| \le R_p\} \quad ; \quad p = 0,\dots,k-1 \quad ; \quad n = 0,1,2,\dots$$

is a sequence of pre value regions corresponding to $\{E_n\}$ when

(2.26) $$R_{kn+p} = R_p = \frac{D-\mu}{2t_p} \quad ; \quad p = 0,\dots,k-1 \quad ; \quad n = 0,1,2,\dots .$$

(C) If $a_n \in E_n$ for all $n \in \mathbb{N}$, then

(I): $f^{(m)} \in V_m \quad ; \quad m = 0,1,2,\dots,$ and

(2.27) (II): $$|f-S_{kn+p}(w_n^{(p)})| \le Q^n \cdot \frac{D-\mu}{t_0} \quad ; \quad p = 1,\dots,k \quad ; \quad n = 0,1,2,\dots ,$$

where

(2.28) $$Q = \prod_{j=1}^{k} \frac{t_j|1+x_j| + t_{j-1}|x_{j-1}| - \mu}{t_j|1+x_j| + t_{j-1}|x_{j-1}| + \mu} ,$$

and $\{\{w_n^{(p)}\}_{n=0}^{\infty}\}_{p=1}^{k}$ is an arbitrary sequence such that $w_n^{(p)} \in V_p$ for all $n \geq 0$ and $p \in \{1,\ldots,k\}$.

(D) The k-periodic sequence $\{E_n^*\}$ given by

$$(2.29)\qquad E_{kn+p}^* = E_p^* = E_{k-p+1} \quad ; \quad p = 1,\ldots,k \ ; \quad n = 0,1,2,\ldots$$

is a sequence of convergence regions. The k-periodic sequence $\{V_n^*\}$ given by

$$(2.30)\qquad V_{kn+p}^* = V_p^* = \{w \in \mathbb{C} \ ; \ |w+1+y_{k-p}| \leq r_{k-p}\} \quad ; \quad p = 0,\ldots,k-1 \ ; \quad n = 0,1,2,\ldots$$

is a sequence of pre-value regions corresponding to $\{E_n^*\}$, when

$$(2.31)\qquad r_{kn+p} = r_p = \frac{D-\mu}{2v_p} \quad ; \quad p = 0,\ldots,k-1 \ ; \quad n = 0,1,2,\ldots .$$

If $a_n \in E_n^*$ for all $n \in \mathbb{N}$, then

$$(2.32)\qquad \begin{aligned} &(I): \ f^{(m)} \in V_m^* \ \text{ for all } m \geq 0 \ , \text{ and} \\ &(II): \ \left|f - S_{kn+p}(w_n^{(p)*})\right| \leq (Q^*)^n \cdot \frac{D-\mu}{v_0} \quad ; \quad p = 1,\ldots,k \ ; \quad n = 0,1,2,\ldots \end{aligned}$$

where

$$(2.33)\qquad Q^* = \prod_{j=1}^{k} \frac{v_j|1+y_j| + v_{j-1}|y_{j-1}| - \mu}{v_j|1+y_j| + v_{j-1}|y_{j-1}| + \mu}$$

and $\{\{w_n^{(p)*}\}_{n=0}^{\infty}\}_{p=1}^{k}$ is an arbitrary sequence such that $w_n^{(p)*} \in V_p^*$ for all $n \geq 0$ and $p \in \{1,\ldots,k\}$.

Remarks.

(1) By Theorem 2.1 (C), $x_n \neq y_n$ for all n whenever $x_{n_0} \neq y_{n_0}$ for one special n_0. Furthermore, by Theorem 2.1 (A), $D > 0$.

(2) $v_p, t_p \neq 0$ for all p, because $x_n \neq 0,-1$ for all n, when $\Gamma_1 \cdots \Gamma_k \neq 0$. By Theorem 2.1 (B), we therefore also have $y_n \neq 0,-1$ for all n.

(3) Condition (i) in (A) is not necessary. If $\Gamma_1 \cdots \Gamma_k = 0$, we let $x_p = \frac{\Gamma_{p+1}}{1} + \cdots + \frac{\Gamma_{p+k}}{1}$ for $p = 0,\ldots,k-1$. Under the condition that $x_p \neq -1$ for all p, we get $D > 0$ (since $x_p = 0$ for at least one $p \in \{0,\ldots,k-1\}$), and the conclusions of Theorem 2.2 are still valid (partly by a proof that follows the same pattern as that of Theorem 2.2, partly by noticing that the disks $E_1,\ldots,E_k$ are contained in some slightly larger disks with centers slightly shifted.)

(4) Conditions (iii) may also be abandoned. But if $D = 0$ (the only other possibility by Theorem 2.1 (A)), the disks $E_1,\ldots,E_k$ will degenerate to one-point sets $\{\Gamma_1\},\ldots,\{\Gamma_k\}$ (when we allow $\mu = 0$). And $K(\Gamma_n/1)$ is known to converge.

(5) The conclusion in (B) is still valid if $\{R_p\}_{p=0}^{\infty}$ is replaced by $\{R'_p\}_{p=0}^{\infty}$ given by

$$(2.34)\qquad R'_{kn+p} = R'_p = \frac{D+\nu}{2t_p} \quad ; \quad p = 0,\ldots,k-1 \ ; \quad n = 0,1,2,\ldots$$

where $-\mu \leq \nu \leq \mu$. Furthermore, we may have $\mu = 0$ without altering the conclusion of (B) . Finally, (B) is still true when K_p is replaced by $4t_p t_{p-1}$; $p = 1,\ldots,k$ in the radii of the disks $E_1,\ldots,E_k$.

(6) The theorem generalizes the k-periodic continued fraction. It considers continued fractions $K(a_n/1)$ that are nearly k-periodic in the sense that the elements $\{a_n\}$ are picked from k disks, periodically.

(7) Part (D) generalizes Galois' theorem [4, Theorem 3.4, p. 56] , because $\{E_n^*\}$ may be looked upon as the dual of $\{E_n\}$.

(8) For $k = 1$ we have $v_n = t_n = 1$. For $k = 2$ we have

$$t_n = |1 + x_{n+1}| + |x_{n+1}| = |y_n| + |1 + y_{n+2}| = v_{n+1}$$

So, in those cases $K_p = 4t_p t_{p-1} = 4v_p v_{p-1}$.

The proof of Theorem 2.2 is based on a lemma:

Lemma 2.3. *Let $\{\tau_n\}$ be a sequence of linear fractional transformations such that*

(i) $|\tau_n(\infty)| \leq q$; $n = 1,2,3,\ldots$ *for some* $q < 1$, *and*

(ii) $\tau_n(U) \subseteq U$; $n = 1,2,3,\ldots$ *where*

(2.35) $$U = \{z \in \mathbb{C} ; |z| \leq 1\} .$$

Let $G_n = \tau_1 \circ \tau_2 \circ \cdots \circ \tau_n$ *for all* $n \in \mathbb{N}$. *Then*

(2.36) $$\lim_{n\to\infty} G_n(z_n) = G \in U$$

for any sequence $\{z_n\}$ *such that* $|z_n| \leq q'$ *for all* $n \in \mathbb{N}$ *and some* $q' < 1$.
(G *is independent of the actual choice of* $\{z_n\}$.)

This lemma is a strengthened form of Lemma 4.38 in [4] . We leave out the proof here, since it is almost identical to that of Lemma 4.38 in [4] .

When we prove Theorem 2.2 , we first prove part (B) , because that result is used in the proof of (A) .

Proof of Theorem 2.2. (B) Conditions (i) and (ii) in Definition 1.1 are clearly satisfied. To prove condition (iii) , let $z = \Gamma_n + re^{i\phi} \in E_n$ and $w = x_n + Re^{i\theta} \in V_n$ for some fixed $n \geq 0$. Then

$$\left|\frac{z}{1+w} - x_{n-1}\right| = \left|\frac{re^{i\phi} - x_{n-1}Re^{i\theta}}{1 + x_n + Re^{i\theta}}\right| \leq \frac{\rho_n + |x_{n-1}|R_n}{|1+x_n| - R_n}$$

$$\leq \frac{D^2-\mu^2+2t_{n-1}|x_{n-1}|(D-\mu)}{2t_{n-1}[2t_n|1+x_n|-(D-\mu)]} = \frac{D-\mu}{2t_{n-1}} = R_{n-1}$$

by using Theorem 2.1 (D) , since $x_{n-1}(1 + x_n) = \Gamma_n$ and

(2.37) $$|1 + x_n| - R_n = \frac{1}{2t_n}[2t_n|1+x_n| - (D-\mu)]$$

$$= \frac{1}{2t_n}[t_n|1+x_n| + t_{n-1}|x_{n-1}| + \mu] > 0 .$$

(A) When $\mu = D$, we have $E_n = \{\Gamma_n\}$ for all n so therefore $\{E_n\}$ is clearly a sequence of convergence regions (since $K(\Gamma_n/1)$ converges). Suppose in the following that $0 < \mu < D$.

Let $K(a_n/1)$ be an arbitrarily chosen continued fraction where $a_n \in E_n$ for all $n \in \mathbb{N}$. In order to prove that $K(a_n/1)$ converges (and thereby that $\{E_n\}$ is a sequence of convergence regions), we shall first use Lemma 2.3 to prove that

(2.38) $$\lim_{n\to\infty} S_{kn}(w_n^{(0)}) = F \in V_0$$

for any sequence $\{w_n^{(0)}\}$ from V_0. (F is independent of the actual choice of $\{w_n^{(0)}\}$).

Let $v(z)$ be a linear fractional transformation such that $v(U) = V_0'$ and $v(\infty) = \xi$ when U is defined by (2.35).

(2.39) $$V_0' = \{w \in \mathbb{C} ; |w-x_0| \le \frac{D-(\mu-\delta)}{2t_0}\}$$

where $0 < \delta < \min\{\mu, \frac{R_{k-1}R_0 2t_0}{|x_{k-1}| - R_{k-1}}\}$, and

(2.40) $$\xi = \begin{cases} \infty & \text{when } 0 \in V_{k-1} \\ x_0 + \frac{1+x_0}{|1+x_0|}\left(R_0 + \frac{R_{k-1}R_0}{|x_{k-1}| - R_{k-1}}\right) & \text{when } 0 \notin V_{k-1} . \end{cases}$$

(The existence of v is insured by the fact that $\xi \notin V_0'$.) By repeated application of (B), we get

(α): $S_k^{(nk)}(V_0) \subseteq V_0$; $n = 0,1,2\ldots$,

and

(β): $S_k^{(nk)}(V_0') \subseteq V_0'$,

since

(2.41) $a_n \in E_n \subseteq E_n' = \{z \in \mathbb{C} ; |z-\Gamma_n| \le \frac{D^2-(\mu-\delta)^2}{K_n}\}$; $n = 1,2,3,\ldots$

and $E_n'/(1+V_n') \subseteq V_{n-1}'$ by (B). ($K_{kn+p} = K_p$; $p = 1,\ldots k$; $n = 0,1,2,\ldots$)

Let $\tau_n = v^{-1} \circ S_k^{(nk)} \circ v$; $n = 0,1,2,\ldots$. Then $\{\tau_n\}$ is a sequence of linear fractional transformations such that $\tau_n(U) \subseteq U$ and $|\tau_n(\infty)| \le q$ for some $q < 1$ $n = 0,1,2,\ldots$. To see that $|\tau_n(\infty)| \le q$, we have to consider the following two possibilities:

<u>Case 1:</u> $0 \in V_{k-1}$. We know that

$$\frac{E_1}{1} + \cdots + \frac{E_{k-1}}{1+V_{k-1}} \subseteq V_0$$

where $\mathrm{dist}(\partial V_0, \partial V_0') = \frac{\delta}{2t_0} > 0$. Therefore, since $S_k^{(nk)}(\xi) = S_{k-1}^{(nk)}(0) \in V_0$, there exists a $q < 1$ such that $|\tau_n(\infty)| \le q$ for all n.

<u>Case 2:</u> $0 \notin V_{k-1}$. Then $\frac{a_{nk}}{1+\xi} \in V_{k-1}$ for all $n \in \mathbb{N}$ because

$$\left|\frac{a_{nk}}{1+\xi} - x_{k-1}\right| \le \frac{\rho_k + R_0|x_{k-1}| + |x_{k-1}|\cdot\frac{R_0R_{k-1}}{|x_{k-1}|-R_{k-1}}}{|1+x_0| + R_0 + \frac{R_0R_{k-1}}{|x_{k-1}|-R_{k-1}}} \le R_{k-1}$$

where the last inequality follows by

$$\rho_k + R_0|x_{k-1}| + |x_{k-1}|\cdot\frac{R_0R_{k-1}}{|x_{k-1}|-R_{k-1}} - R_{k-1}\left(|1+x_0| + R_0 + \frac{R_0R_{k-1}}{|x_{k-1}|-R_{k-1}}\right)$$

$$\le \frac{1}{4t_kt_{k-1}}[(D^2-\mu^2) - 2D(D-\mu) + (D-\mu)^2 - (D-\mu)^2]$$

$$= -\frac{1}{4t_kt_{k-1}}[D-\mu]^2 < 0$$

since $0 < \mu < D$. Therefore, by the same argument as in case 1 , we get that there exists a $q < 1$ such that

$$|\tau_n(\infty)| = |v^{-1}\circ S_k^{(nk)}(\xi)| = \left|v^{-1}\circ S_{k-1}^{(nk)}\left(\frac{a_{nk+k}}{1+\xi}\right)\right| \le q \quad .$$

By Lemma 2.3 we know that there exists a $G \in U$ such that

$$\lim_{n\to\infty}(\tau_0\circ\tau_1\circ\cdots\circ\tau_n(z_n)) = G$$

for any sequence $\{z_n\}$ such that $|z_n| \le q' < 1$ for all n . In particular, this is true when $q' < 1$ is chosen such that $v^{-1}(V_0) \subseteq \{z \in \mathbb{C}\,;\ |z| \le q'\}$. Then, since

$$S_{nk} = S_k \circ S_k^{(k)} \circ \cdots \circ S_k^{(n-1)k}$$

$$= v\circ v^{-1}\circ S_k\circ v\circ v^{-1}\circ S_k^{(k)}\circ v\circ\cdots\circ v^{-1}\circ S_k^{(n-1)k}\circ v\circ v^{-1}$$

$$= v\circ\tau_0\circ\tau_1\circ\cdots\circ\tau_{n-1}\circ v^{-1} \quad ; \quad n = 1,2,3,\ldots$$

we see that

$$\lim_{n\to\infty} S_{nk}(w_n^{(0)}) = v(G) = F$$

when $\{w_n^{(0)}\}$ is any sequence from V_0 . That $F \in V_0$ follows from the fact that $S_{nk}(w_n^{(0)}) \in V_0$ for all n , and V_0 is closed. This proves (2.38) .

This result is also true if we consider a tail $\overset{\infty}{\underset{n=m+1}{K}}(a_n/1)$ of the continued fraction; that is, we get

$$\text{(2.42)} \qquad \lim_{n\to\infty} S_{nk}^{(m)}(w_n^{(m)}) = F^{(m)} \in V_m$$

when $\{w_n^{(m)}\}$ is any sequence from V_m , for $m = 0,1,2,\ldots$. Furthermore, we get

$$\lim_{n\to\infty} S_{nk+p}^{(m)}(w_n^{(m+p)}) = \lim_{n\to\infty} S_{nk}^{(m)}\left(\frac{a_{m+nk+1}}{1} + \cdots + \frac{a_{m+nk+p}}{1+w_n^{(m+p)}}\right)$$

$$= \lim_{n\to\infty} S_{nk}^{(m)}(w_n^*) = \lim_{n\to\infty} S_{nk}^{(m)}(w_n^{(m)}) = F^{(m)} \in V_m$$

for $p = 0,\dots,k-1$ and $m = 0,1,2,\dots$, because

$$w_n^* = \frac{a_{m+nk+1}}{1} + \dots + \frac{a_{m+nk+p}}{1+w_n^{(m+p)}} \in V_m \quad ; \quad m = 0,1,2,\dots \quad .$$

Therefore

$$(2.43) \qquad \lim_{n\to\infty} S_n^{(m)}(w_n^{(m+n)}) = F^{(m)} \in V_m$$

so $\{F^{(m)}\}_{m=0}^{\infty}$ satisfies the condition

$$(2.44) \qquad F^{(m)}(1 + F^{(m+1)}) = a_{m+1} \quad ; \quad m = 0,1,2,\dots .$$

We shall use the results (2.43) , (2.44) to prove that $K(a_n/1)$ converges.

By (2.42) we see that $\{S_{nk}^{(m)}\}_{n=1}^{\infty}$ is a sequence of linear fractional transformations which in the limit maps V_m onto a point $F^{(m)}$. This is what is called the limit point case in Thron's work on convergence of such sequences [6] , which states that when $\{T_n\}$, defined by

$$T_n(z) = \frac{z+\overline{G_n}}{G_n z+1} \quad ; \quad |G_n| < 1 \quad ; \quad n = 1,2,3,\dots \quad ,$$

is such that $\lim_{n\to\infty} T_n(U) = \{c\}$ ($U = \{z \; ; \; |z| \leq 1\}$),then $\lim_{n\to\infty} T_n(z) = c$ for every z that is not a limit point on the sequence $\{-1/G_n\}$.

Transformed to our situation, it says that $\lim_{n\to\infty} S_{nk}^{(m)}(z) = F^{(m)}$ for every z that is not a limit point of $\{-h_{kn}^{(m)}\}_{n=1}^{\infty}$, since

$$(2.45) \qquad S_{nk}^{(m)}(z) = \frac{A_{nk}^{(m)} + A_{nk-1}^{(m)} z}{B_{nk}^{(m)} + B_{nk-1}^{(m)} z}$$

and

$$(2.46) \qquad \frac{B_{kn}^{(m)}}{B_{kn-1}^{(m)}} = h_{kn}^{(m)} = 1 + \frac{a_{m+kn}}{1} + \dots + \frac{a_{m+2}}{1} \quad ; \quad m \geq 0 \quad , \quad n \geq 1 \quad .$$

Since

$$\lim_{n\to\infty} S_{nk+p}^{(m)}(z) = \frac{a_{m+1}}{1} + \dots + \frac{a_{m+p}}{1 + \lim_{n\to\infty} S_{nk}^{(m+p)}(z)}$$

$$= \frac{a_{m+1}}{1} + \dots + \frac{a_{m+p}}{1 + F^{(m+p)}} = F^{(m)}$$

by (2.44) , the following statement is true:

<u>Statement</u>: If $K(c_n/1)$ is a continued fraction such that $c_n \in E_n$ for all $n \in \mathbb{N}$ and $\{h_{kn}^{(m)'}\}_{n=1}^{\infty}$ has no limit point at 0 for any $m \in \{0,\dots,k-1\}$, where

$$(2.47) \qquad h_n^{(m)'} = 1 + \frac{c_{m+n}}{1} + \dots + \frac{c_{m+2}}{1} \quad ,$$

then $K(c_n/1)$ converges to $F' = \lim_{n\to\infty} S_n'(w_n) \in V_0$ where $w_n \in V_n$; $n = 1,2,3,\dots$

and $S_n' = S_n^{(0)'}$ and

(2.48) $$S_n^{(m)'}(w) = \frac{c_{m+1}}{1} + \cdots + \frac{c_{m+n}}{1+w} \; ; \quad n = 1,2,3,\ldots \; ; \quad m = 0,1,2,\ldots \; .$$

By Galois' theorem [4, Theorem 3.4, p. 56] , we know that $\lim_{n\to\infty} h_{kn}^{(m)*} = -y_m$; $m = 0,1,2,\ldots$, where

(2.49) $$h_n^{(m)*} = 1 + \frac{\Gamma_{m+n}}{1} + \frac{\Gamma_{m+n-1}}{1} + \cdots + \frac{\Gamma_{m+2}}{1} \; ; \quad n = 2,3,4,\ldots \; .$$

Therefore, we can find an $N_0 \in \mathbb{N}$ such that

(2.50) $$\left|h_{k(N_0-1)}^{(m)*} + y_m\right| \leq r_m \quad \text{for} \quad m = 0,1,\ldots,k-1 \; ,$$

where r_m is defined by (2.31) . Let $K(c_n/1)$ be the continued fraction given by:

(2.51) $$c_n = \begin{cases} \Gamma_n & ; \quad n \leq N_0 k \; , \\ a_{n-N_0 k} & ; \quad n > N_0 k \; . \end{cases}$$

Then $K(c_n/1)$ is a continued fraction such that

(α) $c_n \in E_n$ for all $n \in \mathbb{N}$, and

(β) $\{h_{kn}^{(m)'}\}$ has no limit point at 0 for any $m \in \{0,\ldots,k-1\}$, when $h_n^{(m)'}$ is defined by (2.47) because

(i) $\left|h_{k(N_0-1)}^{(m)'} + y_m\right| \leq r_m$ for $m = 0,\ldots,k-1$ by (2.50).

(ii) If $\left|h_n^{(m)} + y_{m+n}\right| \leq r_{m+n}$, then

$$\left|h_{n+1}^{(m)} + y_{m+n+1}\right| = \left|1 + \frac{c_{m+n+1}}{h_n^{m'}} + \left(-1 + \frac{\Gamma_{m+n+1}}{y_{m+n}}\right)\right|$$

$$\leq \frac{\rho_{m+n+1} + \left|1 + y_{m+n+1}\right| \cdot \left|h_n^{(m)'} + y_{m+n}\right|}{\left|y_{m+n}\right| - \left|h_n^{(m)'} + y_{m+n}\right|}$$

$$\leq \frac{\rho_{m+n+1} + \left|1 + y_{m+n+1}\right| \cdot r_{m+n}}{\left|y_{m+n}\right| - r_{m+n}}$$

$$\leq \frac{D^2 - \mu^2 + 2v_{m+n+1}\left|1+y_{m+n+1}\right|(D-\mu)}{2v_{m+n+1}[2v_{m+n}\left|y_{m+n}\right| - (D-\mu)]} = r_{m+n+1}$$

by use of Theorem 2.1 , since

(2.52) $$\left|y_n\right| - r_n = \frac{1}{2v_n}[2v_n\left|y_n\right| - D + \mu]$$

$$= \frac{1}{2v_n}[v_n\left|y_n\right| + v_{n+1}\left|1 + y_{n+1}\right| + \mu] > 0 \; .$$

(iii) By induction

(2.53) $$\left|h_n^{(m)'} + y_{m+n}\right| \leq r_{m+n} \quad \text{for all} \quad n \geq k(N_0-1)$$

for $m = 0,\dots,k-1$. So by (2.52) $\{h_n^{(m)'}\}$ will have no limit point at 0 for any $m \in \{0,\dots,k-1\}$.

Therefore, by the previous statement, $K(c_n/1)$ converges. In particular, the kN_0^{th} tail of $K(c_n/1)$, which is $K(a_n/1)$, converges. And, since $a_n \in E_n$ was arbitrarily chosen for all n , this proves (A) .

(C.I) If we continue the last argument in the proof of (A) , we also know that

$$\operatorname*{K}_{n=1}^{\infty} (c_n/1) = F' = \lim_{n\to\infty} S_n^{(0)'}(w_n^{(n)}) \in V_0$$

where $w^{(m+n)} \in V_{n+m}$ for all $n \in \mathbb{N}$. By (2.44) we know that $F^{(m)'}(1 + F^{(m+1)'}) = c_{m+1}$; $m = 0,1,2,\dots$ when $F^{(m)'} = \lim_{n\to\infty} S_n^{(m)'}(w_n^{(m+n)})$; $m = 0,1,2,\dots$. Since the tails $\{f^{(n)'}\}$ of $K(c_n/1)$ satisfy the equations:

$$f^{(0)'} = F'$$

$$f^{(m)'}(1 + f^{(m+1)'}) = c_{m+1} \quad ; \quad m = 0,1,2,\dots$$

we must have $f^{(m)'} = F^{(m)'}$ for all m . And since $f^{(m)} = f^{(m+kN_0)'}$, (C.I) follows.

(C.II) To prove (2.27) , we use the fact that $f \in V_0$ and $S_{kn+p}^{(m)}(w_n^{(p+m)}) \in V_m$ when $w_n^{(p+m)} \in V_{p+m}$. Then

$$|f - S_{nk+p}(w_n^{(p)})| = \left|\frac{a_1}{1+f^{(1)}} - \frac{a_1}{1+S_{kn+p-1}^{(1)}(w_n^{(p)})}\right|$$

$$= \left|\frac{S_{kn+p}(w_n^{(p)})}{1 + f^{(1)}}\right| \cdot |f^{(1)} - S_{kn+p-1}^{(1)}(w_n^{(p)})|$$

$$\leq \frac{|x_0| + \frac{D-\mu}{2t_0}}{|1+x_1| - \frac{D-\mu}{2t_1}} \cdot |f^{(1)} - S_{kn+p-1}^{(1)}(w_n^{(p)})|$$

$$= \frac{t_0}{t_1} \cdot \frac{t_1|1+x_1| + t_0|x_0| - \mu}{t_1|1+x_1| + t_0|x_0| + \mu} \cdot |f^{(1)} - S_{kn+p-1}^{(1)}(w_n^{(p)})|$$

$$\leq Q^n \cdot |f^{(nk)} - S_p^{(nk)}(w_n^{(p)})| \leq Q^n \cdot 2R_0 = Q^n \cdot \frac{D-\mu}{t_0} .$$

(D) Let

(2.54) $$\Gamma_{kn+p}^* = \Gamma_p^* = \Gamma_{k-p+1} \quad ; \quad p = 1,\dots,k \quad ; \quad n = 0,1,2,\dots$$

(that is $K(\Gamma_n^*/1)$ is the dual of $K(\Gamma_n/1)$.). By Galois' theorem [4, Theorem 3.4, p. 56] , we know that the tails of $K(\Gamma_n^*/1)$ are given by

(2.55) $$f^{(n)*} = x_n^* = \operatorname*{K}_{m=n+1}^{\infty} (\Gamma_m^*/1) = -1 - y_{k-n} \quad ; \quad n = 0,1,2,\dots .$$

Since $K(\Gamma_n/1)$ is the dual of $K(\Gamma_n^*/1)$, we also get that the repulsive fixed point y_n^* of

(2.56) $$s_k^{(n)*}(w) = \frac{\Gamma_{n+1}^*}{1} + \cdots + \frac{\Gamma_{n+k}^*}{1+w}$$

is given by

(2.57) $$y_n^* = -1 - x_{k-n} \quad ; \quad n = 0,1,2,\ldots \quad .$$

Furthermore,

(2.58) $$t_{kn+p}^* = t_p^* = \sum_{j=p}^{k+p-1} \left(\prod_{i=p+1}^{j} |1+x_i^*| \cdot \prod_{i=j+1}^{k+p-1} |x_i^*| \right)$$

$$= \sum_{j=p}^{k+p-1} \left(\prod_{i=p+1}^{j} |y_{2k-i}| \cdot \prod_{i=j+1}^{k+p-1} |1+y_{2k-i}| \right) = \sum_{j=p}^{k+p-1} \left(\prod_{i=2k-j}^{2k-p-1} |y_i| \cdot \prod_{i=k-p+1}^{2k-j-1} |1+y_i| \right)$$

$$= \sum_{j=k-p}^{2k-p-1} \left(\prod_{i=k-p+1}^{j} |1+y_i| \cdot \prod_{i=j+1}^{2k-p-1} |y_i| \right) = v_{k-p} \quad .$$

In the same way

(2.59) $$v_{kn+p}^* = v_p^* = \sum_{j=p}^{k+p-1} \left(\prod_{i=p+1}^{j} |1+y_i^*| \cdot \prod_{i=j+1}^{k+p-1} |y_i^*| \right) = t_{k-p} \quad .$$

Finally

(2.60) $$D^* = \prod_{n=1}^{k} |1+x_n^*| - \prod_{n=1}^{k} |x_n^*| = \prod_{n=1}^{k} |y_n| - \prod_{n=1}^{k} |1+y_n| = D \quad .$$

Therefore part (D) follows by using Theorem 2.2 (A) , (B) and (C) on $\{E_n^*\}$.

References

1. Basic definitions and notation, these Lecture Notes
2. Lisa Jacobsen, A method for convergence acceleration of continued fractions $K(a_n/1)$, these Lecture Notes.
3. Lisa Jacobsen and Haakon Waadeland, Some useful formulas involving tails of continued fractions, these Lecture Notes.
4. William B. Jones and W.J. Thron, Continued Fractions: Analytic Theory and Applications; Encyclopedia of Mathematics and Its Applications, vol. 11, Addison Wesley, Reading, Massachusetts, (1980).
5. William B. Jones and W.J. Thron, Twin-convergence regions for continued fractions $K(a_n/1)$, Trans. Amer. Math. Soc. 150 (1970), 93-119.
6. W.J. Thron, Convergence of sequences of linear fractional transformations and of continued fractins, J. Indian Math. Soc. 27, (1963) 103-127.

Lisa Jacobsen

Institutt for Matematikk

Universitetet i Trondheim

7034 Trondheim - NTH

Norway

SOME USEFUL FORMULAS INVOLVING TAILS OF CONTINUED FRACTIONS

Lisa Jacobsen

Haakon Waadeland

In the use of auxiliary continued fractions for the purpose of convergence acceleration [2, 3, 7, 10] or analytic continuation [8, 9, 10], the tails of the auxiliary continued fraction are of vital importance. Since the method heavily depends upon knowledge of the tails, it seems to be a good idea to express other quantities connected with the problem in terms of the known tails. The purpose of the present paper is to list some formulas of this type and in some cases to indicate some previous and present use of them. Although the formulas are very elementary we have not found them in the literature (except in the few cases we refer to). This may partly be due to the fact that tails were less interesting earlier than they are now, partly that some of the results may be folklore among people working with continued fractions.

The proofs are in most of the cases straightforward induction, and shall be omitted. Rather than giving such proofs, we shall devote the space to comments. For definitions and notations we refer to [1], in particular to (DN5), (DN6a), (DN6b) and (DN9). Since in the proofs the only tail-property used will be the recursion relation

(1) $$g^{(n)} = \frac{a_{n+1}}{1+g^{(n+1)}}, \quad n = 0,1,2,\ldots.$$

the formulas will be valid not only for tails, but also for any sequence $\{g^{(n)}\}$ satisfying (1), which also means that we are not restricted to convergent continued fractions. In view of the possible applications this general setting makes sense.

<u>Proposition</u> 1. <u>For a given continued fraction</u> $K(a_n/1)$ <u>let</u> $\{g^{(n)}\}_{n=0}^{\infty}$ <u>be a sequence of complex numbers such that</u>

(1) $$g^{(n)} = \frac{a_{n+1}}{1+g^{(n+1)}}, \quad n=0,1,2,\ldots$$

<u>Then the following formulas hold for</u> $n=0,1,2,\ldots$ n

(2) $$A_n + A_{n-1}g^{(n)} = g^{(0)} \prod_{m=1}^{n} (1+g^{(m)}),$$

(3) $$B_n + B_{n-1}g^{(n)} = \prod_{m=1}^{n} (1+g^{(m)}),$$

(4) $$A_n - B_n g^{(0)} = (-1)^{n-1} \prod_{m=0}^{n} g^{(m)}.$$

The proofs are trivial verifications (inductions using [1,(DN6)] and (1)).

The case one thinks of first is when the continued fraction converges and $g^{(n)} = f^{(n)}$ are the tail values (all assumed to be finite). In the particular case of a T-fraction formulas of the type (2) and (3) were used in [11, 12]. See

[12,(1.11) and (1.12)]. Actually, in that particular investigation $\{g^{(n)}\}$ is the sequence of tails in one part of the plane and a sequence of "wrong tails" in another, leading to analytic continuation of the function represented by the T-fraction. For periodic continued fractions of period one with $g^{(n)}$ = the "right" or "wrong" root the formulas are particularly simple. This case occurs in the paper [9] in connection with analytic continuation of functions represented by continued fractions. The formulas [9,(2.2) and (2.3)] are valid for limit periodic continued fractions, but reduce to the periodic case in the obvious way. Also there they lead to analytic continuation of functions, represented by continued fractions.

The formulas (2), (3), (4) and several of the subsequent formulas are used in [4].

Using the formulas (2) and (3) we easily get

Proposition 2. *With* $g^{(n)}$ *as in Proposition* 1 *we have*

(5)
$$A_n = g^{(0)} \sum_{p=2}^{n+1} (-1)^{n+1-p} \left(\prod_{\nu=p}^{n} g^{(\nu)}\right)\left(\prod_{\nu=1}^{p-1} (1+g^{(\nu)})\right)$$

and

(6)
$$B_n = \sum_{p=1}^{n+1} (-1)^{n+1-p} \left(\prod_{\nu=p}^{n} g^{(\nu)}\right)\left(\prod_{\nu=1}^{p-1} (1+g^{(\nu)})\right)$$

We can of course get similar formulas for $A_n^{(m)}$ and $B_n^{(m)}$. (For definition, see [1,(DN 5,6)] and the definition of $S_n^{(m)}$, also in [1].)

(5')
$$A_n^{(m)} = g^{(m)} \sum_{p=m+2}^{m+n+1} (-1)^{m+n+1-p} \left(\prod_{\nu=p}^{m+n} g^{(\nu)}\right)\cdot\left(\prod_{\nu=m+1}^{p-1} (1+g^{(\nu)})\right)$$

(6')
$$B_n^{(m)} = \sum_{p=m+1}^{m+n+1} (-1)^{m+n+1-p} \left(\prod_{\nu=p}^{m+n} g^{(\nu)}\right)\cdot\left(\prod_{\nu=m+1}^{p-1} (1+g^{(\nu)})\right)$$

And from these equalities, we get the following result:

(7)
$$B_n^{(m)} g^{(m+n+1)} + B_n^{(m+1)}(1+g^{(m+1)}) = \prod_{\nu=m+1}^{m+n+1} (1+g^{(\nu)}) + (-1)^n \prod_{\nu=m+1}^{m+n+1} g^{(\nu)}$$

In the special case when $K(a_n/1)$ is convergent and k-periodic, we get the following proposition:

Proposition 3. *Let* $K(a_n/1)$ *be a* k-*periodic, convergent continued fraction. Let* $\tilde{f}^{(n)}$ *denote the repulsive fixed point of* $S_k^{(n)}(w)$. *Then*

(8)
$$B_{k-1}^{(m)} f^{(m)} + B_{k-1}^{(m+1)}(1+f^{(m+1)}) = \prod_{\nu=1}^{k} (1+f^{(\nu)}) + (-1)^{k-1} \prod_{\nu=1}^{k} f^{(\nu)}$$

(9)
$$B_{k-1}^{(m)} \tilde{f}^{(m)} + B_{k-1}^{(m+1)}(1+\tilde{f}^{(m+1)}) = \prod_{\nu=1}^{k} (1+\tilde{f}^{(\nu)}) + (-1)^{k-1} \prod_{\nu=1}^{k} \tilde{f}^{(\nu)}$$

$$= - \prod_{\nu=1}^{k} (1+f^{(\nu)}) - (-1)^{k-1} \prod_{\nu=1}^{k} f^{(\nu)}$$

(10) $$f^{(m)} - \tilde{f}^{(m)} = \frac{1}{B_{k-1}^{(m)}} \left(\prod_{\nu=1}^{k} (1+f^{(\nu)}) + (-1)^{k-1} \prod_{\nu=1}^{k} f^{(\nu)} \right)$$

(11) $$(A_{k-1}^{(n)} - B_k^{(n)})^2 + 4B_{k-1}^{(n)} A_k^{(n)} = \left[\prod_{\nu=1}^{k} (1+f^{(\nu)}) + (-1)^{k-1} \prod_{\nu=1}^{k} f^{(\nu)} \right]^2$$

<u>independently of</u> n.

Here (8) follows directly from (7), by using the facts that $\{f^{(n)}\}$ is k-periodic and

$$f^{(n)} = \frac{a_{n+1}}{1+f^{(n+1)}} .$$

(9) follows in a similar way, using in addition

(12) $$B_{k-1}^{(n)} f^{(n)} = - B_{k-1}^{(n+1)} (1+\tilde{f}^{(n+1)}) ,$$

(13) $$B_{k-1}^{(n)} \tilde{f}^{(n)} = - B_{k-1}^{(n+1)} (1+f^{(n+1)}) ,$$

by [4,Thm. 2.1] where $\{B_{k-1}^{(n)}\}_{n=0}^{\infty}$ also is k-periodic. (10) follows by combining (8) or (9) with (13) or (12) respectively. (11) is a direct consequence of (10), since $f^{(m)}$ and $\tilde{f}^{(m)}$ are solutions of the quadratic equation

$$x = S_k^{(m)}(x) = \frac{A_k^{(m)} + A_{k-1}^{(m)} x}{B_k^{(m)} + B_{k-1}^{(m)} x} .$$

In the theory of continued fractions, the quantity $h_n = B_n / B_{n-1}$ is of great importance [5, (4.1.3)]. The following proposition is a direct result of (3) and (4):

<u>Proposition 4.</u> <u>Let</u> $K(a_n/1)$ <u>be a convergent continued fraction. Then</u>

(14) $$h_n + f^{(n)} = (-1)^{n-1} \frac{\prod_{m=1}^{n} (1+f^{(m)})}{\prod_{m=0}^{n-1} f^{(m)}} \cdot (f - f_{n-1})$$

$$= f^{(n)} \cdot \frac{f_{n-1} - f_n}{f - f_n} ,$$

(15) $$h_n = - f^{(n)} \cdot \frac{f - f_{n-1}}{f - f_n} .$$

These formulas contain some useful information on h_n, and combining them can give further equalities.

Another useful concept in the theory of continued fractions is the concept of contractions: Let $K(a_n/1)$ be any continued fraction, convergent or not. Furthermore, let $\{n_\nu\}_0^\infty$ be an increasing subsequence of the non-negative integers. Then $b_0^* + K(a_n^*/b_n^*)$ is a contraction of $K(a_n/1)$ with respect to $\{n_\nu\}_0^\infty$ iff $f_\nu^* = f_{n_\nu}$ for all integers $\nu \geq 0$. Such contractions may be used to simplify the study of some continued fractions $K(a_n/1)$. For instance in the special case when $K(a_n/1)$ is k-periodic and $n_\nu = \nu \cdot k$ for all $\nu \geq 0$, we get 1-periodic contractions. Since the behaviour of tails is of growing importance, it is also interesting to see what the tails of a contraction of a continued fraction look like.

Proposition 5. Let $K(a_n/1)$ be a convergent continued fraction. Let $b_0^* + K(a_n^*/b_n^*)$ be the contraction given by [6, p.11]

$$b_0^* = \frac{A_{n_0}}{B_{n_0}}, \quad b_1^* = B_{n_1} \tag{16}$$

$$b_\nu^* = \frac{B^{(n_{\nu-2}+1)}_{n_\nu - n_{\nu-2}-1}}{B^{(n_{\nu-2}+1)}_{n_{\nu-1}-n_{\nu-2}-1}} \; ; \; \nu = 2,3,4,\cdots\cdot \tag{17}$$

$$a_1^* = (-1)^{n_0} \quad a_1 \cdots a_{n_0+1} \frac{B^{(n_0+1)}_{n_1-n_0-1}}{B_{n_0}} \tag{18}$$

$$a_2^* = (-1)^{n_1-n_0-1} \quad a_{n_0+2} \cdots a_{n_1+1} \cdot \frac{B_{n_0} B^{(n_1+1)}_{n_2-n_1-1}}{B^{(n_0+1)}_{n_1-n_0-1}} \tag{19}$$

$$a_\nu^* = (-1)^{n_{\nu-1}-n_{\nu-2}-1} \quad a_{n_{\nu-2}+2} \cdots a_{n_{\nu-1}+1} \frac{B^{(n_{\nu-1}+1)}_{n_\nu-n_{\nu-1}-1}}{B^{(n_{\nu-2}+1)}_{n_{\nu-1}-n_{\nu-2}-1}} \tag{20}$$

where $\{n_\nu\}_{\nu=0}^\infty$ is an increasing subsequence of the non-negative integers. Then

$$f^{(\nu)*} = (-1)^{n_\nu - n_{\nu-1}-1} \prod_{m=n_{\nu-1}+1}^{n_\nu} f^{(m)} \; , \; n \geq 1 \; , \tag{21}$$

$$f^{(0)*} = \frac{(-1)^{n_0}}{B_{n_0}} \prod_{m=0}^{n_0} f^{(m)} \; . \tag{21'}$$

(22) $$b_\nu^* + f^{(\nu)*} = \frac{B^{(n_{\nu-1}+1)}_{n_\nu - n_{\nu-1}-1}}{B^{(n_{\nu-2}+1)}_{n_{\nu-1}-n_{\nu-2}-1}} \cdot \prod_{m=n_{\nu-2}+2}^{n_{\nu-1}+1} (1+f^{(m)})$$

__for__ $\nu = 2, 3, 4, \ldots$

(22') $$b_0^* + f^{(0)*} = f$$

(22") $$b_1^* + f^{(1)*} = B^{(n_0+1)}_{n_1-n_0-1} \prod_{m=1}^{n_0+1} (1+f^{(m)}) \quad .$$

Proof (outline): By [6, Satz 1.6],
$b_0^* + K(a_n^*/b_n^*)$ is a contraction of $K(a_n/1)$ with respect to $\{n_\nu\}_{\nu=0}^\infty$.
Furthermore, it follows from the construction of the coefficients $\{b_n^*\}$, $\{a_n^*\}$, that

(23) $$A_0^* = \frac{A_{n_0}}{B_{n_0}} , \; B_0^* = 1 ,$$

(24) $$A_\nu^* = A_{n_\nu} , \; B_n^* = B_{n_\nu} \; ; \; \nu = 1,2,3,\ldots$$

Since

$$f = S_\nu^* (f^{(\nu)*}) = S_{n_\nu} (f^{(n_\nu)})$$

the proposition follows, by use of a generalized determinant formula, [6, p.1, formula (9)], and (23), (24).

A contraction $b_0^* + K(a_n^*/b_n^*)$ of a continued fraction $K(a_n/1)$ with respect to $\{n_\nu\}_{\nu=0}^\infty$ is not unique. In fact, any continued fraction equivalent to $b_0^* + K(a_n^*/b_n^*)$ will be such a contraction, since two continued fractions are equivalent iff they have the same sequence of approximants, (by definition [5, p. 31]). By [5, Theorem 2.6], $b_0' + K(a_n'/b_n')$ is equivalent to $b_0^* + K(a_n^*/b_n^*)$ iff

$$a_n' = r_n r_{n-1} a_n^* , \; b_n' = r_n b_n^*$$

for all n, some sequence $\{r_n\}$ of non-zero constants with $r_0 = 1$.
But what happens to the tails under equivalence transformations?

__Proposition 6.__ __Let__ $b_0 + K(a_n/b_n)$ __and__ $b_0' + K(a_n'/b_n')$ __be equivalent continued fractions such that__

(25) $$a_n' = r_n r_{n-1} a_n , \; b_n' = r_n b_n , \; r_0 = 1 .$$

__Then__

$$f^{(n)'} = r_n f^{(n)} . \qquad (26)$$

This is essentially stated in [6, Satz 1.2], although the word tail does not enter into that context. And it fully shows the importance of emphasizing which one of the contractions we are looking at.

Since the contraction given in Proposition 5 is the only one with the property that

$$A_\nu^* = A_{n_\nu} , \quad B_\nu^* = B_{n_\nu} \quad ; \nu = 1,2,3,\ldots \qquad (27)$$

we shall call that one the principal contraction of $K(a_n/1)$ with respect to $\{n_\nu\}_{\nu=0}^\infty$. Then Proposition 5 may be rewritten;

Proposition 5'. The principal contraction of a convergent continued fraction $b_0 + K(a_n/b_n)$ with respect to an increasing sequence $\{n_\nu\}_{\nu=0}^\infty$ of non-negative integers, is unique and has tails given by

$$f^{(\nu)*} = (-1)^{n_\nu - n_{\nu-1} - 1} \prod_{m=n_{\nu-1}+1}^{n_\nu} f^{(m)} .$$

Furthermore, if $b_\nu^* + f^{(\nu)*}$ is the quantity of interest, one may study the equivalent continued fraction $b_0' + K(a_n'/b_n')$ given by (25) with

$$r_n = \frac{B^{(n_{\nu-2}+1)}_{n_{\nu-1}-n_{\nu-2}-1}}{B^{(n_{\nu-1}+1)}_{n_\nu - n_{\nu-1}-1}} \quad ; n = 1,2,3,\cdots \qquad (28)$$

since in that case

$$b_n' + f^{(n)'} = \prod_{m=n_{\nu-2}+2}^{n_{\nu-1}+1} (1+f^{(m)}) . \qquad (29)$$

Finally, we mention that for equivalent continued fractions (using the notations from Prop. 6) the following formulas hold

$$A_n^{(m)'} = A_n^{(m)} \cdot \prod_{j=m}^{m+n} r_j ,$$
$$B_n^{(m)'} = B_n^{(m)} \cdot \prod_{j=m+1}^{m+n} r_j . \qquad (30)$$

Observe that for $m = 0$ the r-products are equal, since $r_0 = 1$.

Several other related results could have been included, among them some on dual periodic continued fractions. But hopefully the sample above is sufficient to accomplish the intention of the paper: to present some examples of useful formulas involving tails (or "wrong tails") in the theory of continued fractions.

References

1. BASIC DEFINITIONS AND NOTATIONS, these Lecture Notes.
2. Lisa Jacobsen, Convergence Acceleration for Continued Fractions $K(a_n/1)$, Trans. Amer. Math. Soc., to appear.
3. Lisa Jacobsen, A Method for Convergence Acceleration of Continued Fractions $K(a_n/1)$, these Lecture Notes.
4. Lisa Jacobsen, Some Periodic Sequence of Circular Convergence Regions, these Lecture Notes.
5. William B. Jones and Wolfgang J. Thron, Continued fractions: Analytic theory and applications, ENCYCLOPEDIA OF MATHEMATICS AND ITS APPLICATIONS, No. 11, Addison-Wesley Publishing Company, Reading, Mass. 1980.
6. O. Perron, Die Lehre von den Kettenbrüchen, 3. Auflage, 2. Band. Stuttgart, Teubner 1957.
7. W. J. Thron and Haakon Waadeland, Accelerating convergence of limit periodic continued fractions $K(a_n/1)$, Numer. Math. 34 (1980), 155-170.
8. W. J. Thron and Haakon Waadeland, Analytic continuation of functions defined be means of continued fractions, Math. Scand., 47 (1980), 72-90.
9. W. J. Thron and Haakon Waadeland, Convergence questions for limit periodic continued fractions, Rocky Mountain J. Math. 11, (1981), Number 4.
10. W. J. Thron and Haakon Waadeland, Modifications of Continued Fractions, a Survey, these Lecture Notes.
11. Haakon Waadeland, On T-fractions of functions, holomorphic and bounded in a circular disk, Norske Vid. Selsk. Skr. (Trondheim) (1964) No. 8, 1-19.
12. Haakon Waadeland, A convergence property of certain T-fraction expansions, Norske Vid. Selsk. Skr. (Trondheim) (1966) No. 9, 1-22.

Lisa Jacobsen
Institutt for Matematikk NTH
Universitetet i Trondheim
7034 Trondheim -- NTH
Norway

Haakon Waadeland
Institutt for Matematikk
og Statistikk NLHT
Universitetet i Trondheim
7055 Dragvoll
Norway

UNIFORM TWIN-CONVERGENCE REGIONS FOR CONTINUED FRACTIONS $K(a_n/1)$

William B. Jones

Walter M. Reid

1. Introduction. A sequence $\{E_n\}$ of subsets of $\mathbb{C}$ is called a sequence of convergence regions for continued fractions of the form

$$(1.1) \qquad \mathop{K}_{n=1}^{\infty} (a_n/1) = \frac{a_1}{1} + \frac{a_2}{1} + \frac{a_3}{1} + \cdots$$

if

$$(1.2) \qquad 0 \neq a_n \in E_n \quad , \quad n = 1,2,3,\ldots$$

insures the convergence of $K(a_n/1)$ to a finite value. Let f_n denote the nth approximant of (1.1) and let $f = \lim_{n\to\infty} f_n$, when that limit exists. A sequence of convergence regions $\{E_n\}$ is called a uniform sequence of convergence regions for continued fractions of the form (1.1) , if there exists a sequence of positive numbers $\{\varepsilon_n\}$ (depending only on $\{E_n\}$) with $\lim_{n\to\infty} \varepsilon_n = 0$ such that (1.2) implies that

$$(1.3) \qquad |f - f_n| \leq \varepsilon_n \quad , \quad n = 1,2,3,\ldots \quad .$$

In the special case where $E_{2n-1} = E_1$ and $E_{2n} = E_2$, $n = 1,2,3,\ldots$, one speaks of twin-convergence regions and uniform twin-convergence regions.

In 1959 Thron [10] proved that if $w = |w|e^{i\theta}$ and $\rho^* > 1$, then

$$(1.4) \qquad E_1^* = [w : |w| \leq \rho^*] \text{ and } E_2^* = [w : |w| \geq 2(\rho^* - \cos\theta)]$$

are twin-convergence regions for continued fractions $K(a_n/1)$. By means of a counter example he showed that (E_1^*, E_2^*) in (1.4) are not uniform twin-convergence regions. In the same paper it was shown that

$$(1.5a) \qquad E_1^*(\eta) = [w : |w| \leq \rho^*(1-\eta)] \text{ and } E_2^* = [w : |w| \geq 2(\rho^* - \cos\theta)]$$

are uniform twin-convergence regions for $K(a_n/1)$, provided that

$$(1.5b) \qquad 0 < \eta < \frac{1}{4}\left(\frac{\rho^*-1}{\rho^*}\right) \text{ and } \rho^* > 1 \quad .$$

Since $E_1^*(\eta) \subset E_1^*$, we say that $E_1^*(\eta)$ is a contraction of E_1^* .

By taking a different approach to convergence theory, Jones and Thron [6, Corollary 5.7] were able to extend the result of Thron for (1.4) by proving that, if Γ is a complex number and ρ a positive number satisfying

$$(1.6a) \qquad |\Gamma| < |1 + \Gamma| < \rho$$

then

$$(1.6b) \qquad E_1 = [w : |\bar{\Gamma}w - \Gamma(|1+\Gamma|^2 - |\Gamma|^2)| + |1+\Gamma||w| \leq \rho(|1+\Gamma|^2 - |\Gamma|^2)]$$

and

$$(1.6c)\qquad E_2 = [w : \rho|w| - |w(1+\bar{\Gamma}) - (1+\Gamma)(\rho^2 - |1+\Gamma|^2)| \geq |1+\Gamma|(\rho^2 - |1+\Gamma|^2)]$$

are twin-convergence regions for $K(a_n/1)$. We note that the first inequality in (1.6a) is equivalent to $\text{Re}(\Gamma) > -1/2$. It is easily seen that, when $\Gamma = 0$, the regions in (1.6) reduce to the corresponding regions in (1.4) (i.e. , $E_1 = E_1^*$ and $E_2 = E_2^*$). A disadvantage of the method used in [6] is that it gives no information about uniform convergence or speed of convergence of the continued fractions.

This paper is concerned with a sequence of convergence regions $\{E_n(\delta_n)\}$ for continued fractions $K(a_n/1)$ where each region $E_n(\delta_n)$ depends on a parameter δ_n (with $0 < \delta_n \leq 1$) and where

$$(1.7)\qquad E_n(\delta_n) \subset E_n(\delta_n') \quad \text{if} \quad \delta_n < \delta_n' \leq 1$$

and

$$(1.8)\qquad E_{2n-1}(1) = E_1 \quad \text{and} \quad E_{2n}(1) = E_2 \;, \quad n = 1,2,3,\ldots \;.$$

Here E_1 and E_2 are defined by (1.6). Thus $E_{2n-1}(\delta_{2n-1})$ and $E_{2n}(\delta_{2n})$ are contractions of E_1 and E_2, respectively. In our main result (Theorem 1) it is shown that $\{E_n(\delta_n)\}$ is a uniform sequence of convergence regions if the infinite product $\Pi\delta_n$ diverges to zero. Moreover, it is shown that, for continued fractions $K(a_n/1)$ with $0 \neq a_n \in E(\delta_n)$, the truncation error $|f - f_n|$ of the nth approximant f_n is bounded above by $2\rho\prod_2^n \delta_j$. This theorem therefore provides an extension of Thron's uniform twin convergence regions (1.5).

The methods used in [6], [10], and in the present paper are based on a study of sequences of nested circular disks $\{\mathcal{D}_n\}$ in $\mathbb{C}$ having the property that, for $n = 1,2,3,\ldots$,

$$(1.9)\qquad f_{n+m} \in \mathcal{D}_n \;, \quad m = 0,1,2,\ldots \;,$$

so that

$$(1.10)\qquad |f_{n+m} - f_n| < 2R_n \;, \quad m = 0,1,2,\ldots$$

where R_n is the radius of $\mathcal{D}_n$. The approach taken in [6] was to show that a continued fraction $K(a_n/1)$ converges even if $\lim R_n \neq 0$ (limit circle case). Hence nothing can be said about the speed of convergence. In [10] and in the present paper, estimates are obtained for R_n which insure that $\lim R_n = 0$ (limit point case). Therefore uniform convergence regions and speed-of-convergence estimates are obtained. This approach was employed first by Thron [9] in the investigation of parabolic convergence regions for continued fractions $K(a_n/1)$. It has subsequently been used here and in [1], [2], [3], [4], [5], [7], [8], and [10].

One of the main steps in Thron's method is obtaining suitable parametric representations for the element regions $E_n(\delta_n)$. In terms of the parameters defining $E_n(\delta_n)$, we derive an expression for R_n. This expression is then maximized for all $a_n \in E_n(\delta_n)$, thus giving an estimate of R_n and hence of (1.10).

Questions on best twin value regions are dealt with in Section 3. It is shown (Theorem 2) that some of the twin-convergence regions (1.6) are not best. In Theorem 3 it is established that for every pair of twin-convergence regions (1.4), there is no better pair of twin-convergence regions within the larger family (1.6).

We conclude this introduction by defining some special symbols that are used. Let A and B denote subsets of $\mathbb{C}$. Then Int(A) and ∂A denote the interior of A and boundary of A, respectively. B-A denotes the complement of A with respect to B. If f is a function of A into $\mathbb{C}$, then $f(A) = [f(x) : x \in A]$.

2. Uniform Convergence Regions. Here and throughout this paper, Arg z denotes the principal value of arg z, where $-\pi < \text{Arg } z \leq \pi$.

Theorem 1. *Let Γ be a complex number and ρ be a positive real number such that*

$$|\Gamma| < |1+\Gamma| < \rho \, . \tag{2.1}$$

Let $\{\delta_n\}$ be a sequence of positive numbers satisfying

$$|1+\Gamma|/\rho < \delta_{2n-1} \leq 1 \; , \quad 0 < \delta_{2n} \leq 1 \; , \quad n = 1,2,3,\ldots \, . \tag{2.2}$$

For each $n = 1,2,3,\ldots$, let $E_n(\delta_n) \subset \mathbb{C}$ be defined by

$$E_{2n-1}(\delta_{2n-1}) = [w = re^{i\theta} : 0 \leq r \leq t(\delta_{2n-1},\theta) - \sqrt{t^2(\delta_{2n-1},\theta) - K(\delta_{2n-1})}] \, , \tag{2.3a}$$

where

$$t(\delta_{2n-1},\theta) = \delta_{2n-1}\rho|1+\Gamma| - |\Gamma|^2\cos(\theta - 2\text{Arg}\Gamma) \, , \tag{2.3b}$$

$$K(\delta_{2n-1}) = (\delta_{2n-1}^2\rho^2 - |\Gamma|^2)(|1+\Gamma|^2 - |\Gamma|^2) \, , \tag{2.3c}$$

and

$$E_{2n}(\delta_{2n}) = [w = re^{i\theta} : r \geq 2|1+\Gamma|(\rho/\delta_{2n} - |1+\Gamma|\cos(\theta - 2\text{Arg}(1+\Gamma)))] \, . \tag{2.4}$$

Then: (A) *$\{E_n(\delta_n)\}$ is a sequence of convergence regions for continued fractions of the form $K(a_n/1)$.*

(B) *If the infinite product $\Pi\delta_n$ diverges to zero, then $\{E_n(\delta_n)\}$ is a uniform sequence of convergence regions for continued fractions $K(a_n/1)$.*

(C) *If $K(a_n/1)$ is a continued fraction with elements satisfying*

$$0 \neq a_n \in E_n(\delta_n) \; , \quad n = 1,2,3,\ldots \tag{2.5}$$

and with nth approximant f_n and value $f = \lim f_n$, then

$$|f - f_n| < 2\rho \prod_{j=2}^{n} \delta_j \; , \quad n = 2,3,4,\ldots \, . \tag{2.6}$$

Our proof of Theorem 1 is based on several lemmas which will now be given. Throughout this section the symbols $\Gamma, \rho, \{\delta_n\}$ and $\{E_n(\delta_n)\}$ are defined as in

Theorem 1. The first five lemmas describe useful properties of the element regions $E_n(\delta_n)$.

Lemma 1. (A) $E_{2n-1}(\delta_{2n-1})$ is a closed, bounded, convex region, symmetric with respect to the line (axis) determined by the ray $\arg w = \theta = \mathrm{Arg}\Gamma$.
(B) Let g be the function

(2.7) $$g(\delta,\theta) = t(\delta,\theta) - \sqrt{t^2(\delta,\theta)-K(\delta)} \quad , \quad |1+\Gamma|/\rho < \delta \leq 1 \quad ,$$

where t and K are defined by (2.3b,c). Then

(2.8) $$\max_{0\leq\theta<2\pi} g(\delta,\theta) = g(\delta,\theta_2) = (\delta\rho + |\Gamma|)(|1+\Gamma|-|\Gamma|)$$

where $\theta_2 = 2\mathrm{Arg}\Gamma$ and

(2.9) $$\min_{0\leq\theta<2\pi} g(\delta,\theta) = g(\delta,\theta_1) = (\delta\rho-|\Gamma|)(|1+\Gamma|-|\Gamma|) \quad ,$$

where $\theta_1 = 2\mathrm{Arg}\Gamma + \pi$.

Proof: Except for convexity, (A) is an immediate consequence of (2.3). The proof of convexity will be given after we have proved Lemma 4.
(B): Since $\delta > |1+\Gamma|/\rho > |\Gamma|/\rho$, we have $K(\delta) > 0$ and hence

$$\sqrt{t^2(\delta,\theta)-K(\delta)} \neq t(\delta,\theta) \quad .$$

It follows that $g_\theta(\delta,\theta) = 0$ if and only if $t_\theta(\delta,\theta) = |\Gamma|^2\sin(\theta-2\mathrm{Arg}\Gamma) = 0$; that is, if and only if

$$\theta = \theta_2 = 2\mathrm{Arg}\Gamma \quad \text{or} \quad \theta = \theta_1 = 2\mathrm{Arg}\Gamma + \pi \quad .$$

By an elementary calculation one can verify (2.8) and (2.9). ■

Lemma 2. (A) $\partial E_{2n}(\delta_{2n})$ is a cardioid with axis of symmetry passing through the ray $\arg w = \theta = 2\mathrm{Arg}(1+\Gamma)$. $E_{2n}(\delta_{2n})$ is the unbounded region consisting of $\partial E_{2n}(\delta_{2n})$ and its exterior. (B) Let h be the function defined by

(2.10) $$h(\delta,\theta) = 2|1+\Gamma|[\rho/\delta-|1+\Gamma|\cos(\theta-2\mathrm{Arg}(1+\Gamma))] \quad , \quad 0 < \delta \leq 1 \quad .$$

Then

(2.11) $$\max_{0\leq\theta<2\pi} h(\delta,\theta) = h(\delta,\theta_2^*) = 2|1+\Gamma|(\rho/\delta+1)$$

where $\theta_2^* = 2\mathrm{Arg}(1+\Gamma) + \pi$ and

(2.12) $$\min_{0\leq\theta<2\pi} h(\delta,\theta) = h(\delta,\theta_1^*) = 2|1+\Gamma|(\rho/\delta-1) \quad ,$$

where $\theta_1^* = 2\mathrm{Arg}(1+\Gamma)$.

The proof of Lemma 2 follows readily from (2.4) and hence is omitted.

Lemma 3. (A)

(2.13) $$E_{2n-1}(\delta') \subset E_{2n-1}(\delta) \quad \text{if} \quad |1+\Gamma|/\rho < \delta' < \delta \leq 1 \quad .$$

(B)

(2.14) $$E_{2n}(\delta') \subset E_{2n}(\delta) \quad \text{if} \quad 0 < \delta' < \delta \leq 1 \quad .$$

Proof. (A): Let θ be fixed and define $f(\delta) = g(\delta,\theta)$ where g is defined by (2.7). It suffices to show that f is an increasing function of δ. If $\Gamma = 0$, then it can be seen that $f(\delta) = \rho\delta$ which is clearly increasing. Suppose that $\Gamma \neq 0$. Then by (2.3b), $t(\delta,\theta) \geq \rho\delta|1+\Gamma|-|\Gamma|^2 > 0$. Thus

$$(2.15)\qquad \begin{aligned} t^2(\delta,\theta) - K(\delta) &= [\delta\rho|1+\Gamma|-|\Gamma|^2\cos(\theta-2\mathrm{Arg}\Gamma)]^2-(\delta^2\rho^2-|\Gamma|^2)(|1+\Gamma|^2-|\Gamma|^2) \\ &\geq [\delta\rho|1+\Gamma|-|\Gamma|^2]^2-(\delta^2\rho^2-|\Gamma|^2)(|1+\Gamma|^2-|\Gamma|^2) \\ &= |\Gamma|^2(|1+\Gamma|-\rho\delta)^2 > 0 \ . \end{aligned}$$

It follows that $f'(\delta) > 0$ if and only if

$$(2.16)\qquad 2t_\delta(\delta,\theta)\sqrt{t^2(\delta,\theta)-K(\delta)} > 2t(\delta,\theta)t_\delta(\delta,\theta) - K'(\delta) \ .$$

Since

$$(2.17)\qquad t_\delta(\delta,\theta) = \rho|1+\Gamma| \quad \text{and} \quad K'(\delta) = 2\delta\rho^2(|1+\Gamma|^2-|\Gamma|^2) \ ,$$

the right side of (2.16) becomes

$$(2.18)\qquad \begin{aligned} &2t(\delta,\theta)t_\delta(\delta,\theta) - K'(\delta) \\ &= 2\rho|1+\Gamma|[\rho\delta|1+\Gamma|-|\Gamma|^2\cos(\theta-2\mathrm{Arg}\Gamma)] - 2\delta\rho^2(|1+\Gamma|^2-|\Gamma|^2) \\ &\geq 2\rho|1+\Gamma|[\rho\delta|1+\Gamma|-|\Gamma|^2] - 2\delta\rho^2(|1+\Gamma|^2-|\Gamma|^2) \\ &= 2\rho|\Gamma|^2(\delta\rho-|1+\Gamma|) > 0 \ . \end{aligned}$$

Therefore if we square both sides of (2.16), rearrange terms and simplify, we obtain the equivalent inequality

$$(2.19)\qquad 4t(\delta,\theta)t_\delta(\delta,\theta)K'(\delta) - 4[t_\delta(\delta,\theta)]^2K(\delta) - [K'(\delta)]^2 > 0 \ .$$

Now substituting from (2.3) and (2.17) into (2.19) and dividing throughout by $4\rho^2(|1+\Gamma|^2-|\Gamma|^2)$ yields the equivalent inequality

$$\begin{aligned} &2\delta\rho|1+\Gamma|[\rho\delta|1+\Gamma|-|\Gamma|^2\cos(\theta-2\mathrm{Arg}\Gamma)] - |1+\Gamma|^2(\rho^2\delta^2-|\Gamma|^2) \\ &\quad - \delta^2\rho^2(|1+\Gamma|^2-|\Gamma|^2) > 0 \ . \end{aligned}$$

This inequality will surely hold (for all θ) if the new inequality, obtained by replacing the cosine by one, holds. That inequality can be shown to be equivalent to

$$|\Gamma|^2(|1+\Gamma|^2-\rho\delta)^2 > 0$$

which clearly holds. We conclude that $f'(\delta) > 0$, which proves (A). To prove (B) it suffices to show that $h(\delta,\theta)$ (defined by (2.10)) is a decreasing function of δ. Since this is readily done, the proof is complete. ■

Lemma 4. Let E_1 and E_2 be defined by (1.6) and let $E_n(1)$ be defined by (2.3) and (2.4). Then

$$(2.20)\qquad E_{2n-1}(1) = E_1 \quad \text{and} \quad E_{2n}(1) = E_2 \ , \quad n = 1,2,3\ldots \ .$$

Proof: It suffices to prove that $E_1(1) = E_1$ and $E_2(1) = E_2$. We begin with $E_2(1) = E_2$. The inequality in (1.6c) is equivalent to

(2.21) $$\rho|w| - |1+\Gamma|(\rho^2-|1+\Gamma|^2) \geq |w(1+\bar{\Gamma})-(1+\Gamma)(\rho^2-|1+\Gamma|^2)| ,$$

which implies that

(2.22) $$r = |w| \geq \frac{|1+\Gamma|(\rho^2-|1+\Gamma|^2)}{\rho} > 0 .$$

Squaring both sides of (2.21) , collecting like terms in r and dividing throughout by the positive quantity $r(\rho^2-|1+\Gamma|^2)$ yields

(2.23) $$r \geq 2|1+\Gamma|[\rho-|1+\Gamma|\cos(\theta-2\mathrm{Arg}(1+\Gamma)] ,$$

which is the inequality in (2.4) . We have shown that (2.21) implies both (2.22) and (2.23) and hence that $E_2 \subset E_2(1)$. It can be seen that (2.23) implies (2.21) provided that (2.23) implies (2.22) . To prove that (2.23) implies (2.22) it suffices to show that

(2.24) $$2|1+\Gamma|[\rho-|1+\Gamma|\cos(\theta-2\mathrm{Arg}(1+\Gamma)] \geq |1+\Gamma|(\rho^2-|1+\Gamma|^2)/\rho .$$

But (2.24) is equivalent to

$$2\rho[\rho-|1+\Gamma|\cos(\theta-2\mathrm{Arg}(1+\Gamma))] \geq \rho^2 - |1+\Gamma|^2 = (\rho+|1+\Gamma|)(\rho-|1+\Gamma|) ,$$

which follows immediately from the fact that $\rho > |1+\Gamma|$, so that $2\rho > \rho + |1+\Gamma|$. This proves that $E_2(1) = E_2$.

To prove that $E_1(1) = E_1$ it suffices to show that the inequalities

(2.25) $$|\bar{\Gamma}w-\Gamma(|1+\Gamma|^2-|\Gamma|^2)| \leq \rho(|1+\Gamma|^2-|\Gamma|^2) - |1+\Gamma||w|$$

and

(2.26) $$0 \leq r \leq t(\theta) - \sqrt{t^2(\theta)-K} \qquad (\text{where } w = re^{i\theta})$$

are equivalent. Here

$$t(\theta) = t(1,\theta) = \rho|1+\Gamma| - |\Gamma|^2\cos(\theta-2\mathrm{Arg}\Gamma) ,$$

and

$$K = K(1) = (\rho^2-|\Gamma|^2)(|1+\Gamma|^2-|\Gamma|^2) .$$

For that purpose we establish the following

(2.27) $$0 \leq t(\theta) - \sqrt{t^2(\theta)-K} \leq \frac{\rho(|1+\Gamma|^2-|\Gamma|^2)}{|1+\Gamma|} \leq t(\theta) + \sqrt{t^2(\theta)-K} .$$

The first inequality in (2.27) follows from the fact that $t(\theta) \geq \rho|1+\Gamma| - |\Gamma|^2 > 0$ and $0 < K \leq t^2(\theta)$, since

$$\begin{aligned} t^2(\theta) - K &= [\rho|1+\Gamma|-|\Gamma|^2\cos(\theta-2\mathrm{Arg}\Gamma)]^2 - (\rho^2-|\Gamma|^2)(|1+\Gamma|^2-|\Gamma|^2) \\ &\geq (\rho|1+\Gamma|-|\Gamma|^2) - (\rho^2-|\Gamma|^2)(|1+\Gamma|^2-|\Gamma|^2) \\ &= |\Gamma|^2(\rho-|1+\Gamma|)^2 \geq 0 . \end{aligned}$$

To prove the second inequality in (2.27) we note that by (2.7) and (2.8)

$$t(\theta) - \sqrt{t^2(\theta)-K} = g(1,\theta) \leq (\rho+|\Gamma|)(|1+\Gamma|-|\Gamma|)$$
$$\leq \frac{\rho(|1+\Gamma|^2-|\Gamma|^2)}{|1+\Gamma|} \ .$$

The third inequality in (2.27) is equivalent to

$$|1+\Gamma|\sqrt{t^2(\theta)-K} + |\Gamma|^2(\rho-|1+\Gamma|\cos(\theta-2\text{Arg}\Gamma)) \geq 0 \ ,$$

which is readily verified. Now suppose that (2.25) holds. This implies that

(2.28) $$r = |w| \leq \frac{\rho(|1+\Gamma|^2-|\Gamma|^2)}{|1+\Gamma|} \ .$$

Squaring both sides of (2.25), collecting like terms in $|w| = r$, and dividing throughout by $(|1+\Gamma|^2-|\Gamma|^2)$ yields

(2.29) $$r^2 - 2rt(\theta) + K \geq 0 \ .$$

Completing the square on the left side gives the equivalent relation

(2.30) $$(r-t(\theta))^2 \geq t^2(\theta) - K \ .$$

Since $t^2(\theta) - K \geq 0$ for all real θ, (2.30) is equivalent to

(2.31) $$|r-t(\theta)| \geq \sqrt{t^2(\theta)-K} \ ,$$

where we take the positive square root. Clearly (2.31) implies that either

(2.32a) $$r \leq t(\theta) - \sqrt{t^2(\theta)-K} \ , \quad (\text{when } r < t(\theta))$$

or

(2.32b) $$r \geq t(\theta) + \sqrt{t^2(\theta)-K} \ , \quad (\text{when } r \geq t(\theta)) \ .$$

Combining (2.28) with the third inequality in (2.27) shows that (2.32b) cannot hold. Hence (2.32a) holds, which proves (2.26).

Conversely, suppose that (2.23) is satisfied. Then by the second inequality in (2.27), it follows that (2.28) holds. Hence the above steps can be reversed to show that (2.21) is valid. ■

<u>Remarks.</u> We shall now prove the assertion made in Lemma 1(A) that $E_{2n-1}(\delta_{2n-1})$ is a convex set. It suffices to show that $E_1(\delta)$ is convex if $|1+\Gamma|/\rho < \delta < 1$. In view of Lemma 4 it can be seen that, if ρ is replaced by $\rho\delta$ in (1.6b), then the resulting inequality is equivalent to the inequality in (2.3a) (with $\delta_{2n-1} = \delta$ and $n = 1$) defining $E_1(\delta)$. Therefore we can write

$$E_1(\delta) = [w : |\bar{\Gamma}w-\Gamma(|1+\Gamma|^2-|\Gamma|^2)|+|1+\Gamma||w| \leq \rho\delta(|1+\Gamma|^2-|\Gamma|^2)] \ .$$

To show that $E_1(\delta)$ is convex, it suffices to show that, if A, B and C are arbitrary positive numbers and $w_0 \in \mathbb{C}$, then

$$E = [w : A|w| + B|w-w_0| \leq C]$$

is a convex set. Let w_1 and w_2 be any two points in E. Then we shall show that, for all $0 \leq \alpha \leq 1$, $w(\alpha)$ E, where $w(\alpha) = \alpha w_1 + (1-\alpha)w$.

Since $w_1, w_2 \in E$,

$$A|w_k| + B|w_k - w_0| \le C \quad , \quad k = 1,2 \ .$$

Now

$$A|\alpha w_1+(1-\alpha)w_2| \le \alpha A|w_1| + (1-\alpha)A|w_2|$$

and

$$B|\alpha w_1+(1-\alpha)w_2-w_0| = B|\alpha(w_1-w_0)+(1-\alpha)(w_2-w_0)|$$
$$\le \alpha B|w_1-w_0| + (1-\alpha)B|w_2-w_0| \ .$$

Thus by using the preceding three inequalities, we obtain, for $0 \le \alpha \le 1$,

$$A|w(\alpha)| + B|w(\alpha)-w_0| = A|\alpha w_1+(1-\alpha)w_2| + B|\alpha w_1+(1-\alpha)w_2-w_0|$$
$$\le \alpha A|w_1| + (1-\alpha)A|w_2| + \alpha B|w_1-w_0| + (1-\alpha)B|w_2-w_0|$$
$$= \alpha[A|w_1|+B|w_1-w_0|] + (1-\alpha)[A|w_2|+B|w_2-w_0|]$$
$$\le \alpha C + (1-\alpha)C = C \ .$$

It follows that E is convex and hence also that $E_1(\delta)$ is convex as asserted in Lemma 1(A) .

<u>Lemma</u> 5. <u>Let Γ,ρ, and $\{E_n(\delta_n)\}_{n=1}^{\infty}$ be as in Theorem 1 and let $\{V_n\}_{n=0}^{\infty}$ be a sequence of subsets of $\hat{\mathbb{C}}$ defined by</u>

(2.33a) $$V_{2n} = [v \in \hat{\mathbb{C}} : |v-\Gamma| \le \rho] \quad , \quad n = 0,1,2,\ldots \ ,$$

(2.33b) $$V_{2n+1} = [v \in \hat{\mathbb{C}} : |v+(1+\Gamma)| \ge |1+\Gamma|] \quad , \quad n = 0,1,2,\ldots \ .$$

<u>Let</u>

$$s(w,v) = \frac{w}{1+v} \ .$$

<u>Then</u>

(2.34a) $$s(E_{2n}(\delta_{2n}),V_{2n}) \subseteq V_{2n-1} \quad , \quad n = 1,2,3,\ldots \ ,$$

(2.34b) $$s(E_{2n+1}(\delta_{2n+1}),V_{2n+1}) \subseteq V_{2n} \quad , \quad n = 0,1,2,\ldots \ .$$

<u>Proof</u>: Let E_1 and E_2 be defined as in (1.6) . Then by [6, 1970, Lemma 5.5] we have

$$s(E_1,V_1) \subseteq V_0 \quad \text{and} \quad s(E_2,V_2) \subseteq V_1 \ .$$

Hence (2.34) is an immediate consequence of Lemmas 3 and 4 . ■

<u>Proof of Theorem 1</u>. (A) follows from Lemmas 3 and 4 and [6, 1970, Corollary 5.7]. (B) is an immediate consequence of (C) . To prove (C) let $K(a_n/1)$ be a continued fraction with elements a_n satisfying (2.5) and with nth approximant f_n and value f . Let $\{s_n\}$ and $\{S_n\}$ be sequences of linear fractional transformations (l.f.t.'s) defined by (DN1) . Let $\{V_n\}$ be defined by (2.33) . It follows then from Lemma 5 that

(2.35a) $$s_n(V_n) \subseteq V_{n-1} \quad , \quad n = 0,1,2,\ldots$$

and hence that

(2.35b) $$S_n(V_n) \subseteq S_{n-1}(V_{n-1}) \subseteq V_0 \ , \quad n = 1,2,3\ldots \ .$$

Therefore $\{S_n(V_n)\}$ is a nested sequence of closed circular disks. Let R_n denote the radius of $S_n(V_n)$. Since $f_n = S_n(0)$ and $0 \in V_n$, for all $n = 1,2,3,\ldots$ it follows from (2.35b) that $f_{n+m} \in S_n(V_n)$ for $m = 1,2,\ldots$ and hence that

(2.36) $$\left|f_{n+m} - f_n\right| < 2R_n \ , \quad n = 2,3,4,\ldots \ , \quad m = 0,1,2,\ldots \ .$$

Thus to prove (2.6) , it suffices to show that

$$R_n \leq \rho \prod_{j=2}^{n} \delta_j \ , \quad n = 2,3,4,\ldots \ .$$

For that purpose we let

$$\Gamma_{2n} = \Gamma \ , \ \rho_{2n} = \rho \ , \ \Gamma_{2n+1} = -(1+\Gamma) \ , \ \rho_{2n+1} = \left|1+\Gamma\right| \ , \quad n = 0,1,2,\ldots \ .$$

Then for each n , ∂V_n is a circle with center Γ_n and radius ρ_n . Let β_n denote a point on ∂V_n with its exact location to be determined. Let z_n be a point in $\mathrm{Int}(V_n)$ such that $S_n(z_n)$ is the center of $S_n(V_n)$. Let A_n and B_n denote the nth numerator and nth denominator of $K(a_n/1)$, respectively. Then by use of the determinant formula

$$A_n B_{n-1} - A_{n-1} B_n = (-1)^n \prod_{k=1}^{n} a_k \ , \quad n = 1,2,3,\ldots$$

and (DN5) , one can show that

(2.37) $$R_n = \left|S_n(\beta_n) - S_n(z_n)\right| = \frac{\left|\beta_n - z_n\right| \prod_{k=1}^{n} \left|a_k\right|}{\left|B_{n-1}\right|^2 \left|h_n + \beta_n\right| \left|h_n + z_n\right|} \ ,$$

where $h_n = B_n/B_{n-1}$. Formula (2.37) can be simplified as follows. We recall that inverses with respect to a circle are preserved under l.f.t.'s . Moreover, ∞ is the inverse of $S_n(z_n)$ with respect to the circle $\partial S_n(V_n)$. Since $S_n(-h_n) = \infty$, it follows that the points z_n and $-h_n$ are inverses with respect to the circle ∂V_n (see Figure 1 for a schematic diagram). Now we choose β_n to be the point of intersection of the circle ∂V_n and the line segment passing from Γ_n through z_n and $-h_n$; that is,

$$\beta_n = \Gamma_n + \rho_n e^{i\tau_n} \ .$$

It follows that (2.37) can be reduced to

(2.38) $$R_n = \frac{\rho_n \prod_{k=1}^{n} \left|a_k\right|}{\left|B_{n-1}\right|^2 \left|\left|h_n + \Gamma_n\right|^2 - \rho_n^2\right|} \ .$$

Since $h_n = B_n/B_{n-1}$ can be estimated more readily than B_{n-1} , we shall consider the ratio

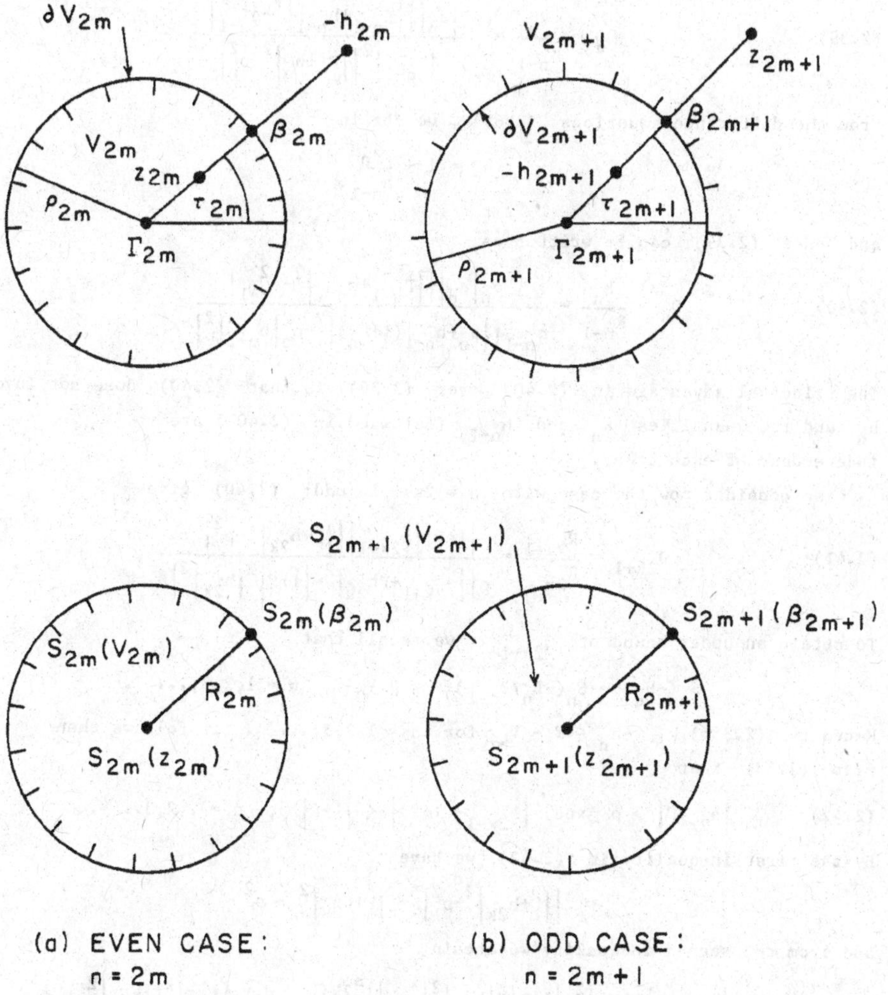

∂V_{2m}
$-h_{2m}$
V_{2m}
β_{2m}
z_{2m}
ρ_{2m}
τ_{2m}
Γ_{2m}
V_{2m+1}
z_{2m+1}
β_{2m+1}
∂V_{2m+1}
$-h_{2m+1}$
τ_{2m+1}
Γ_{2m+1}
ρ_{2m+1}
$S_{2m+1}(V_{2m+1})$
$S_{2m}(\beta_{2m})$
$S_{2m}(V_{2m})$
R_{2m}
$S_{2m}(z_{2m})$
$S_{2m+1}(\beta_{2m+1})$
R_{2m+1}
$S_{2m+1}(z_{2m+1})$
(a) EVEN CASE: $n = 2m$
(b) ODD CASE: $n = 2m+1$

Figure 1

(2.39) $$Q_n = \frac{R_n}{R_{n-1}} = \frac{\rho_n|a_n|\,\big||\Gamma_{n-1}+h_{n-1}|^2-\rho_{n-1}^2\big|}{|h_{n-1}|^2\,\big||\Gamma_n+h_n|^2-\rho_n^2\big|} \; .$$

From the difference equations (DN6) , we obtain

$$h_n = 1 + \frac{a_n}{h_{n-1}}$$

and hence (2.39) can be written as

(2.40) $$Q_n = \frac{R_n}{R_{n-1}} = \frac{\rho_n|a_n|\,\big||\Gamma_{n-1}+h_{n-1}|^2-\rho_{n-1}^2\big|}{\rho_{n-1}\big||a_n+h_{n-1}(1+\Gamma_n)|^2-\rho_n^2|h_{n-1}|^2\big|} \; .$$

The principal advantage in (2.40) over (2.39) is that (2.40) does not involve h_n and the quantities a_n and h_{n-1} (involved in (2.40)) are independent of each other.

We consider now the case with $n = 2k + 1$ odd; (2.40) gives

(2.41) $$Q_{2k+1} = \frac{R_{2k+1}}{R_{2k}} = \frac{|1+\Gamma|\,|a_{2k+1}|\,\big||\Gamma+h_{2k}|^2-\rho^2\big|}{\rho\big||a_{2k+1}+\Gamma h_{2k}|^2-|1+\Gamma|^2|h_{2k}|^2\big|} \, .$$

To obtain an upper bound of Q_{2k+1} , we recall that

$$S_n(-h_n) = S_n(-B_n/B_{n-1}) = \infty \notin V_0 \quad , \quad n = 1,2,3,\ldots \; .$$

Hence by (2.35b) , $-h_n \in \hat{\mathbb{C}} - V_n$ for $n = 1,2,3,\ldots$. It follows then from (2.33) that

(2.42) $$|h_{2k}+\Gamma| > \rho \quad \text{and} \quad |h_{2k-1}-(1+\Gamma)| < |1+\Gamma| \quad , \quad n = 1,2,3.. \; .$$

By the first inequality in (2.42) we have

$$\big||\Gamma+h_{2k}|^2-\rho^2\big| = |\Gamma+h_{2k}|^2 - \rho^2$$

and from the second inequality we obtain

$$\big||a_{2k+1}-\Gamma h_{2k}|^2-|1+\Gamma|^2|h_{2k}|^2\big| = \left|\left|\frac{a_{2k+1}}{h_{2k}} - \Gamma\right|^2-|1+\Gamma|^2\right||h_{2k}|^2$$

$$= \big||h_{2k+1}-(1+\Gamma)|^2-|1+\Gamma|^2\big||h_{2k}|^2 \quad , \quad (\text{since } h_n = 1 + a_n/h_{n-1})$$

$$= [|1+\Gamma|^2-|h_{2k+1}-(1+\Gamma)|^2]|h_{2k}|^2$$

$$= |1+\Gamma|^2|h_{2k}|^2 - |a_{2k+1}-\Gamma h_{2k}|^2 \; .$$

Therefore certain absolute value signs in (2.41) can be removed and we obtain

(2.43) $$Q_{2k+1} = \frac{|1+\Gamma|\,|a_{2k+1}|\,[|\Gamma+h_{2k}|^2-\rho^2]}{\rho[|1+\Gamma|^2|h_{2k}|^2-|a_{2k+1}-\Gamma h_{2k}|^2]} \; .$$

Next we introduce, by means of (2.5) and (2.42) , the parametric representations of a_{2k+1} and h_{2k} given by

(2.44a) $$a_{2k+1} = \varepsilon u e^{i\theta} \quad , \quad 0 < \varepsilon \leq 1 \ , \quad 0 \leq \theta < 2\pi \ ,$$

where

(2.44b) $$u = g(\delta,\theta) = t(\delta,\theta) - \sqrt{t^2(\delta,\theta)-K(\delta)} \quad , \quad (\text{see} \quad (2.7))$$

and

(2.44c) $$h_{2k} = -\Gamma + \frac{\rho}{\gamma} e^{i\lambda} \quad , \quad 0 < \gamma < 1 \ , \quad 0 \leq \lambda < 2\pi$$

It is understood that the parameters $\varepsilon, u, \theta, \delta, \gamma,$ and λ all depend on k . Substituting (2.44a) into (2.43) and expanding the denominator yields

$$Q_{2k+1} = \frac{u|1+\Gamma|[|\Gamma+h_{2k}|^2-\rho^2]}{\rho} F(\varepsilon,u,\theta,h_{2k})$$

where

$$F(\varepsilon,u,\theta,h_{2k}) = \frac{\varepsilon}{D_0+2D_1\varepsilon-D_2\varepsilon^2} \ ,$$

$$D_0 = (|1+\Gamma|^2-|\Gamma|^2)|h_{2k}|^2 \quad , \quad D_1 = uRe(\Gamma h_{2k}e^{-i\theta}) \quad , \quad D_2 = u^2 \ .$$

Since

$$\frac{\partial F}{\partial \varepsilon} = \frac{D_0+D_2\varepsilon^2}{[D_0+2D_1\varepsilon-D_2\varepsilon^2]^2} > 0 \ ,$$

we have

$$\max_{0<\varepsilon\leq 1} F(\varepsilon,u,\theta,h_{2k}) = F(1,u,\theta,h_{2k}) \ ,$$

and hence

(2.45) $$\max_{\varepsilon} Q_{2k+1} = \frac{u|1+\Gamma|(|\Gamma+h_{2k}|^2-\rho^2)}{\rho[(|1+\Gamma|^2-|\Gamma|^2)|h_{2k}|^2+2uRe(\Gamma h_{2k}e^{-i\theta})-u^2]} \ .$$

Now we substitute (2.44c) into the numerator of (2.45) and obtain

(2.46) $$\max_{\varepsilon} Q_{2k+1} = \frac{u\rho|1+\Gamma|(1-\gamma^2)}{\gamma^2[(|1+\Gamma|^2-|\Gamma|^2)|h_{2k}|^2+2uRe(\Gamma h_{2k}e^{-i\theta})-u^2]} \ .$$

We see that fortunately the parameter λ appears by way of h_{2k} only in the denominator of (2.46) . Therefore we can find the maximum of (2.46) with respect to λ by finding the minimum of its denominator. Letting D denote the denominator of (2.46) and substituting (2.44c) into D yields after some rearrangement of terms

(2.47a) $$D = (|1+\Gamma|^2-|\Gamma|^2)(\rho^2+\gamma^2|\Gamma|^2) - \gamma^2 u[2|\Gamma|^2\cos(\theta-2\operatorname{Arg}\Gamma)+u] \\ -2\rho\gamma|\Gamma|J(\lambda) \ ,$$

where

(2.47b) $$J(\lambda) = (|1+\Gamma|^2-|\Gamma|^2)\cos(\lambda-\mathrm{Arg}\Gamma) - u\cos(\lambda-\theta+\mathrm{Arg}\Gamma) \quad .$$

Therefore to minimize D with respect to λ, it suffices to maximize $J(\lambda)$. For that purpose we write

$$J(\lambda) = e_1\cos\lambda + e_2\sin\lambda = \sqrt{e_1^2+e_2^2}\,\cos(\lambda-\Lambda) \quad ,$$

where

$$e_1 = (|1+\Gamma|^2-|\Gamma|^2)\cos(\mathrm{Arg}\Gamma) - u\cos(\theta-\mathrm{Arg}\Gamma) \quad ,$$

$$e_2 = (|1+\Gamma|^2-|\Gamma|^2)\sin(\mathrm{Arg}\Gamma) - u\sin(\theta-\mathrm{Arg}\Gamma) \quad ,$$

$$\cos\Lambda = e_1/\sqrt{e_1^2+e_2^2} \quad , \quad \sin\Lambda = e_2/\sqrt{e_1^2+e_2^2} \quad .$$

Consequently

(2.48) $$\max_\lambda J(\lambda) = \sqrt{e_1^2+e_2^2} = |(|1+\Gamma|^2-|\Gamma|^2)-ue^{i(\theta-2\mathrm{Arg}\Gamma)}| \quad .$$

Equation (2.48) can be further simplified by recalling from (2.44b) and (2.3) that $u \in \partial E_{2k+1}(\delta)$ and hence from (1.6b) with ρ replaced by $\delta\rho$, we conclude that

$$|\bar{\Gamma}u-\Gamma(|1+\Gamma|^2-|\Gamma|^2)| + |1+\Gamma|u = \rho(|1+\Gamma|^2-|\Gamma|^2) \quad .$$

Therefore

$$\max_\lambda J(\lambda) = \frac{1}{|\Gamma|}[\delta\rho(|1+\Gamma|^2-|\Gamma|^2)-u|1+\Gamma|] \quad .$$

The minimum of D with respect to λ can now be written, after some rearrangement of terms, as

(2.49) $$\min_\lambda D = -\gamma^2u^2-2\gamma u[\gamma|\Gamma|^2\cos(\theta-2\mathrm{Arg}\Gamma)-\rho|1+\Gamma|]$$
$$+ (|1+\Gamma|^2-|\Gamma|^2)(\rho^2+\gamma^2|\Gamma|^2-2\gamma\delta\rho^2) \quad .$$

From (2.44b) it follows that $u^2 = 2ut(\delta,\theta) - K(\delta)$. Substituting this into (2.49) and then rearranging and cancelling some terms yields

$$\min_\lambda D = \rho(1-\gamma\delta)[\rho(|1+\Gamma|^2-|\Gamma|^2)(1-\gamma\delta)+2\gamma u|1+\Gamma|] \quad .$$

Using this in (2.46) we arrive at

(2.50) $$\max_{\varepsilon,\lambda} Q_{2k+1} = \frac{u(1-\gamma^2)|1+\Gamma|}{(1-\gamma\delta)[\rho(|1+\Gamma|^2-|\Gamma|^2)(1-\gamma\delta)+2\gamma u|1+\Gamma|]} \quad .$$

Our next step is to maximize (2.50) with respect to θ. For that purpose we write

(2.51a) $$\max_{\varepsilon,\lambda} Q_{2k+1} = \frac{|1+\Gamma|(1-\gamma^2)}{(1-\gamma\delta)}G(u) \quad ,$$

where

(2.51b) $$G(u) = \frac{u}{\rho(|1+\Gamma|^2-|\Gamma|^2)(1-\gamma\delta)+2|1+\Gamma|\gamma u} \quad ,$$

(2.51c) $$u = u(x) = T(x) - \sqrt{T^2(x)-K(\delta)} \quad ,$$

(2.51d) $$T(x) = t(\delta,\theta) \quad , \quad x = \cos(\theta - 2\mathrm{Arg}\Gamma) \quad .$$

It is easily seen that $G'(u) \geq 0$. From (2.51c) we have $u^2(x) = 2T(x)u(x) - K(\delta)$; hence implicit differentiation with respect to x gives

$$u'(x) = \frac{T'(x)u(x)}{u(x)-T(x)} = \frac{|\Gamma|^2 u(x)}{\sqrt{T^2(x)-K(\delta)}} \geq 0 \quad .$$

Therefore, by the chain rule $\frac{dG}{dx} = \frac{dG}{du}\frac{du}{dx} \geq 0$; hence $G(u)$ attains its maximum value when $x = 1$, that is, when $\theta = 2\mathrm{Arg}\Gamma$. We conclude from this and Lemma 1 that

$$\max_{\theta} G(u) = (\delta\rho + |\Gamma|)(|1+\Gamma| - |\Gamma|) \quad .$$

Combining this with (2.51) gives

(2.52a) $$H(\delta,\gamma) = \max_{\varepsilon,\lambda,\theta} Q_{2k+1} = \frac{C_3(1-\gamma^2)}{C_0 - 2C_1\gamma - C_2\gamma^2} \quad ,$$

where

(2.52b) $$C_0 = \rho(|1+\Gamma| + |\Gamma|) > 0 \quad ,$$

(2.52c) $$C_1 = |\Gamma|(\rho\delta - |1+\Gamma|) \geq 0 \quad ,$$

(2.52d) $$C_2 = \delta[\rho\delta(|1+\Gamma| - |\Gamma|) + 2|\Gamma||1+\Gamma|] > 0 \quad ,$$

(2.52e) $$C_3 = |1+\Gamma|(\rho\delta + |\Gamma|) > 0 \quad .$$

It remains to show that

(2.53) $$H(\delta,\gamma) \leq \delta \quad , \text{ for all } 0 < \gamma < 1 \quad .$$

It is readily seen that the denominator on the right side of (2.52a) is positive. Hence (2.53) is equivalent to

$$C_3(1-\gamma^2) \leq \delta C_0 - 2\delta C_1\gamma - \delta C_2\gamma^2 \quad ,$$

which can be shown to be equivalent to

$$|\Gamma|(1-\delta\gamma)^2(\rho\delta - |1+\Gamma|) + \gamma^2(1-\delta^2)|1+\Gamma|(\rho\delta + |\Gamma|) \geq 0 \quad .$$

Since this inequality clearly holds, we conclude that

(2.54) $$Q_{2k+1} \leq \max_{\varepsilon,\lambda,\theta,\gamma} Q_{2k+1} \leq \delta = \delta_{2k+1} \quad .$$

We turn now to the case with $n = 2k$ even; (2.40) then gives

(2.55) $$Q_{2k} = \frac{R_{2k}}{R_{2k-1}} = \frac{\rho|a_{2k}|\,\left||1+\Gamma|^2 - |h_{2k-1} - (1+\Gamma)|^2\right|}{|1+\Gamma|\,\left||(1+\Gamma)h_{2k-1} + a_{2k}|^2 - \rho^2|h_{2k-1}|^2\right|} \quad .$$

We shall establish that

(2.56) $$Q_{2k} \leq \delta_{2k} \quad , \quad k = 1,2,3,\ldots \quad .$$

Since most of the arguments involved parallel those used in the proof of (2.54), we shall give only an outline of the steps used for (2.56). Two sets of absolute value signs in (2.55) may be removed since, by (2.42),

$$|1+\Gamma|^2 - |h_{2k-1}-(1+\Gamma)|^2 > 0$$

and

$$|(1+a_{2k}/h_{2k-1})+\Gamma| - \rho > 0 \quad .$$

We note that $h_{2k-1} \neq 0$, since $-h_{2k-1} \notin V_{2k-1}$ and $0 \in V_{2k-1}$. Consequently one obtains

$$|(1+\Gamma)h_{2k-1}+a_{2k-1}|^2 - \rho^2|h_{2k-1}|^2 > 0$$

and hence

(2.57) $$Q_{2k} = \frac{\rho|a_{2k}|[|1+\Gamma|^2-|h_{2k-1}-(1+\Gamma)|^2]}{|1+\Gamma|[|(1+\Gamma)h_{2k-1}+|a_{2k}|^2-\rho^2|h_{2k-1}|^2]} ,$$

From (2.4) it is seen that the element a_{2k} has the parametric representation

(2.58a) $$a_{2k} = |1+\Gamma|\varepsilon\beta e^{i\theta}$$

where

(2.58b) $$\varepsilon \geq 1 \quad , \quad 0 \leq \theta < 2\pi \quad ,$$

and

(2.58c) $$\beta = 2[\rho/\delta-|1+\Gamma|\cos(\theta-2\mathrm{Arg}(1+\Gamma))] \quad .$$

Here ε,θ,β and δ all depend upon $2k$; a subscript of $2k$ has been suppressed for simplicity. Substituting (2.58) into (2.57) and cancelling the common factor $|1+\Gamma|$, we may express

(2.59a) $$Q_{2k} = \frac{D_3\varepsilon}{D_2\varepsilon^2+D_1\varepsilon-D_0} ,$$

where

(2.59b) $$D_0 = (\rho^2-|1+\Gamma|^2)|h_{2k-1}|^2 > 0 \quad ,$$

(2.59c) $$D_1 = 2|1+\Gamma|\beta\mathrm{Re}[(1+\Gamma)h_{2k-1}e^{-\alpha}] \quad ,$$

(2.59d) $$D_2 = |1+\Gamma|\beta^2 > 0 \quad ,$$

(2.59e) $$D_3 = \rho\beta[|1+\Gamma|^2-|h_{2k-1}-(1+\Gamma)|^2] \quad .$$

An argument similar to that used to maximize Q_{2k+1} with respect to ε shows that Q_{2k} attains its maximum with respect to ε at $\varepsilon = 1$; that is, Q_{2k} has its maximum value when a_{2k} lies on the boundary of $E_{2k}(\delta_{2k})$. Thus we find that

(2.60) $$\max_{\varepsilon} Q_{2k} = \frac{\rho\beta[|1+\Gamma|^2-|h_{2k-1}-(1+\Gamma)|^2]}{|1+\Gamma|^2\beta^2+2|1+\Gamma|\beta\mathrm{Re}[(1+\Gamma)h_{2k-1}e^{-i\theta}]-(\rho^2-|1+\Gamma|^2)|h_{2k-1}|^2} .$$

By (2.42) , h_{2k-1} can be written in the parametric form

(2.61a) $$h_{2k-1} = (1+\Gamma) + \gamma|1+\Gamma|e^{i\lambda} ,$$

where

(2.61b) $$0 < \gamma < 1 \quad \text{and} \quad 0 \leq \lambda < 2\pi .$$

Substituting (2.61) in (2.60) and simplifying, the resulting expression yields

(2.62a) $$\max_{\varepsilon} Q_{2k} = \frac{\rho\beta(1-\gamma^2)}{\Delta} ,$$

where

(2.62b) $$\Delta = \beta^2 + 2|1+\Gamma|\beta\cos(\theta-2\mathrm{Arg}(1+\Gamma)) - (\rho^2-|1+\Gamma|^2)(1+\gamma^2) + 2\gamma J(\lambda) ,$$

(2.62c) $$J(\lambda) = a \sin\lambda + b \cos\lambda ,$$

(2.62d) $$a = |1+\Gamma|\beta\cos(\theta-2\mathrm{Arg}(1+\Gamma)) - (\rho^2-|1+\Gamma|^2)\cos(\mathrm{Arg}(1+\Gamma)) ,$$

(2.62e) $$b = |1+\Gamma|\beta\sin(\theta-2\mathrm{Arg}(1+\Gamma)) - (\rho^2-|1+\Gamma|^2)\sin(\mathrm{Arg}(1+\Gamma)) .$$

By an argument similar to that used to maximize $\max_{\varepsilon} Q_{2k+1}$ with respect to λ , we find that (2.62a) attains its maximum with respect to λ at $\lambda = \Lambda$, where

$$\Lambda = \cos^{-1}(b/\sqrt{a^2+b^2}) .$$

After simplification we obtain

$$\min_{\lambda} J(\lambda) = J(\Lambda) = -|(\rho^2-|1+\Gamma|^2)-|1+\Gamma|\beta e^{i(\theta-2\mathrm{Arg}(1+\Gamma)}| .$$

Let $x = \cos(\theta-2\mathrm{Arg}(1+\Gamma))$. Then in (2.62b) one has the simplification

$$\beta^2 + 2|1+\Gamma|\beta\cos(\theta-2\mathrm{Arg}(1+\Gamma)) = \beta[\beta+2x|1+\Gamma|]$$
$$= 2\rho\beta/\delta .$$

Consequently one now has

(2.63a) $$\max_{\varepsilon,\lambda} Q_{2k} = \frac{\rho\beta(1-\gamma^2)}{2\rho\beta/\delta-(\rho^2-|1+\Gamma|^2)(1+\gamma^2)-2\gamma\sqrt{M(x)}} ,$$

where

(2.63b) $$\sqrt{M(x)} = |\rho^2-|1+\Gamma|^2-|1+\Gamma|\beta e^{i(\theta-2\mathrm{Arg}(1+\Gamma))}| .$$

$M(x)$ may be expanded in the form

$$M(x) = 4|1+\Gamma|^2\rho^2x^2 - 4\rho|1+\Gamma|(\rho^2+|1+\Gamma|^2)x/\delta + (\rho^2-|1+\Gamma|^2)^2 + 4|1+\Gamma|^2\rho^2/\delta .$$

This can be simplified somewhat by writing

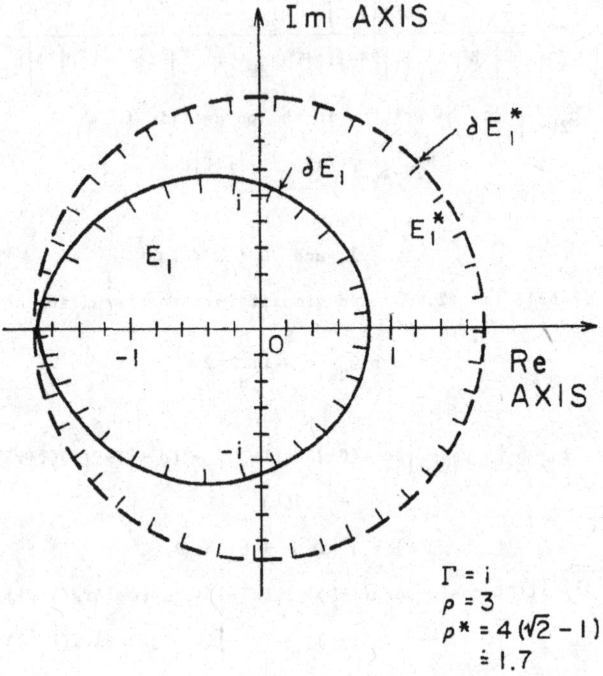
Im AXIS
∂E$_1^*$
∂E$_1$
i
E$_1^*$
E$_1$
-1
0
1
Re AXIS
-i
Γ = i
ρ = 3
ρ* = 4(√2 - 1)
≐ 1.7

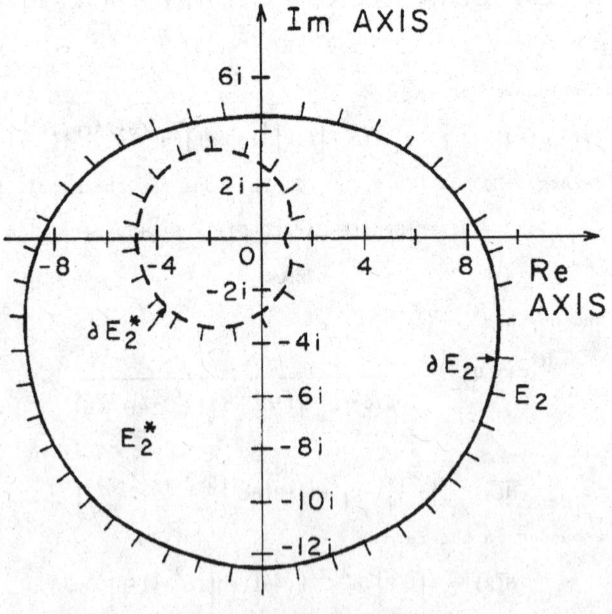
Im AXIS
6i
2i
-8
-4
0
4
8
Re AXIS
-2i
∂E$_2^*$
-4i
∂E$_2$
-6i
E$_2$
E$_2^*$
-8i
-10i
-12i

Figure 2

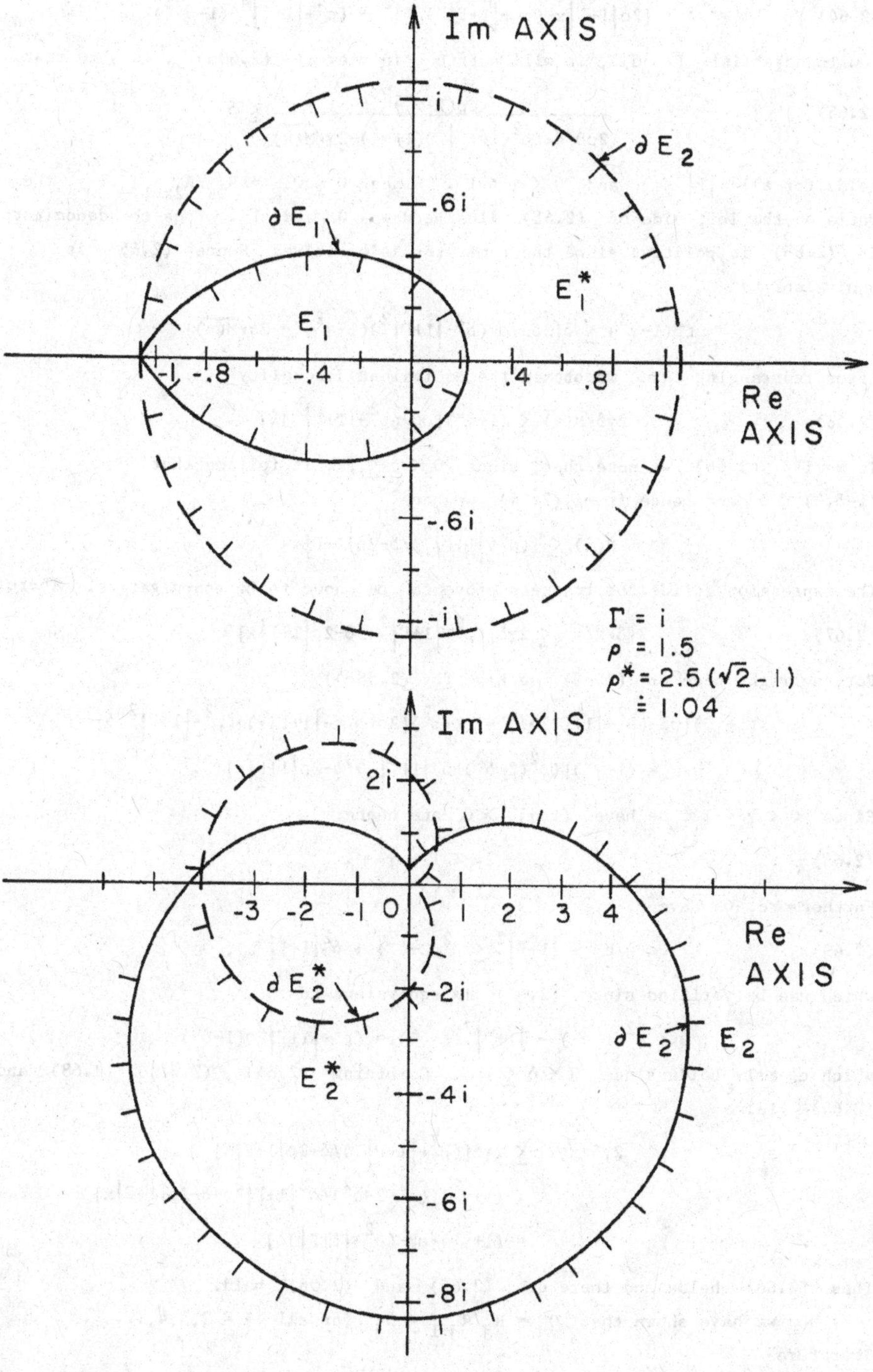
Im AXIS
i
∂E₂
.6i
∂E₁
E₁*
E₁
-1
-.8
-.4
0
.4
.8
1
Re AXIS
-.6i
-i
Γ = i
ρ = 1.5
ρ* = 2.5(√2 - 1)
≐ 1.04
Im AXIS
2i
-3
-2
-1
0
1
2
3
4
Re AXIS
∂E₂*
-2i
∂E₂
E₂
E₂*
-4i
-6i
-8i

Figure 3

(2.64) $$M(x) = [2\rho|1+\Gamma|x-(\rho^2+|1+\Gamma|^2)/\delta]^2 + (\rho^4+|1+\Gamma|^4)(1-\delta^{-1}) \quad .$$

Now to establish (2.56), it will suffice, in view of (2.63a), to show that

(2.65) $$\frac{\rho\beta(1-\gamma^2)}{2\rho\beta/\delta-(\rho^2-|1+\Gamma|^2)(1+\gamma^2)-2\gamma\sqrt{M(x)}} \leq \delta$$

holds for all $|x| \leq 1$ and $0 < \gamma < 1$. Since $0 \leq Q_{2k} = R_{2k}/R_{2k-1} \leq 1$, the ratio on the left side of (2.65) lies between 0 and 1. Thus the denominator in (2.65) is positive since the numerator is positive. Hence (2.65) is equivalent to

$$\rho\beta(1-\gamma^2) \leq \delta[2\rho\beta/\delta-(\rho^2-|1+\Gamma|^2)(1+\gamma^2) - 2\gamma\sqrt{M(x)}] \quad .$$

After rearranging terms we obtain the equivalent inequality

(2.66) $$2\gamma\delta\sqrt{M(x)} \leq (1+\gamma^2)[\rho\beta-(\rho^2-|1+\Gamma|^2)\delta] \quad .$$

To verify (2.66) we note that, since $0 < \delta \leq 1$, it follows that $(1-\delta^{-1}) \leq 0$ and hence from (2.64) we have

$$M(x) \leq [(\rho^2+|1+\Gamma|^2)/\delta-2\rho|1+\Gamma|x]^2 \quad .$$

The expression inside the brackets above can be shown to be non-negative. Therefore

(2.67) $$2\gamma\delta\sqrt{M(x)} \leq 2\gamma\delta[(\rho^2+|1+\Gamma|^2)/\delta-2\rho|1+\Gamma|x] \quad .$$

For the right side of (2.66) we have (by (2.58c))

$$\begin{aligned}(1+\gamma^2)[\rho\beta-(\rho^2-|1+\Gamma|^2)\delta] &= (1+\gamma^2)[2\rho(\rho/\delta-|1+\Gamma|x)-(\rho^2-|1+\Gamma|^2)\delta] \\ &= (1+\gamma^2)[(\rho^2(2-\delta^2)+\delta^2|1+\Gamma|^2)/\delta-2\rho|1+\Gamma|x] \quad .\end{aligned}$$

Since $0 < \gamma < 1$, we have $(1-\gamma)^2 > 0$ and hence

(2.68) $$2\gamma < 1 + \gamma^2 \quad .$$

Furthermore, we have

(2.69) $$\rho^2 + |1+\Gamma|^2 \leq \rho^2(2-\delta^2) + \delta^2|1+\Gamma|^2 \quad ,$$

which can be verified since (2.69) is equivalent to

$$0 \leq \rho^2(1-\delta^2) - |1+\Gamma|^2(1-\delta^2) = (\rho^2-|1+\Gamma|^2)(1-\delta^2) \quad ,$$

which clearly holds since $0 < \delta \leq 1$. Combining (2.66), (2.67), (2.68) and (2.69) yields

$$\begin{aligned}2\gamma\delta\sqrt{M(x)} &\leq 2\gamma\delta[(\rho^2+|1+\Gamma|^2)/\delta-2\rho|1+\Gamma|x] \\ &< (1+\gamma^2)[(\rho^2(2-\delta^2)+\delta^2|1+\Gamma|^2)/\delta-2\rho|1+\Gamma|x] \\ &= (1+\gamma^2)[\rho\beta-(\rho^2-|1+\Gamma|)\delta] \quad .\end{aligned}$$

Thus (2.66) holds and therefore (2.65) and (2.56) hold.

Now we have shown that $Q_j = R_j/R_{j-1} \leq \delta_j$ for all $j = 2,3,4,\ldots$.

Therefore

$$R_n \le R_1 \prod_{j=2}^{n} Q_j \le R_1 \prod_{j=2}^{n} \delta_j \ , \quad n = 2,3,4,\ldots \ . \tag{2.70}$$

Since R_1 is the radius of $S_1(V_1)$ and since $S_1(V_1) \subseteq V_0$ it follows that $R_1 \le \rho$. Combining this with (2.70) and (2.36) proves (2.6) . This completes the proof of Theorem 1 . ■

3. Elimination of Twin-Convergence Regions Which Are Not Best. A pair of twin-convergence regions (E_1,E_2) for continued fractions $K(a_n/1)$ is said to be best if there does not exist another pair of twin-convergence regions (E_1',E_2') such that

$$E_k \subseteq E_k' \ , \quad k = 1,2$$

where proper containment holds for $k = 1$ or 2 or both. In our next result it is shown that there exist pairs of twin-convergence regions (E_1,E_2) of the form (1.6) which are not best.

Theorem 2. *Let Γ be a complex number and ρ be a positive real number such that*

$$0 < |\Gamma| < |1+\Gamma| < \rho \ . \tag{3.1}$$

Let (E_1,E_2) denote the pair of twin-convergence regions for continued fractions $K(a_n/1)$ defined by (1.6) . Let

$$\rho^* = (\rho+|\Gamma|)(|1+\Gamma|-|\Gamma|) \tag{3.2}$$

and let (E_1^*,E_2^*) denote the pair of twin-convergence regions (1.4) . Then

$$E_1 \subseteq E_1^* \ . \quad \text{(proper containment)} \tag{3.3a}$$

Moreover

$$E_2 \subseteq E_2^* \tag{3.3b}$$

if and only if

$$\rho \ge |1+\Gamma| + |2+\Gamma| - |\Gamma| \ . \tag{3.4}$$

Proof. By Lemma 4 , $E_1 = E_1(1)$ and $E_2 = E_2(1)$ where $E_1(1)$ and $E_2(1)$ are given by (2.3) and (2.4) . Let $g(\theta) = g(1,\theta)$ and $h(\theta) = h(1,\theta)$ where $g(\delta,\theta)$ and $h(\delta,\theta)$ are defined by (2.7) and (2.10) , respectively. Then

$$E_1 = E_1(1) = [w = re^{i\theta} : 0 \le r \le g(\theta) \ , \ 0 \le \theta < 2\pi] \tag{3.5a}$$

and

$$E_2 = E_2(1) = [w = re^{i\theta} : r \ge h(\theta) \ , \ 0 \le \theta < 2\pi] \ . \tag{3.5b}$$

Here $r = |w| \ge 0$. It follows immediately from Lemma 1 that $E_1 \subset E_1^*$ and the containment is clearly proper. It can be seen that (3.3b) holds if and only if $h(\theta) \ge 2(\rho^*-\cos\theta)$ for all θ ; that is, if and only if

$$2|1+\Gamma|[\rho-|1+\Gamma|\cos(\theta-2\mathrm{Arg}(1+\Gamma))] \ge 2(\rho^*-\cos\theta) \ , \quad 0 \le \theta < 2\pi \ . \tag{3.6}$$

Upon substituting (3.2) in (3.6) , cancelling terms and rearranging, we obtain the equivalent inequality

$$(3.7)\qquad |\Gamma|(\rho-|1+\Gamma|+|\Gamma|) - |1+\Gamma|^2\cos(\theta-2\mathrm{Arg}(1+\Gamma)) + \cos\theta \geq 0 \quad .$$

Let

$$D = |\Gamma|(\rho-|1+\Gamma|+|\Gamma|) \quad \text{and} \quad \phi = 2\mathrm{Arg}(1+\Gamma) \quad .$$

Then (3.7) can be written as

$$(3.8)\qquad D - |1+\Gamma|^2\cos(\theta-\phi) - \cos\theta \geq 0 \quad .$$

This is easily seen to be equivalent to

$$(3.9a)\qquad D + A\sin\theta + B\cos\theta \geq 0$$

where

$$(3.9b)\qquad A = -|1+\Gamma|^2\sin\phi \quad \text{and} \quad B = 1 - |1+\Gamma|^2\cos\phi \quad .$$

Inequality (3.9b) is equivalent to

$$(3.10a)\qquad D + C\cos(\theta-\psi) > 0 \quad ,$$

where

$$(3.10b)\qquad C = \sqrt{A^2+B^2} \quad , \quad \sin\psi = A/C \quad , \quad \cos\psi = B/C \quad .$$

It can be seen that

$$(3.11)\qquad C = \sqrt{1-2|1+\Gamma|^2\cos\phi+|1+\Gamma|^4} > \big|1-|1+\Gamma|^2\big| > 0 \quad ,$$

since $\Gamma \neq 0$. Now (3.10a) will hold for all θ if and only if $D \geq C$; that is, if and only if

$$(3.12)\qquad |\Gamma|(\rho-|1+\Gamma|+|\Gamma|) \geq \sqrt{1-2|1+\Gamma|^2\cos\phi+|1+\Gamma|^4} \quad .$$

To investigate (3.12) we note that

$$\begin{aligned} 1 - 2|1+\Gamma|^2\cos\phi + |1+\Gamma|^4 &= 1 - 2\mathrm{Re}(1+\Gamma)^2 + |1+\Gamma|^4 \\ &= 1 - (1+\Gamma)^2 - (1+\bar{\Gamma})^2 + |1+\Gamma|^4 \\ &= |1-(1+\Gamma)^2|^2 \\ &= |\Gamma|^2|2+\Gamma|^2 \quad . \end{aligned}$$

Using this in (3.12) , we obtain the equivalent inequality

$$|\Gamma|(\rho-|1+\Gamma|+|\Gamma|) \geq |\Gamma||2+\Gamma| \quad ,$$

which is equivalent to (3.4) since $|\Gamma| \neq 0$. This completes the proof. ■

<u>Remarks</u>. It can be seen that there will exist a ρ satisfying (3.1) and not satisfying (3.4) if and only if $|2+\Gamma| > |\Gamma|$; that is, if and only if $\mathrm{Re}(\Gamma) > -1$. But the requirement $|\Gamma| < |1+\Gamma|$ in (3.1) is equivalent to $\mathrm{Re}(\Gamma) > -1/2$. Thus for every Γ satisfying $0 < |\Gamma| < |1+\Gamma|$, there exists a $\rho > |1+\Gamma|$ such that in Theorem 2 $E_2 \not\subset E_2^*$.

Two examples are considered to illustrate the situations that can occur.

Example 1. Let $\Gamma = i$ and $\rho = 3$. Then $\rho^* = 4(\sqrt{2} - 1) \doteq 1.7$ and

$$\rho = 3 > 2.65 \doteq \sqrt{2} + \sqrt{5} - 1 = |1+\Gamma| + |2+\Gamma| - |\Gamma| .$$

Thus by Theorem 1 we will have $E_1 \subset E_1^*$ and $E_2 \subseteq E_2^*$. The boundaries of these regions are shown in Figure 2, where it is seen that E_2 is properly contained in E_2^*.

Example 2. Let $\Gamma = i$ and $\rho = 1.5$. Then $\rho^* = 2.5(\sqrt{2} - 1) \doteq 1.04$ and

$$\rho = 1.5 < 2.65 \doteq \sqrt{2} + \sqrt{5} - 1 = |1+\Gamma| + |2+\Gamma| - |\Gamma| .$$

Hence by Theorem 2 we will have $E_1 \subset E_1^*$ and $E_2 \not\subseteq E_2^*$. The boundaries of these regions are shown in Figure 3.

In our final result (Theorem 3) we shall show that, given any pair of twin-convergence regions (E_1^*,E_2^*) of the form (1.4), there does not exist a better pair of twin-convergence regions (E_1,E_2) in the larger family defined by (1.6).

Theorem 3. Let $\rho^* > 0$ be given and let (E_1^*,E_2^*) be defined by (1.4). Let Γ be a complex number and ρ a positive number such that

$$(3.13) \qquad 0 < |\Gamma| < |1+\Gamma| < \rho$$

and let (E_1,E_2) be defined by (1.6). Then: (A)

$$(3.14) \qquad E_1^* \subseteq E_1$$

if and only if

$$(3.15) \qquad \rho^* = (\rho-|\Gamma|)(|1+\Gamma|-|\Gamma|) .$$

(B) If (3.15) is satisfied, then

$$(3.16) \qquad E_2^* \not\subseteq E_2 .$$

Proof. (A) is an immediate consequence of (1.4), Lemma 1 and Lemma 4. (B) Suppose that (3.15) holds and assume that

$$(3.17) \qquad E_2^* \subseteq E_2 .$$

Then by Lemma 4, (3.17) is equivalent to

$$(3.18) \qquad 2|1+\Gamma|[\rho-|1+\Gamma|\cos(\theta-2\mathrm{Arg}(1+\Gamma))] < 2(\rho^*-\cos\theta) , \quad 0 \leq \theta < 2\pi .$$

Substituting (3.15) in (3.18), cancelling terms, rearranging and setting $\phi = 2\mathrm{Arg}(1+\Gamma)$ yields the equivalent inequality

$$(3.19) \qquad |\Gamma|(\rho+|1+\Gamma|-|\Gamma|) - |1+\Gamma|^2\cos(\theta-\phi) + \cos\theta \leq 0 , \quad 0 \leq \theta < 2\pi .$$

By a method similar to that used in the proof of Theorem 2, we set

$$D = \rho + |1+\Gamma| - |\Gamma| , \quad A = |1+\Gamma|^2\sin\phi , \quad B = |1+\Gamma|^2\cos\phi - 1 ,$$

$$C = \sqrt{A^2+B^2} , \quad \sin\psi = A/C , \quad \cos\psi = B/C .$$

Then it is readily shown that (3.19) is equivalent to

(3.20) $$C\cos(\theta-\psi) \geq D \ , \quad 0 \leq \theta < 2\pi \ .$$

Clearly (3.20) holds for all θ if and only if $-C \geq D$. But this is impossible since $C \geq 0$ and $D > 0$. Hence the assumption (3.17) is false. ∎

References

1. Field, David A., Estimates of the speed of convergence of continued fraction expansions of functions, Math. of Comp. 31, No. 138 (April 1977), 495-502.
2. Field, David A., Error bounds for elliptic convergence regions for continued fractions, SIAM J. Numerical Anal. 15 (1978a), 444-449.
3. Field, David A., Error bounds for continued fractions $K(1/b_n)$, Numer. Math. 29 (1978b), 261-267.
4. Field, David A. and William B. Jones, A priori estimates for truncation error of continued fractions $K(1/b_n)$, Numer. Math. 19 (1972), 283-302.
5. Jones, William B., and R.I. Snell, Truncation error bounds for continued fractions, SIAM J. Numer. Anal. 6, No. 2 (June 1969), 210-221.
6. Jones, William B., and W.J. Thron, Twin-convergence regions for continued fractions $K(a_n/1)$, Trans. Amer. Math. Soc. 150 (1970), 93-119.
7. Lange, L.J., On a family of twin convergence regions for continued fractions, Illinois J. Math. 10 (1966), 97-108.
8. Sweezy, W.B. and W.J. Thron, Estimates of the speed of convergence of certain continued fractions, SIAM J. Numer. Anal. 4, No. 2 (1967), 254-270.
9. Thron, W.J., On parabolic convergence regions for continued fractions, Math. Zeitschr. 69 (1958), 173-182.
10. Thron, W.J., Zwillingskonvergenzgebiet für Kettenbrüche $1 + K(a_n/1)$, deren eines die Kreisscheibe $|a_{2n-1}| < \rho^2$ ist, Math. Zeitschr. 70 (1959), 310-344.

William B. Jones
Department of Mathematics
University of Colorado
Boulder, Colorado 80309

Walter M. Reid
Department of Mathematics
University of Wisconsin - Eau Claire
Eau Claire, Wisconsin 54701

DIGITAL FILTERS AND CONTINUED FRACTIONS

William B. Jones

Allan Steinhardt

1. Introduction. Digital filtering is a discipline, born of the computer revolution, that is concerned with the extraction and/or enhancement of information contained in a sequence of measurements of a continuous waveform phenomenon. Digital filtering has been used in such diverse fields as astronomy, economics, electrical power planning, medicine, radar, seismology and speech processing. The most active area of digital filter applications is speech. Although a continuous phenomenon, speech is so redundant that a sequence of sampled values, if taken frequently, can contain all of the desired information in the original message. By processing such a sequence one can filter the static noise from a telephone message or transform the message to a form recognizable by a computer [9]. Recent studies have been made [7] in computer processing, recognition, and classification of picture images by digital filtering. In seismology, digital filtering is being used to process seismographs for detecting the presence of gas, oil or precious minerals in a given geological formation [1].

The purpose of this paper is to investigate connections between digital filters and continued fractions. Digital filters implemented by ladder-directed graphs (Figure 1) are used in machine analysis and synthesis of speech [9], and are known to perform well in the presence of arithmetical roundoff [12]. It is shown (Theorem 2.1(A)) that the transfer function of such a ladder filter can be represented by a terminating continued fraction $R_N(z;c_{N+1})$ (see (2.1)). This continued fraction is seen to be equivalent to a modified approximant of a Schur fraction and it follows that all such filters are stable. Conversely, it is shown (Theorem 2.1(B)) that the transfer function of an arbitrary stable digital filter can be represented either by a continued fraction of the form $R_N(z;c_{N+1})$, or as the limit of such continued fractions. The question of stability is also studied in Section 3. We state and prove (Theorem 3.1) that, for a given polynomial to have all of its zeros located in the interior of the unit disk, it is necessary and sufficient for a certain test function to be represented by a continued fraction of the form $R_N(z;c_{n+1})$. In a sense this result is the analogue of a theorem of Wall [19] and Frank [2] for stable (Hurwitz) polynomials (i.e. polynomials whose zeros all lie in the left half-plane). We obtain as a consequence of Theorem 3.1 the well-known Schur-Cohn test (see Algorithm 1 and Corollary 3.8). In Section 4 we describe a numerical method for computing the poles of transfer functions of the form $R_N(z;c_{N+1})$, even in the case (of most interest here) when two or more poles have equal moduli. The method is based on the FC-algorithm of McCabe and Murphy [10]. In the applications considered, the coefficients c_j of the continued fraction $R_N(z;c_{N+1})$ are measured parameters called reflection coefficients. Two examples are presented, one to speech analysis and the other to signal detection in noise [18]. In the latter example the FG-algorithm was found to be faster for this application than Laguerre's global convergence method with polynomial deflation which is known to be cubically convergent. An explanation for this phenomenon is given in Section 4.

Connections between digital filters and continued fractions have been investigated previously by Mitra and Sherwood [11]. However, they did not consider continued fractions of the forms dealt with here. We conclude this introduction with a summary of basic concepts, definitions and theorems about digital filters, directed graphs and Schur fractions that are subsequently used. More on digital filters and directed graphs can be found in [4], [6], and [13].

The set ℓ of all sequences $\{x(n)\}_{n=0}^{\infty}$ of complex numbers forms a linear space over the field $\mathbb{C}$ with respect to the usual operations of addition and scalar multiplication. Two other operations on ℓ are of special interest. The <u>convolution</u> $\{h(n)\} * \{x(n)\}$ of sequences $\{h(n)\}$ and $\{x(n)\}$ is defined by

$$(1.1) \qquad \{h(n)\} * \{x(n)\} = \{\sum_{k=0}^{n} h(k)x(n-k)\} .$$

The <u>unit delay</u> $D\{x(n)\}$ of a sequence $\{x(n)\}$ is defined by

$$(1.2) \qquad D\{x(0),x(1),x(2),\ldots\} = \{0,x(0),x(1),x(2),\ldots\} .$$

For $j = 1,2,3,\ldots$, the <u>jth order delay</u> is

$$(1.3) \qquad D^j\{x(n)\} = \{y(n)\} , \text{ where } y(n) = \begin{cases} 0 , & \text{if } 0 \leq n \leq j-1 \\ x(n-j) & \text{if } n \geq j \end{cases} .$$

Two subspaces of ℓ are particularly notable. ℓ^0 denotes the subspace consisting of all sequences $\{x(n)\}$ such that

$$(1.4) \qquad \limsup_{n\to\infty} |x(n)|^{1/n} < \infty .$$

ℓ_∞ denotes the subspace of ℓ (and of ℓ^0) consisting of all bounded sequences $\{x(n)\}$; that is, $\{x(n)\} \in \ell_\infty$ if and only if there exists a constant B such that

$$(1.5) \qquad |x(n)| \leq B , \quad n = 0,1,2,\ldots .$$

A <u>digital filter</u> F is a mapping of sequences $\{x(n)\}$ in ℓ into sequences $\{y(n)\}$ according to equations of the form

$$(1.6) \qquad y(n) + \sum_{k=1}^{N} b_k y(n-k) = \sum_{k=0}^{M} a_k x(n-k) , \quad n = 0,1,2,\ldots ,$$

where $a_0,a_1,\ldots,a_M$ and $b_1,b_2,\ldots,b_N$ are given complex constants with

$$(1.7) \qquad a_M \neq 0 , \quad b_N \neq 0 \text{ and } x_m = y_m = 0 \text{ if } m < 0 .$$

If $N = 0$ then the sum on the left side of (1.7) is interpreted to be zero, in which case the filter is called <u>nonrecursive</u>. If $N > 0$ then the filter is called <u>recursive</u>. The sequences $\{x(n)\}$ and $\{y(n)\}$ are called the <u>input</u> and <u>output</u>, respectively, of the filter F and we write

$$\{y(n)\} = F(\{x(n)\}) \quad .$$

It is easily seen from (1.6) that a digital filter F is a linear transformation of ℓ into ℓ ; that is, if $\{x(n)\}$, $\{z(n)\} \in \ell$ and $a,b \in \mathbb{C}$, then

$$F(a\{x(n)\} + b\{z(n)\}) = aF(\{x(n)\}) + bF(\{z(n)\}) \quad .$$

It can also be shown that a digital filter F maps ℓ^0 into ℓ^0 .

Directed graphs are used to describe specific procedures for implementing a digital filter. A <u>directed graph</u> is a geometric configuration consisting of two types of elements. There are points (called <u>nodes</u>) and simple directed curves (called <u>branches</u>). Each branch connects two nodes and the direction of the branch is indicated by an arrow. The nodes will be labeled from 1 to 4N . Each node has associated with it a node sequence; $\{x_i(n)\}_{n=0}^{\infty}$ denotes the <u>node sequence</u> of the ith node. Each branch has associated with it a <u>transmittance sequence</u> of the form

$$(1.8) \qquad \{t_{ij}(n)\}_{n=0}^{\infty} = \{a_{ij},b_{ij},0,0,0,\cdots\} \quad , \quad \begin{matrix} a_{ij},b_{ij} \in \mathbb{C} \ , \\ a_{ij} = 0 \ \text{ or } \ b_{ij} = 0 \ . \end{matrix}$$

The first index i denotes the node from which the branch originates and the second index j denotes the node at which the branch terminates. The effect of $\{t_{ij}(n)\}$ is a scalar multiplication by a_{ij} if $b_{ij} \neq 0$; it is a scalar multiplication by b_{ij} and a unit shift if $a_{ij} = 0$. Every directed graph has two special nodes called the <u>sink</u> node and the <u>source</u> node. As the names imply, the source node has no branch terminating at it and the sink node has no branch originating at it. The node sequences are interrelated by means of the <u>fundamental equations</u>

$$(1.9) \qquad \{x_i(n)\} = \sum_{\substack{j=1 \\ j \neq i}}^{4N} \{t_{ji}(n)\} * \{x_j(n)\} \quad , \ i = 2,3,\ldots,4N \quad .$$

If there is no branch from node j to node i , then $t_{ji}(n) = 0$ for all $n = 0,1,2,\ldots$. If node 1 is the source node and node 4N is the sink node, then (1.9) defines a digital filter in which $\{x_1(n)\}$ is the input and $\{x_{4N}(n)\}$ is the output. It is well known that every digital filter can be described by equations of the form (1.9) and hence by means of a directed graph.

It is sometimes useful to decompose a directed graph into subgraphs. A directed graph is a <u>subgraph</u> of a given directed graph if it has the property that exactly one branch of the parent graph enters the subgraph (at the subgraph source node) and exactly one branch of the parent graph leaves the subgraph (at the subgraph sink node). The following theorem governs the role of a subgraph in the parent graph.

<u>Theorem</u> 1.1. <u>For a subgraph of a given directed graph</u>, let $\{x_{SO}(n)\}$ <u>denote the source nodal sequence and</u> $\{x_{SI}(n)\}$ <u>denote the sink nodal sequence</u>.

Then there exists a sequence $\{t(n)\}$ *with the property that if the subgraph is replaced by the subgraph source and sink nodes connected by a single branch (from source to sink) with transmittance sequence* $\{t(n)\}$ *, then the nodal sequences at all nodes on the parent graph exterior to the subgraph remain the same. In other words, a subgraph can be replaced by a single branch and the remainder of the parent graph is unchanged. Moreover,* $\{t(n)\}$ *can be obtained by setting* $\{x_{SO}(n)\} = \{1,0,0,0,\cdots\}$ *and solving the fundamental equations* (1.9) *for* $\{t(n)\} = \{x_{SI}(n)\}$.

We note that in this case the transmittance sequence $\{t(n)\}$ will not (in general) be a sequence of the special form (1.8) . Some examples of directed graphs and subgraphs will be considered in Section 2.

The z-transform is very useful in work with digital filters and directed graphs. The *z-transform* $X(z)$ of a sequence $\{x(n)\}$ in ℓ is defined by

$$X(z) = \sum_{n=0}^{\infty} x(n)z^{-n} \tag{1.10}$$

Following Henrici [6] , we shall use the symbol

$$X(z) \overset{z}{\bullet\!\!-\!\!\!-\!\!\circ} \{x(n)\}$$

to indicate that $X(z)$ is the z-transform of $\{x(n)\}$. The z-transform provides a one-to-one mapping of ℓ onto the set of all formal power series of the form (1.10) . If $\{x(n)\} \in \ell^0$, then the z-transform of $\{x(n)\}$ defines a function $X(z)$ analytic at $z = \infty$. The following are some elementary properties of the z-transforms that are used here. If $X(z) \overset{z}{\bullet\!\!-\!\!\!-\!\!\circ} \{x(n)\} \in \ell^0$ and $Y(z) \overset{z}{\bullet\!\!-\!\!\!-\!\!\circ} \{y(n)\} \in \ell^0$, then

$$aX(z) + bY(z) \overset{z}{\bullet\!\!-\!\!\!-\!\!\circ} a\{x(n)\} + b\{y(n)\} \quad , \quad a,b \in \mathbb{C} \tag{1.11}$$

$$1 \overset{z}{\bullet\!\!-\!\!\!-\!\!\circ} \{1,0,0,0,\cdots\} \quad , \tag{1.12}$$

$$Y(z)X(z) \overset{z}{\bullet\!\!-\!\!\!-\!\!\circ} \{y(n)\} * \{x(n)\} \quad , \tag{1.13}$$

$$z^{-1}X(z) \overset{z}{\bullet\!\!-\!\!\!-\!\!\circ} D\{x(n)\} \quad , \tag{1.14}$$

$$\left(\sum_{j=0}^{m} a_j z^{-j}\right)X(z) \overset{z}{\bullet\!\!-\!\!\!-\!\!\circ} \sum_{j=0}^{m} a_j D^j\{x(n)\} \quad , \quad a_j \in \mathbb{C} \quad , \quad m \geq 0 \ . \tag{1.15}$$

We shall now summarize some further properties of the z-transform and its relation to digital filters that are needed.

Theorem 1.2. *Let* f *be a digital filter defined by* (1.7) *and let* $H(z)$ *be the corresponding rational function*

$$H(z) = \frac{\sum_{k=0}^{M} a_k z^{-k}}{\sum_{k=0}^{N} b_k z^{-k}} \quad , \quad b_0 = 1 \ . \tag{1.16}$$

If $X(z) \bullet\!\!\overset{z}{-\!\!-}\!\!\circ \{x(n)\} \in \ell^0$, then

$$H(z)X(z) \bullet\!\!\overset{z}{-\!\!-}\!\!\circ F(\{x(n)\}) \quad . \tag{1.17}$$

The function $H(z)$ is called the transfer function of the digital filter F . Theorem 1.2 shows that the z-transform of the output of a digital filter is the product of the transfer function and the z-transform of the input. It can be seen that $H(z)$ is analytic at $z = \infty$, since $b_0 = 1$. The sequence $\{h(n)\}_{n=0}^{\infty}$ defined by

$$H(z) \bullet\!\!\overset{z}{-\!\!-}\!\!\circ \{h(n)\}$$

is called the shock response of the filter. The next property states that the output of a digital filter is the convolution of the shock response and the input.

Theorem 1.3. Let F be a digital filter with shock response $\{h(n)\}$. If $\{x(n)\} \in \ell^0$, then

$$F(\{x(n)\}) = \{h(n)\} * \{x(n)\} \quad . \tag{1.18}$$

A digital filter F is said to be stable if $\{x(n)\} \in \ell_\infty$ implies that $F(\{x(n)\} \in \ell_\infty$.

The following result provides conditions for determining whether or not a given filter is stable.

Theorem 1.4. Let F be a digital filter with transfer function $H(z)$ and shock response $\{h(n)\}$. Then the following three statements are equivalent:

(i) F is stable.

(ii) All poles of $H(z)$ are in the interior of the unit disk.

(iii) $\sum_{n=0}^{\infty} |h(n)| < \infty$.

A directed graph and its fundamental equations (1.9) can be mapped by the z-transform into a corresponding directed graph and set of fundamental equations involving functions of the complex variable z . The simplification induced by this mapping is mainly due to the fact that functions of z can be manipulated more easily than infinite sequences. Moreover, the theory of analytic functions can sometimes be used to deal with stability and other related problems. The nodal function $X_i(z)$ associated with the ith node and the transmission function $T_{ij}(z)$ associated with the branch connecting node i to node j are defined by

$$X_i(z) \bullet\!\!\overset{z}{-\!\!-}\!\!\circ \{x_i(n)\} \quad , \quad T_{ij}(z) \bullet\!\!\overset{z}{-\!\!-}\!\!\circ \{t_{ij}(n)\} \quad . \tag{1.19}$$

If $\{t_{ij}(n)\}$ has the form (1.8) , then

$$T_{ij}(z) = a_{ij} + b_{ij}z^{-1} \quad . \tag{1.20}$$

By applying the z-transform to both sides of (1.9) and using (1.11) and (1.13) , we obtain the fundamental equations in the transform domain

(1.21) $$X_i(z) = \sum_{\substack{j=1 \\ j\neq i}}^{4N} T_{ji}(z)X_j(z) \quad , \quad i = 2,3,\ldots,4N \quad .$$

Now let node 1 be the source node and node 4N be the sink node of a directed graph. Then $\{x_1(n)\} \in \ell^0$ and $\{x_{4N}(n)\}$ are the input and output respectively of the associated digital filter F, so that

$$F(\{x_1(n)\}) = \{x_{4N}(n)\} \quad .$$

Thus by (1.17), $F(\{x_1(n)\}) \circ\!\!\xrightarrow{\;z\;}\!\!\bullet\, H(z)X_1(z)$, where $H(z)$ is the transfer function of F. It follows that

(1.22) $$X_{4N}(z) = H(z)X_1(z) \quad .$$

Equation (1.22) can be obtained by eliminating $X_2(z), X_3(z), \ldots, X_{4N-1}(z)$ from (1.21). Therefore one can obtain the transfer function $H(z)$ of the filter defined by (1.21) simply by setting $X_1(z) \equiv 1$ and solving (1.21) for $H(z) = X_{4N}(z)$. This method of finding $H(z)$ will be employed in Section 2.

We conclude this introduction by stating a well known theorem about Schur continued fractions for functions $G(\zeta)$ analytic for $|\zeta| < 1$ and such that

$$M(G) = \underset{|\zeta|<1}{\text{l.u.b.}} |G(\zeta)| \leq 1 \quad .$$

The theorem is based on methods of Schur [15] and was first proved by Wall [20, Theorem 77.1]. Although Schur did not consider continued fractions, Hamel [3] gives a theorem on continued fractions closely related to the Schur problem and to the result of Wall. Connections with two-point Padé tables and general T-fractions are given by Thron [17].

<u>Theorem</u> 1.5. (A) <u>Let $\{c_k\}$ be an infinite sequence of complex numbers with moduli less than unity,</u>

(1.23) $$|c_k| < 1 \quad , \quad k = 1,2,3,\ldots \quad ,$$

<u>and let</u>

(1.24) $$S_n(\zeta;t) = c_1 + \frac{(1-|c_1|^2)\zeta}{\bar{c}_1\zeta} - \frac{1}{\bar{c}_2} + \frac{(1-|c_2|^2)\zeta}{\bar{c}_2\zeta} - \cdots \frac{1}{-\bar{c}_n} + \frac{(1-|c_n|^2)\zeta}{\bar{c}_n\zeta} - \frac{1}{t} \quad .$$

<u>Then</u>

(1.25) $$|S_n(\zeta;t)| < 1 \quad , \text{ for } |t| \leq 1 \quad , \quad |\zeta| < 1 \quad , \quad n = 0,1,2,\ldots \quad .$$

<u>Let $\{t_k\}$ be a sequence of independent variables. There exists a function $G(\zeta)$, analytic for $|\zeta| < 1$, such that $M(G) \leq 1$, and such that, for every positive number $r < 1$,</u>

(1.26) $$\lim_{n\to\infty} S_n(\zeta;t_n) = G(\zeta) \quad ,$$

uniformly for $|\zeta| \le r$, $|t_n| \le 1$, $n = 1,2,3,\ldots$.

(B) Conversely, if $G(\zeta)$ is a given function, analytic for $|\zeta| < 1$, such that $M(G) \le 1$, then one of the following three statements holds.

(a) $G(\zeta) \equiv c_1$, where c_1 is a constant with $|c_1| = 1$.

(b) There exists uniquely a finite sequence $c_1, c_2, \ldots, c_{n+1}$, $n \ge 1$, such that

(1.27) $$|c_k| < 1 \quad \text{for} \quad k = 1,2,\ldots,n \quad \text{and} \quad |c_{n+1}| = 1 ,$$

and such that

(1.28) $$G(\zeta) = S_n(\zeta; c_{n+1}) .$$

(c) There exists uniquely an infinite sequence $\{c_k\}$ satisfying (1.23) such that (1.26) holds.

A continued fraction of the form

(1.29) $$c_1 + \frac{(1-|c_1|^2)\zeta}{\bar{c}_1\zeta} \, \frac{1}{-\,c_2} + \frac{(1-|c_2|^2)\zeta}{\bar{c}_2\zeta} \, \frac{1}{-\,c_3} + \frac{(1-|c_3|^2)\zeta}{\bar{c}_3\zeta} + \cdots ,$$

where the c_k satisfy (1.23) will be called a Schur fraction. The successive approximants of (1.29) starting with the zeroth approximant are as follows:

$$c_1 , \; c_1 + \frac{(1-|c_1|^2)\zeta}{\bar{c}_1\zeta} , \; c_1 + \frac{(1-|c_1|^2)\zeta}{\bar{c}_1\zeta} \, \frac{1}{-\,c_2} , \; \ldots .$$

Therefore

$$S_n(\zeta;0) = c_1 + \frac{(1-|c_1|^2)\zeta}{\bar{c}_1\zeta} \, \frac{1}{-\,c_2} + \frac{(1-|c_2|^2)\zeta}{\bar{c}_2\zeta} \, \cdots \, - \, \frac{1}{-\,c_n}$$

is the $(2n-2)$th approximant of (1.29) . Thus Theorem 1.5 deals with even order approximants of (1.29) and more generally with modified approximants $S_n(\zeta, t_n)$.

2. A Family of Ladder Directed Graphs.

In this section we are concerned with digital filters that can be implemented by ladder directed graphs of the form shown in Figure 1 .

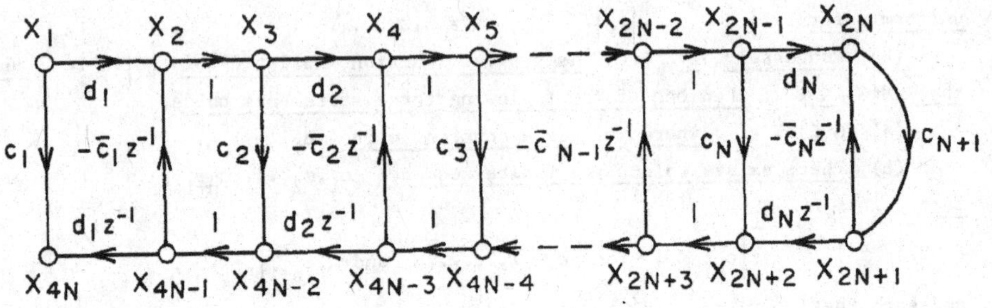

Figure 1. A ladder directed graph in the transform domain with $4N$ nodes. The X_k are the nodal functions of the complex variable z. Next to each branch is the associated transmittance function of z, where the c_k are complex constants with $|c_k| < 1$, $k = 1,2,\ldots,N$, $|c_{N+1}| = 1$ and $d_k = \sqrt{1-|c_k|^2}$. Node 1 is the source node and node $4N$ is the sink node.

These filters are useful in machine analysis and synthesis of speech [Gray and Markel]. The physical mechanism for speech can be modeled mathematically as a ladder digital filter with an input which is a fixed periodic sequence (for vowels) and a white noise sequence (for consonants). The type of input and the parameters c_k can be obtained from the speech waveform in a time period so small as to be imperceptible to the naked ear. From the parametric representation it is possible to reconstruct the speech by using the appropriate ladder directed graph and input.

Mullis and Roberts [12] have shown recently that a directed graph of the type in Figure 1 provides a nearly optimal implementation of the digital filter in the sense that the influence on the output due to arithmetic roundoff error occuring at each node sequence is comparatively small. The directed graphs known to be optimal in this regard are impractical, since the number of nodes involved is of the order of N^2. Hence the ladder directed graphs in Figure 1 offer a good compromise between the requirements of speed and good roundoff behavior. We shall show that every digital filter represented by a directed graph of the type in Figure 1 is a stable filter. Moreover, every stable filter can (up to a constant factor) either be represented by a graph of this type or can be approximately represented by such a graph where the error of approximation can be made arbitrarily small but positive. For this purpose we consider continued fractions of the form

$$(2.1)\quad R_N(z;t) = c_1 + \frac{(1-|c_1|^2)z^{-1}}{\bar{c}_1 z^{-1}} + \frac{1}{c_2} + \frac{(1-|c_2|^2)z^{-1}}{\bar{c}_2 z^{-1}} + \cdots + \frac{1}{c_N} + \frac{(1-|c_N|^2)z^{-1}}{\bar{c}_N z^{-1}} + \frac{1}{t} .$$

Theorem 2.1. (A) Let F be a digital filter represented by a directed graph of the form in Figure 1. Let $H_N(z)$ denote the transfer function of F. Then F is a stable filter and $H_N(z)$ can be expressed as a continued fraction

$$(2.2)\qquad H_N(z) = R_N(z;c_{N+1}) .$$

(B) Let F be an arbitrary stable digital filter and let $H(z)$ denote its transfer function. (We assume that $H(z)$ is not identically constant.) Then one of the following statements holds.

(B1) There exists uniquely a finite sequence $c_1,c_2,\ldots,c_{N+1}$, with $|c_k| < 1$ for $k = 1,2,\ldots,N$ and $|c_{N+1}| = 1$, and a positive constant $\beta > 0$, such that

$$(2.3)\qquad H(z) = \beta R_N(z;c_{N+1}) .$$

(B2) There exists uniquely a sequence $\{c_k\}$ with $|c_k| < 1$ for $k = 1,2,3,\ldots$ and a positive constant β such that

$$(2.4)\qquad H(z) = \lim_{N\to\infty} \beta R_N(z;1) \quad \text{for} \quad |z| > 1 .$$

For each r with $0 < r < 1$, the convergence is uniform on $|z| \geq 1/r$.

Proof. (A): First we will show that $H_N(z)$ has the form (2.2). Our proof is by induction. Suppose that $N = 1$. Then by the fundamental equation (1.21) we have

$$(2.5)\qquad \begin{aligned} X_2(z) &= \sqrt{1-|c_1|^2}X_1(z) - \bar{c}_1 z^{-1}X_3(z) , \\ X_3(z) &= c_2 X_2(z) , \\ X_4(z) &= c_1 X_1(z) + \sqrt{1-|c_1|^2}z^{-1}X_3(z) . \end{aligned}$$

Setting $X_1(z) \equiv 1$ and solving for $H_1(z) = X_4(z)$ we obtain

$$H_1(z) = c_1 + \frac{(1-|c_1|^2)z^{-1}}{\bar{c}_1 z^{-1}} + \frac{1}{c_2}$$

as asserted. Next we assume that the assertion (2.2) holds for $N-1$ and consider the subgraph shown in Figure 2. By the induction hypothesis the transfer function $T(z)$ for the subgraph in Figure 2 has the form

$$(2.6)\quad T(z) = c_2 + \frac{(1-|c_2|^2)z^{-1}}{\bar{c}_2 z^{-1}} + \frac{1}{c_3} + \cdots + \frac{1}{c_N} + \frac{(1-|c_N|^2)z^{-1}}{\bar{c}_N z^{-1}} + \frac{1}{c_{n+1}} .$$

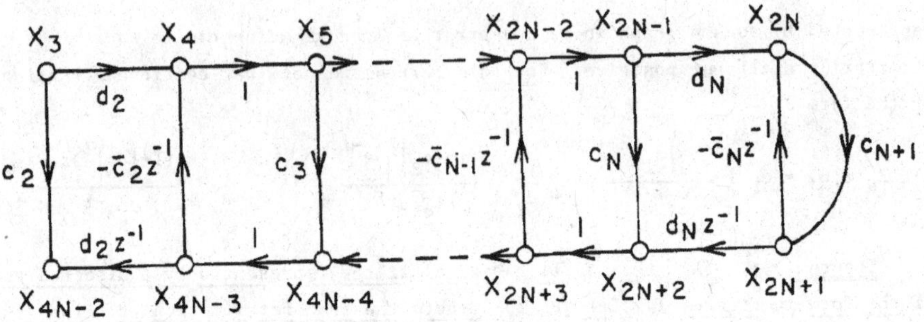

Figure 2. Subgraph of directed graph in Figure 1.

By Theorem 1.1 (interpreted in the transform domain) all nodes on the parent graph exterior to the subgraph remain the same if the subgraph is replaced by a single branch with transmission function T(z) (see Figure 3). The fundamental

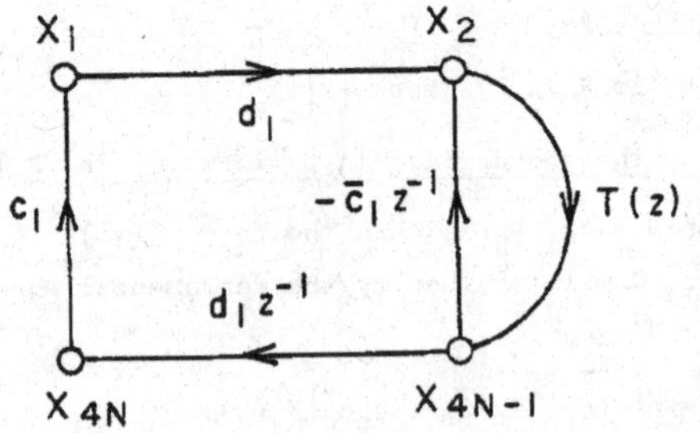

Figure 3. Directed graph (of Figure 1) with subgraph replaced by a single branch having transmission function T(z) .

equations in the transform domain for the directed graph in Figure 3 are the following:

$$(2.7)\qquad \begin{aligned} X_2(z) &= \sqrt{1-|c_1|^2}\,X_1(z) - \bar{c}_1 z^{-1} X_{4N-1}(z) \\ X_{4N-1}(z) &= T(z) X_2(z) \\ X_{4N}(z) &= c_1 X_1(z) + \sqrt{1-|c_1|^2}\, z^{-1} X_{4N-1}(z) \quad . \end{aligned}$$

Setting $X_1(z) \equiv 1$ and solving for $H_N(z) = X_{4N}(z)$, we obtain

$$H_N(z) = c_1 + \cfrac{(1-|c_1|^2)z^{-1}}{\bar{c}_1 z^{-1} + \cfrac{1}{T(z)}} \quad .$$

An application of (2.6) gives (2.2) . Now to prove that F is stable, we note that by (2.2) and (1.24)

$$H_N(z) = R_N(z;c_{N+1}) = S_N(-z^{-1};c_{N+1}) \quad .$$

Therefore by Theorem 1.5(A)

$$|H_N(z)| < 1 \quad \text{for} \quad |z| > 1 \quad .$$

Since $H_N(z)$ is a rational function, all of its poles lie in $|z| < 1$ and hence by Theorem 1.4 F is stable. This proves (A) .

To prove (B) we suppose that F is a stable filter so that by Theorem 1.4 all poles of its transfer function H(z) lie in $|z| < 1$. Thus $G^*(\zeta) = H(1/\zeta)$ is analytic for $|\zeta| \leq 1$. Let

$$\beta = \max_{|\zeta|\leq 1} |G^*(\zeta)|$$

Then $G(\zeta) = G^*(\zeta)/\beta$ is analytic for $|\zeta| \leq 1$ and $M(G) \leq 1$. Theorem 1.5(B) can then be applied to $G(\zeta)$. If case (b) holds then $G(\zeta)$ can be expressed in the form of $S_N(\zeta;c_{n+1})$, where $|c_k| < 1$ for $k = 1,2,\ldots,N$ and $|c_{n+1}| = 1$. Hence

$$H(z) = \beta S_N(z^{-1};c_{N+1}) = \beta R_N(z;c_{N+1}) \quad .$$

Thus (B1) holds.

The only other situation that can occur is case (c) of Theorem 1.5(B) . In that case one can easily verify that (B2) holds. ∎

3. A Stability Test.

A polynomial Q(z) with complex coefficients is called stable (or Hurwitz) if all of its zeros lie in the left half of the complex plane. Wall [19] and Frank [2] have shown that Q(z) is stable if and only if a certain test rational function (with denominator Q(z)) can be expressed in the form of a particular real J-fraction (see, for example, [6, Section 12.7], [8, Theorem 7.32] and [20, Theorem 47.1]).

A polynomial will be called <u>D-stable</u> if all of its zeros lie in the interior of the unit disk $|z| < 1$. The connection between D-stable polynomials and stable digital filters is clear. In this section we shall prove a result for

D-stable polynomials (Theorem 3.1) similar to the result of Wall and Frank for stable polynomials. The well-known Schur-Cohn algorithm for testing D-stability will be derived as a consequence of our proof.

Theorem 3.1. *Let*

$$Q(z) = a_0 + a_1 z + \cdots + a_n z^n \tag{3.1}$$

be given, where $n \geq 1$, $a_n \neq 0$, $a_j \in \mathbb{C}$, $j = 0,1,\ldots,n$. *Let* $P(z)$ *and* $H(z)$ *be defined by*

$$P(z) = z^n \overline{Q_n(1/\bar{z})} = \bar{a}_n + \bar{a}_{n-1} z + \cdots + \bar{a}_0 z^n , \tag{3.2}$$

$$H(z) = \frac{P(z)}{Q(z)} . \tag{3.3}$$

Then $Q(z)$ *is* D-*stable if and only if the test function* $H(z)$ *can be represented by a continued fraction of the form*

$$H(z) = c_1 + \frac{(1-|c_1|^2)}{\bar{c}_1} \, \frac{}{+} \, \frac{z}{c_2} \, \frac{}{+} \, \frac{(1-|c_2|^2)}{\bar{c}_2} \, \frac{}{+} \cdots \frac{}{+} \, \frac{z}{c_n} \, \frac{}{+} \, \frac{(1-|c_n|^2)}{\bar{c}_n} \, \frac{}{+} \, \frac{z}{c_{n+1}} , \tag{3.4a}$$

where

$$|c_j| < 1 , \quad c_j \in \mathbb{C} \quad \text{for } j = 1,2,\ldots,n \quad \text{and} \quad |c_{n+1}| = 1 , \quad c_{n+1} \in \mathbb{C} . \tag{3.4b}$$

Remarks: We note that the continued fraction (3.4a) is equivalent to $R_N(z;c_{n+1})$ (see (2.1) and (2.2)).

Our proof of Theorem 3.1 will be given by means of several lemmas, some of which are of interest for their own sake. In each of these lemmas $P(z)$, $Q(z)$ and $H(z)$ have the meaning as in Theorem 3.1 .

Lemma 3.2. *If* $Q(z)$ *is* D-*stable then every pole of* $H(z)$ *lies in* $|z| < 1$

Proof. $H(\infty) = \bar{a}_0/a_n \neq \infty$ so that all poles of $H(z)$ lie in $\mathbb{C}$. If $H(z_k) = \infty$ for $z_k \in \mathbb{C}$, then clearly $Q(z_k) = 0$ and hence $|z_k| < 1$. ■

Remark: The converse of Lemma 3.2 is not true as can be seen by the following example. Let $Q(z) = z^2 - z$ so that $P(z) = 1-z$ and

$$H(z) = \left(-\frac{1}{z}\right)\left(\frac{z-1}{z-1}\right) .$$

Every pole of $H(z)$ lies in $|z| < 1$ although $Q(z)$ is not D-stable. The reason that $Q(z)$ fails to be D-stable is that $P(z)$ and $Q(z)$ have a common zero at $z = 1$ which is not in $|z| < 1$.

Lemma 3.3. *If* $P(z)$ *and* $Q(z)$ *have no common zeros and if* $H(z)$ *has all of its poles inside* $|z| < 1$, *then* $Q(z)$ *is* D-*stable*.

Proof. Suppose that $Q(z_k) = 0$ for some $z_k \in \mathbb{C}$. Then by hypothesis $P(z_k) \neq 0$ so that $H(z_k) = \infty$. Hence $|z_k| < 1$. ■

<u>Lemma</u> 3.4. <u>If</u> $Q(z)$ <u>is</u> D-<u>stable</u>, <u>then</u> $P(z)$ <u>and</u> $Q(z)$ <u>have no zeros in common</u>.

<u>Proof</u>. Let $z_1, z_2, \ldots, z_k$ denote the distinct zeros of $Q(z)$ and write

(3.5) $$Q(z) = a_n(z-z_1)^{m_1}(z-z_2)^{m_2} \cdots (z-z_k)^{m_k} .$$

Then by (3.2)

(3.6) $$P(z) = \bar{a}_n(1-\bar{z}_1 z)^{m_1}(1-\bar{z}_2 z)^{m_2} \cdots (1-\bar{z}_k z)^{m_k} .$$

Now assume that $P(z_j) = Q(z_j) = 0$ for some z_j, $1 \leq j \leq k$. Then $z_j \neq 0$, since $a_n \neq 0$. It follows then from (3.6) that $z_j = 1/\bar{z}_i$ for some i with $1 \leq i \leq k$ and $z_i \neq 0$. Therefore $z_i = 1/\bar{z}_j$ and hence $Q(1/\bar{z}_j) = Q(z_i) = 0$. Since $Q(z)$ is D-stable, $|1/\bar{z}_j| < 1$. This contradicts the fact that $|z_j| < 1$ so that the above assumption must be false. ∎

<u>Lemma</u> 3.5. <u>If</u> $Q(z)$ <u>is</u> D-<u>stable, then</u>:

(A)

(3.7a) $$H(z) = 1 \quad \underline{\text{for}} \quad |z| = 1 .$$

(B)

(3.7b) $$|H(z)| < 1 \quad \underline{\text{for}} \quad |z| > 1 .$$

(C)

(3.7c) $$|H(z)| > 1 \quad \underline{\text{for}} \quad |z| < 1 .$$

<u>Proof</u>. (A): If $z = e^{i\theta}$, $\theta \in \mathbb{R}$, then by (3.2)

$$H(z) = \left| \frac{e^{in\theta}\overline{Q(e^{i\theta})}}{Q(e^{i\theta})} \right| = 1 .$$

(B): Let $G(\zeta) = H(1/\zeta)$, $\zeta = 1/z$. By (A) we know that $|G(\zeta)| = 1$ if $|\zeta| = 1$ and by Lemma 3.2 we know that every pole of $|G(\zeta)|$ lies in $|z| > 1$. Hence by the maximum principle (and the fact that $G(\zeta)$ is not identically constant) it follows that $|G(\zeta)| < 1$ for $|\zeta| < 1$. This implies (3.7b).

(C) can be proved by an application of the maximum principle to $F(z) = 1/H(z) = Q(z)/P(z)$. In fact, by (3.6) it can be seen that every pole of $F(z)$ lies in $|z| > 1$. Since $|F(z)| = 1$ for $|z| = 1$, it follows from the maximum principle that $|F(z)| < 1$ for $|z| < 1$. This proves (3.7c). ∎

For our next lemma we introduce the notation

(3.8a) $$Q_n(z) = Q(z) = a_0^{(n)} + a_1^{(n)} z + \cdots + a_n^{(n)} z^n ,$$

(3.8b) $$P_n(z) = P(z) \quad \text{and} \quad H_n(z) = H(z) ,$$

so that

(3.8c) $$a_j^{(n)} = a_j , \quad j = 0,1,\ldots,n .$$

Lemma 3.6. Let $Q_n(z) = Q(z)$ be D-stable and let H_{n-1}, P_{n-1}, Q_{n-1} be defined by

(3.9a) $$H_{n-1}(z) = \frac{P_{n-1}(z)}{Q_{n-1}(z)} ,$$

where

(3.9b) $$P_{n-1}(z) = \frac{P_n(z)-c_1 Q_n(z)}{1-|c_1|^2} , \quad Q_{n-1}(z) = \frac{Q_n(z)-\overline{c_1}P_n(z)}{(1-|c_1|^2)z} ,$$

and

(3.9c) $$c_1 = \frac{\overline{a_0^{(n)}}}{a_n^{(n)}} .$$

Then: (A)

(3.10) $$|c_1| < 1 .$$

(B)

(3.11a) $$Q_{n-1}(z) = a_0^{(n-1)} + a_1^{(n-1)}z + \cdots + a_{n-1}^{(n-1)}z^{n-1} ,$$

and

(3.11b) $$P_{n-1}(z) = z^{n-1}\overline{Q_{n-1}(1/\bar{z})} = \overline{a_{n-1}^{(n-1)}} + \overline{a_{n-2}^{(n-1)}}z + \cdots + \overline{a_0^{(n-1)}}z^{n-1} ,$$

where

(3.11c) $$a_j^{(n-1)} = \frac{a_{j+1}^{(n)}-\bar{c}_1 a_{n-j-1}}{1-|c_1|^2} , \quad j = 0,1,\ldots,n-1 .$$

(C)

(3.12) $$\overline{a_{n-1}^{(n-1)}} \neq 0 .$$

(D) If $n = 1$, then

(3.13) $$H_0(z) \equiv \frac{\overline{a_0^{(0)}}}{a_0^{(0)}} , \text{ so that } |H_0(z)| \equiv 1 .$$

Proof. (A): If $a_0^{(n)} = 0$, then $|c_1| = 0 < 1$. Suppose that $a_0^{(n)} \neq 0$. Then $P_n(z)$ has n zeros. Let $z_1, z_2, \ldots, z_k$ denote the distinct zeros of $Q_n(z)$. Then $P_n(z) = P(z)$ can be written in the form (3.6). It follows that the leading coefficient of $P_n(z)$ is

$$\overline{a_0^{(n)}} = \overline{a_n^{(n)}}(-\bar{z}_1)^{m_1}(-\bar{z}_2)^{m_2}\cdots(-\bar{z}_k)^{m_k} .$$

Hence

$$|c_1| = \left|\frac{a_0^{(n)}}{\overline{a_n^{(n)}}}\right| = |z_1^{m_1} z_2^{m_2} \cdots z_k^{m_k}| < 1 ,$$

since $Q_n(z)$ is assumed to be D-stable.

(B): By (3.9b) and (3.9c) we obtain

$$P_{n-1}(z) = \frac{(\overline{a_n^{(n)}} - c_1 a_0^{(n)}) + (\overline{a_{n-1}^{(n)}} - c_1 a_1^{(n)})z + \cdots + (\overline{a_1^{(n)}} - c_1 a_{n-1}^{(n)})z^{n-1}}{1-|c_1|^2} ,$$

which gives (3.11b) with (3.11c). Similarly by the second equation in (3.9b) and (3.9c) we obtain (3.11a) with (3.11c).

(C): By (3.11c), $a_{n-1}^{(n-1)} = (a_n^{(n)} - \bar{c}_1 a_0^{(n)})/(1-|c_1|^2)$. Hence $a_{n-1}^{(n-1)} = 0$ if and only if $c_1 = \overline{a_n^{(n)}}/a_0^{(n)} = 1/\bar{c}_1$. The latter equation does not hold (by (3.10)) and hence (3.12) holds.

(D) follows immediately from (3.9), (3.11) and (3.12). ■

<u>Remarks:</u> From (3.9) we obtain

(3.14) $$P_n(z) = P_{n-1}(z) + c_1 z Q_{n-1}(z) , \quad Q_n(z) = \bar{c}_1 P_{n-1}(z) + z Q_{n-1}(z) ,$$

and hence

$$H_n(z) = \frac{P_{n-1}(z) + c_1 z Q_{n-1}(z)}{\bar{c}_1 P_{n-1}(z) + z Q_{n-1}(z)} = \frac{H_{n-1}(z) + c_1 z}{\bar{c}_1 H_{n-1}(z) + z} .$$

It follows that

(3.15) $$H_n(z) = c_1 + \cfrac{(1-|c_1|^2)}{\bar{c}_1 + \cfrac{z}{H_{n-1}(z)}}$$

which gives the initial part of the continued fraction (3.4a). We wish to show that the procedure described above to obtain (3.15) can now be applied to obtain a similar continued fraction for $H_{n-1}(z)$. That this can be done is a consequence of our next lemma.

<u>Lemma</u> 3.7. <u>Suppose that</u> $Q_n(z) = Q(z)$ <u>is</u> <u>D-stable</u>. <u>Let</u> $P_{n-1}(z)$, $Q_{n-1}(z)$ <u>and</u> $H_{n-1}(z)$ <u>be defined as in Lemma</u> 3.6. <u>Then:</u>

(A) $P_{n-1}(z)$ <u>and</u> $Q_{n-1}(z)$ <u>have no common zeros</u>.

(B) <u>Every pole of</u> $H_{n-1}(z)$ <u>lies in</u> $|z| < 1$.

(C) $Q_{n-1}(z)$ <u>is</u> <u>D-stable</u>.

<u>Proof</u>. (A): Assume that $P_{n-1}(u) = Q_{n-1}(u) = 0$ for some $u \in \mathbb{C}$. Then by (3.14), $P_n(u) = Q_n(u) = 0$ which contradicts the result of Lemma 3.4. Hence the above assumption must be false.

(B): Suppose that u is a pole of $H_{n-1}(z)$. Then $u \neq \infty$ since $a_{n-1}^{(n-1)} \neq 0$. Therefore by (3.15) $H(u) = H_n(u) = 1/\bar{c}_1$, so that $|H(u)| > 1$. It follows from Lemma 3.5 that $|u| < 1$.

(C) is an immediate consequence of (A) , (B) and Lemma 3.3 ■

<u>Proof of Theorem 3.1.</u> Combining the results of the preceding lemmas, we see that if $Q(z)$ is D-stable, then $H(z)$ can be represented by a continued fraction of the form (3.4) . It remains then to prove the converse. For a given $Q(z)$ suppose that $H(z)$ is represented by (3.4) . Then with the notation of (2.1) and (1.24) we have

$$H(z) = R_n(z;c_{n+1}) = S_n(z^{-1};c_{n+1}) \quad .$$

Therefore by Theorem 1.5(A) , $|H(z)| < 1$ for $|z| > 1$ and it follows that all of the poles of $H(z)$ are in $|z| < 1$. Thus Lemma 3.3 implies that $Q(z)$ is D-stable, provided that $P(z)$ and $Q(z)$ have no common zeros. To verify this last condition, we let $\{A_m(z)\}$ and $\{B_m(z)\}$ be sequences of polynomials defined by

$$A_{-1}(z) = 1 \ , \quad B_{-1}(z) = 0 \ , \quad A_0(z) = \beta_0 \ , \quad B_0(z) = 1 \quad ,$$

$$A_m(z) = \beta_m A_{m-1}(z) + \alpha_m A_{m-2}(z) \quad , \quad m \geq 1 \quad ,$$

$$B_m(z) = \beta_m B_{m-1}(z) + \alpha_m B_{m-2}(z) \quad , \quad m \geq 1 \quad ,$$

where

$$\alpha_{2j-1} = 1 - |c_j|^2 \quad , \quad \alpha_{2j} = z \quad , \quad j = 1,2,\ldots,n$$

$$\beta_0 = c_1 \ , \quad \beta_{2j-1} = \bar{c}_j \ , \quad \beta_{2j} = c_{j+1} \ , \quad j = 1,2,\ldots,n \quad .$$

Then $A_m(z)$ is the mth numerator and $B_m(z)$ the mth denominator of the continued fraction (3.4) and $H(z) = A_{2n}(z)/B_{2n}(z)$. Moreover, it is readily seen that $A_{2n}(z)$ is a polynomial in z of degree at most n and $B_{2n}(z)$ is a polynomial in z of degree exactly n with leading coefficient equal 1 . Hence

$$P(z) = a_n A_{2n}(z) \quad \text{and} \quad Q(z) = a_n B_{2n}(z) \quad .$$

By the determinant formula for continued fractions

$$A_{2n}(z)B_{2n-1}(z) - A_{2n-1}(z)B_{2n}(z) = (-1)^{2n-1} \prod_{j=1}^{2n} \alpha_j$$

$$= -z^n \prod_{j=1}^{n} (1-|c_j|^2) \quad .$$

Therefore

(3.16) $$P(z)B_{2n-1}(z) - A_{2n-1}(z)Q(z) = -a_n z^n \prod_{j=1}^{n} (1-|c_j|^2) \quad .$$

Now assume that $P(u) = Q(u) = 0$ for some u C . Then by (3.16) , $u = 0$. But this contradicts the fact that $a_n \neq 0$ in (3.2) . Hence the assumption is false. ■

Lemma 3.6 contains the key to a simple numerical procedure for computing the c_j in the continued fraction (3.4) and hence for determining whether or not a given polynomial $Q(z)$ is D-stable. The procedure is given here.

Algorithm 1 (Schur-Cohn). Let $a_0, a_1, \ldots, a_n$ be given, $a_j \in \mathbb{C}$ for $j = 1,2,\ldots,n$, $a_n \neq 0$. Set

$$a_j^{(n)} = a_j \,, \quad j = 0,1,\ldots,n \;.$$

For $m = n,n-1,\ldots,1$, compute

$$\left\lfloor \begin{array}{l} c_{n-m+1} = \overline{a_0^{(m)}}/a_m^{(m)} \\ a_j^{(m-1)} = \dfrac{a_{j+1}^{(m)} - c_{n-m+1}\, \overline{a_{m-j+1}^{(m)}}}{1-|c_{n-m+1}|^2} \,, \quad j = 0,1,\ldots,m-1 \;. \end{array}\right.$$

Set $c_{n+1} = \overline{a_0^{(0)}}/a_0^{(0)}$.

Corollary 3.8. Let $c_1, c_2, \ldots, c_{n+1}$ be computed as in Algorithm 1 . Then $Q(z) = a_0 + a_1 z + \cdots + a_n z^n$ is D-stable if and only if

$$|c_j| < 1 \text{ for } j = 1,2,\ldots,n \;, \text{ and } |c_{n+1}| = 1 \;.$$

4. Computation of Poles for Speech Filters and Signal Detectors.

In Section 2 it was pointed out that ladder digital filters of the type described in Figure 1 provide a mathematical model of the physical mechanism of speech and that the associated transfer function $H(z)$ can be expressed as a continued fraction $R_n(z;1)$ (see (2.1)). The c_k in (2.1) are the reflection coefficients associated with particular vowel sounds and these parameters can be computed directly from the sampled speech. The parametric representation of the speech waveform can be processed and the utterance identified by computer. The useable utterance information is contained in the location of the poles of the transfer function $H(z)$ [16] . Hence the computation of pole locations is a task that must be performed with much frequency.

Another application requiring pole computation is in signal detection [18] . The example considered is the detection of a sinusoidal wave of a single unknown frequency ω . The signal input is given by a finite sequence $\{x(k)\}$, where

$$x(k) = \sin(\omega k) + \eta_\sigma(k) \;, \quad k = 1,2,\ldots,N \;,$$

and where the $\eta_\sigma(k)$ are numbers chosen at random from a standard normal distribution with variance σ^2 (white noise). Ladder filters have recently been applied to this problem [14] . Again reflection coefficients are computed from the input data $\{x(n)\}$. In this case $H(z)$ will have a single conjugate pair of poles located near the unit circle with the remaining poles located nearer to the origin. The argument (phase) of the poles with greatest magnitude is the estimate of ω .

In this section we describe a method for computing the poles of $H(z)$ directly from the c_k without first having to form the denominator polynomial $Q(z)$ of $H(z)$. The method is based on an algorithm of McCabe and Murphy [10] and is similar to the qd-alrogithm of Rutishauser. Proof of convergence of the iterations is omitted, but follows along the lines of the proof for the qd-algorithm in the case when the poles have equal moduli (see [5, Section 7.9]).

If $c_k \neq 0$ for $k = 2,3,\ldots,n+1$, then the continued fraction $R_n(z;c_{n+1})$ has an even part (see, for example, [8, Section 2.4.2]). The even part of $R_n(\zeta^{-1},c_{n+1})$ can be expressed in the form of a terminating general T-fraction

$$(4.1a) \qquad T_n(\zeta) = R_n(\zeta^{-1};c_{n+1}) = c_1 + \frac{F_1\zeta}{1+G_1\zeta} + \frac{F_2\zeta}{1+G_2\zeta} + \cdots + \frac{F_n\zeta}{1+G_n\zeta},$$

where

$$(4.1b) \qquad F_1 = (1-|c_1|^2)c_2, \quad F_j = -(1-|c_j|^2)\frac{c_{j+1}}{c_j}, \quad j = 2,3,\ldots,n,$$

$$(4.1c) \qquad G_1 = \bar{c}_1 c_2, \quad G_j = \frac{c_{j+1}}{c_j}, \quad j = 2,3,\ldots,n.$$

The following algorithm is now applicable [8, Section 7.3.2].

Algorithm 2. (FG-algorithm)
Let F_j, G_j for $j = 1,2,\ldots,n$ be given and let m_0 be given.
Set $F_1^{(m)} = F_{n+1}^{(m)} = 0$ for $m = 0,1,\ldots,m_0$ and
$G_1^{(0)} = G_1$, $F_k^{(0)} = F_k$ and $G_k^{(0)} = G_k$ for $m = 2,3,\ldots,n$.
Then for each $m = 0,1,\ldots,m_0-1$, compute

$$G_1^{(m+1)} = G_1^{(m)} + F_2^{(m)},$$

and for $k = 2,3,\ldots,n$

$$F_k^{(m+1)} = \frac{F_k^{(m)}(G_k^{(m)}+F_{k+1}^{(m)})}{G_{k-1}^{(m)}+F_k^{(m)}},$$

$$G_k^{(m+1)} = G_k^{(m)} + F_{k+1}^{(m)} - F_k^{(m+1)}.$$

An application of Algorithm 2 produces a table (the FG-table) of the form

$$\begin{array}{cccccccc}
0 = F_1^{(0)} & G_1^{(0)} & F_2^{(0)} & G_2^{(0)} & \cdots & F_n^{(0)} & G_n^{(0)} & F_{n+1}^{(0)} = 0 \\
0 = F_1^{(1)} & G_1^{(1)} & F_2^{(1)} & G_2^{(1)} & \cdots & F_n^{(1)} & G_n^{(1)} & F_{n+1}^{(1)} = 0 \\
\vdots & \vdots & \vdots & \vdots & & \vdots & \vdots & \vdots \\
0 = F_1^{(m_0)} & G_1^{(m_0)} & F_2^{(m_0)} & G_2^{(m_0)} & & F_n^{(m_0)} & G_n^{(m_0)} & F_{n+1}^{(m_0)} = 0.
\end{array}$$

It is known that the FG-table associated with (4.1a) with the $F_j, a_j \neq 0$ can be computed for any positive m_0. $T_n(\zeta)$ has n poles; we shall denote them by $\zeta_1, \zeta_2, \ldots, \zeta_n$, where

$$0 < |\zeta_1| \leq |\zeta_2| \leq \cdots \leq |\zeta_n| .$$

It is well known [8, Theorem 7.21] that if

$$|\zeta_{k-1}| < |\zeta_k| < |\zeta_{k+1}| \tag{4.2}$$

then

$$\lim_{m\to\infty} G_k^{(m)} = -\frac{1}{\zeta_k} . \tag{4.3}$$

If, on the other hand, two or more of the poles have the same modulus, then the situation is somewhat more difficult. That situation is discussed below.

An index k is called a <u>critical index</u> of the function $T_n(\zeta)$ if

$$|\zeta_k| < |\zeta_{k+1}| .$$

It can be shown that if k is a critical index then

$$\lim_{m\to\infty} F_k^{(m)} = 0 .$$

Thus the FG-table is divided into subtables by the F-columns tending to zero. If a subtable contains j G-columns, then this indicates the presence of j poles of equal modulus. When $j = 1$, the single G-column converges to the negative reciprocal of the pole as in (4.3). Suppose that k and $k + j$ are two consecutive critical indices of $T_n(\zeta)$. Let polynomials $p_i^{(m)}(u)$ be defined by

$$p_0^{(m)}(u) = 1 \tag{4.4a}$$

$$p_{i+1}^{(m)}(u) = u p_i^{(m+1)}(u) - G_{k+i+1}^{(m)} p_i^{(m)}(u) , \tag{4.4b}$$

$$m = 0,1,2,\ldots \; ; \quad i = 0,1,\ldots,j-1 .$$

If condition (H) given in [5, Theorem 7.9a] holds, then there exists an infinite set of positive integers M such that

$$\lim_{\substack{m\to\infty \\ m\in M}} p_j^{(m)}(u) = p_j(u) ,$$

where

$$p_j(u) = (u+z_k)(u+z_{k+1})\cdots(u+z_{k+j-1}) \tag{4.5}$$

$$z_i = 1/\zeta_i , \quad i = k,k+1,\cdots,k+j-1 . \tag{4.6}$$

Here the z_i are poles of $H(z) = T_n(z^{-1})$.

For illustration suppose that for some k, we have k and $k+2$ as consecutive critical indexes (i.e. $j = 2$). Then

$$p_2^{(m)}(u) = u^2 - (G_{k+1}^{(m+1)} + G_{k+2}^{(m)})u + G_{k+2}^{(m)} G_{k+1}^{(m)} . \tag{4.7}$$

Now suppose that the limits

$$(4.8) \qquad a_k = \lim_{m\to\infty}(G_{k+1}^{(m+1)} + G_{k+2}^{(m)}) \quad \text{and} \quad b_k = \lim_{m\to\infty} G_{k+2}^{(m)}G_{k+1}^{(m)}$$

both exist, so that

$$(4.9) \qquad p_2(u) = u^2 - a_k u + b_k \quad .$$

The zeros $(a_k \pm \sqrt{a_k^2-4b_k})/2$ of $p_2(u)$ are then poles of $H(z)$ with equal modulus. It will be seen that this situation occurs in the following examples. For convenience we shall let $-z_1^{(m)}, -z_2^{(m)}$ denote the zeros of $p_2^{(m)}(u)$, so that $z_i = \lim_{m\to\infty} z_i^{(m)}$, $i = 1,2$.

Example 1. Speech Analysis. We consider now the reflection coefficients given in Table 1. We wish to compute the poles of the transfer function

Table 1. Reflection coefficients c_j for the Russian vowel /e/ taken from [9, p. 87].

j	c_j	j	c_j
1	0.500	5	-0.189
2	0.456	6	-0.212
3	0.178	7	-0.200
4	-0.044	8	0.381

$H(z) = R_8$ (see (2.1)) associated with the coefficients in Table 1. After using (4.1) to obtain F_j and G_j for $j = 1,2,\ldots,8$, we have applied Algorithm 2 to form an FG-table. The elements $F_k^{(m)}$ of the FG-table for $k = 1,2,\ldots,8$ and $m = 100,200,\ldots,700$ are given in Table 2. As can be seen, $F_3^{(m)}, F_5^{(m)}$ and $F_7^{(m)}$ appear to approach zero as $m \to \infty$. Therefore the indication is that the poles of $H(z)$ occur in pairs having the same modulus. In Table 3 we show the values of the sums and products in (4.8) for $k = 1$ and the resulting $\mathrm{Re}(z_1^{(m)})$ and $\mathrm{Im}(z_1^{(m)})$ for $m = 100,200,\ldots,700$. z_1 and z_2 are complex conjugates. The remaining poles of $H(z)$ are given in Table 4 and all occur in conjugate pairs.

Table 2. Selected elements $F_k^{(m)}$ from the FG-table associated with reflection coefficients c_j in Table 1.

m	$F_2^{(m)}$	$F_3^{(m)}$	$F_4^{(m)}$	$F_5^{(m)}$	$F_6^{(m)}$	$F_7^{(m)}$	$F_8^{(m)}$
100	2.34449	-.10956	-5.11980	.01943	1.65777	-.53872	-.06133
200	.12296	.01990	2.51598	-.00009	2.93129	.02196	-.08759
300	.59788	-.00105	1.99876	.00000	-1.86633	-.00529	-.13034
400	.14277	-.00005	-4.58580	.00000	- .72617	.00051	-.24783
500	.33239	.00001	-2.35639	.00000	-1.38386	-.00002	-.86929
600	.17947	.00000	3.70904	.00000	3.37503	-.00001	4.31542
700	.22212	.00000	1.78708	.00000	-9.09978	.00000	4.36749

Table 3. Sums $G_1^{(m+1)} + G_2^{(m)}$, products $G_1^{(m)} G_2^{(m)}$ and real and imaginary parts of $z_1^{(m)}$ obtained from the FG-table and (4.9) associated with the reflection coefficients c_j in Table 1.

m	$G_1^{(m+1)} + G_2^{(m)}$	$G_1^{(m)} G_2^{(m)}$	$Re(z_1^{(m)})$	$Im(z_1^{(m)})$
100	-1.358154	.767111	.679077	-.553141
200	-1.819877	.924506	.909938	-.310674
300	-1.840151	.945378	.920076	.314386
400	-1.840669	.945794	.920335	.314290
500	-1.840696	.945830	.920348	.314308
600	-1.840695	.945829	.920347	.314309
700	-1.840695	.945829	.920347	.314309

Table 4. Poles z_k of the transfer function $H(z)$ associated with the reflection coefficients c_j in Table 1.

k	$Re(z_k)$	$Im(z_k)$	$\lvert z_k \rvert$
1	.920347	.314309	.972538
2	.920347	-.314309	.972538
3	.00954835	.934707	.939571
4	.00954835	-.934707	.939571
5	-.527641	.715288	.888843
6	-.527641	-.715288	.888843
7	-.836649	.240790	.870610
8	-.836649	-.240790	.870610

For comparison we have also computed the eight poles in Table 4 by use of Laguerre's global convergence method with polynomial deflation. The time required (by a CDC Cyber) to compute the eight poles with at least six significant digits was 0.069 seconds by Laguerres' method and 0.203 seconds by the FG-algorithm. The slowness of the FG-algorithm here is, in part, due to the fact that the magnitudes of the poles are all nearly equal. Thus about 1000 iterations are needed to obtain six significant digits for all of the poles.

Example 2. Signal Detection. We consider now the problem described near the beginning of this section; that is, to determine an unknown frequency ω from a signal consisting of a sinusoidal wave superimposed with white noise. For this example we have taken $N = 200$, $\omega = \pi/4$ radians and $\sigma^2 = 0.02$. (This value of σ^2 corresponds to a signal power to noise power ratio of 30 decibels.) The reflection coefficients c_j for the given input data are given in Table 5 .

Table 5. Reflection coefficients c_j for the sinusoid in noise detection problem.

j	c_j	j	c_j
1	.0138023	5	-.0478801
2	.0376839	6	.177988
3	-.0348238	7	.987017
4	.0187238	8	-.706543

Again we constructed the FG-table; the $F_2^{(m)}, F_3^{(m)}$ columns are given in Table 6 together with the sums and products of G's in (4.7) . Also given in Table 6 are the resulting pole approximations $z_1^{(m)} = \mathrm{Re}(z_1^{(m)}) + i\mathrm{Im}(z_1^{(m)})$, $m = 15,20,\ldots,45$. z_1 is the pole of principal interest and $z_1^{(45)}$ approximates z_1 with six significant digits. Thus as an approximation of $\omega = \pi/4 \doteq .7854$, we obtain $\omega = \mathrm{Arg}\, z_1^{(45)} \doteq .7856$ which has "approximately three" significant digits. The computing time (on the CDC Cyber) to compute z_1 with six significant digits was .044 seconds by the Laguerre method and .010 seconds by the FG-algorithm. The FG-algorithm was favorable in this application because we were searching a single pair of conjugate poles located near the unit circle and somewhat isolated from the other poles. These results suggest that the FG-algorithm may provide an efficient method for signal detection problems.

Table 6. Intermediate output from FG-algorithm approximations $z_1^m = \mathrm{Re}(z_1^{(m)}) + i\mathrm{Im}(z_1^{(m)})$ of the pole z_1 of transfer function $H(z)$ in signal detection problem with reflection coefficients in Table 4 .

m	$F_2^{(m)}$	$F_3^{(m)}$	$G_1^{(m+1)}+G_2^{(m)}$	$G_1^{(m)}G_2^{(m)}$	$\mathrm{Re}(z_1^{(m)})$	$\mathrm{Re}(z_2^{(m)})$
15	1.1747	.00000	27.385539	-24.825296	.878340	.000000
20	-11.38749	.00000	-1.540688	1.028264	.770344	.659420
25	8.02913	.00000	-1.408984	.984263	.704492	.698537
30	.65483	.00000	-1.406301	.989196	.703150	.703403
35	.77308	.00000	-1.406648	.989847	.703324	.703692
40	- 9.16121	.00000	-1.406690	.989883	.703345	.703697
45	7.63801	.00000	-1.406692	.989884	.703346	.703696

References

1. Claerbout, J.F., Fundamentals of Geophysical Data Processing, McGraw Hill, New York (1976).

2. Frank, Evelyn, On the zeros of polynomials with complex coefficients, Bull. Amer. Math. Soc., 52 (1946), 144-157.

3. Hamel, Georg, Eine charakteristische Eigenschaft beschränkter analytischer Funktionen, Math. Annal. 78 (1918), 257-269.

4. Hamming, R.W., Digital Filters, Prentice Hall, New Jersey (1977).

5. Henrici, P., Applied and Computational Complex Analysis, Vol. 1, Power Series, Integration, Conformal Mapping and Location of Zeros, John Wiley and Sons, New York, (1974).

6. Henrici, P., Applied and Computational Complex Analysis, Vol. 2, Special Functions, Integral Transforms, Asymptotics and Continued Fractions, John Wiley and Sons, New York (1977).

7. Huang, T.S. (ed.), Picture Processing and Digital Filtering, Topics in Applied Physics, Vol. 6, 2nd ed., Springer-Verlag, New York (1979).

8. Jones, William B. and Thron, W.J., Continued Fractions: Analytic Theory and Applications, Encyclopedia of Mathematics and Its Applications, No. 11, Addison-Wesley Publishing Company, Reading, Mass. (1980).

9. Markel, J.D. and Gray, A.H., Jr., Linear Prediction of Speech, Springer-Verlag, New York (1976).

10. McCabe, J.H. and J.A. Murphy, Continued fractions which correspond to power series expansions at two points, J. Inst. Maths. Applics 17 (1976), 233-247.

11. Mitra, S.K. and Sherwood, R.J., Canonic realizations of digital filters using the continued fraction expansion, IEEE Transactions on Audio and Electro-acoustics, Vol. AU-20, No. 3 (August, 1972), 185-197.

12. Mullis, C.T. and Roberts, R.A., Roundoff noise in digital filters: frequency transformations and invariants, IEEE Trans. Acoust., Speech Signal Processing, Vol. ASSP-24 (December, 1976), 538-550.

13. Oppenheim, A.V. and Shafer, R.W., Digital Signal Processing, Prentice Hall, New Jersey (1975).

14. Parker, S.R. and Griffiths, L.J., (ed.) Special Issue on Adoptive Signal Processing, IEEE Trans. Acoust. Speech Signal Processing, vol. ASSP-29, No. 3 (June 1981).

15. Schur, I., Über Potenzreihen, die im Innern des Einheitskreises beschränkt sind, J. fur die reine und angewandte Mathematik 147 (1917) and 148 (1918).

16. Tanaka, K., A dynamic processing approach to phoneme recognition (Part I)-Feature extraction, IEEE Trans. Acoust. Speech Signal Processing, vol. ASSP-27 (Dec. 1979), 596-607.

17. Thron, W.J., Two-point Padé tables, T-fractions and sequences of Schur, Padé and Rational Approximation (ed. E.B. Saff and R.S. Varga), Academic Press, Inc., New York (1977), 215-226.

18. Van Trees, H.L., Detection, Estimation, and Modulation Theory, John Wiley and Sons, New York (1968).

19. Wall, H.S., Polynomials whose zeros have negative real parts, Amer. Math. Monthly, 52 (1945), 308-322.

20. Wall, H.S., Analytic Theory of Continued Fractions, D. Van Nostrand Co., New York (1948).

William B. Jones
Department of Mathematics
University of Colorado
Boulder, Colorado 80309
U.S.A.

Allan Steinhardt
Department of Electrical Engineering
University of Colorado
Boulder, Colorado 80309
U.S.A.

δ-FRACTION EXPANSIONS OF ANALYTIC FUNCTIONS

L.J. Lange[1]

1. Introduction. By a δ-fraction we mean a finite or infinite continued fraction of the form

$$(1.1)\qquad b_0 - \delta_0 z + \frac{d_1 z}{1-\delta_1 z} + \frac{d_2 z}{1-\delta_2 z} + \frac{d_3 z}{1-\delta_3 z} + \cdots ,$$

where b_0 and d_n are complex constants, $d_n \neq 0$ for $n \geqq 1$, and the δ_n are real constants restricted to the values 0 or 1. We adopt the convention that the δ-fraction (1.1), and all of its approximants, have value b_0 at $z = 0$. We say that the δ-fraction (1.1) is regular if $d_{k+1} = 1$ for each k such that $\delta_k = 1$. We choose the name δ-fraction for the continued fraction (1.1) because of the binary "impulse" nature of the sequence $\{\delta_n\}$ and the analogies, therefore, with the δ's in the Dirac delta function and the Kronecker delta symbol. We are led to this investigation of δ-fractions, and their connections with analytic functions, in our quest to find an answer to the following question: Is there a class D of "simple" continued fractions

$$(1.2)\qquad b_0(z) + \frac{a_1(z)}{b_1(z)} + \frac{a_2(z)}{b_2(z)} + \frac{a_3(z)}{b_3(z)} + \cdots$$

having the following desirable properties?

(a) the elements $a_n(z)$ and $b_n(z)$ are polynomials in z of degree $\leqq 1$.

(b) D contains the class of regular C-fractions

$$c_0 + \frac{c_1 z}{1} + \frac{c_2 z}{1} + \frac{c_3 z}{1} + \cdots , \quad c_n \in \mathbb{C} , \quad c_n \neq 0 \text{ if } n \geqq 1 .$$

(c) Given the power series

$$L_0 \equiv c_0 + c_1 z + c_2 z^2 + c_3 z^3 + \cdots , \quad c_n \in \mathbb{C} , \quad c_0 \neq 0 ,$$

there exists a unique member $K_0 \in D$ such that K_0 corresponds to L_0, i.e., the Maclaurin series of the n-th approximant of K_0 agrees termwise with L_0 up to and including the term $c_{k(n)} z^{k(n)}$, where $k(n) \to \infty$ as $n \to \infty$.

(d) If L_0 represents a rational function in a neighborhood of $z = 0$, then its corresponding $K_0 \in D$ terminates.

(e) Let us say that $K \in D$ corresponds to a function $f(z)$, analytic at $z = 0$, if K corresponds to the Maclaurin series of f. Then for many classical functions, analytic in a neighborhood of the origin, explicit and useful formulas can be obtained for the elements $a_n(z)$ and $b_n(z)$ of the continued fractions in D corresponding to these functions.

---

[1] This research was funded in part by a grant from the Research Council of the Graduate School, University of Missouri, Columbia.

(f) Much information can be given about the convergence of the continued fractions in D that correspond to functions which are analytic at the origin.

(g) In many cases, at least "some" of the approximants of the continued fraction $K \in D$ corresponding to a power series L_0 are in the Padé table for L_0

The C-fractions of Leighton and Scott [13], the T-fractions of Thron [26], and the P-fractions of Magnus [14,15] all essentially meet requirement (c), but each of these classes fails to meet one or more of the remaining requirements. The C-fractions in general do not meet requirement (a), and the regular C-fractions above do not meet requirement (c). For example, it is known that there is no regular C-fraction corresponding to $1 + z^2$. The T-fractions essentially meet requirements (a) and (f), in addition to (c), but they do not meet (b), (d), and (g). The P-fractions have a close connection with the Padé table of a given power series L_0, but they fail to meet requirements (a) and (b), among others. The class of general T-fractions, studied by Waadeland [31,32] and others, essentially contains our class of δ-fractions, but, to date, this general class has been studied more from the interpolation point of view, i.e., more from the point of view of their connections with two-point Padé tables for meromorphic functions. Closely related to the general T-fractions are the M-fractions studied by a number of authors [2], [17], [18]. For thorough treatments of the subject of correspondence and the properties of many kinds of corresponding continued fractions, including the ones just mentioned, we refer the reader to the recent book on continued fractions by Jones and Thron [9].

We offer the class of regular δ-fractions as our candidate for an answer to the question posed above. It is easily seen that these continued fractions satisfy (a) and (b). In Section 2 we give five basic theorems dealing with the correspondence between δ-fractions and power series. It follows from Theorems 2.1 and 2.4, respectively, that requirements (c) and (d) are satisfied if D denotes the class of regular δ-fractions. The proof of Theorem 2.4 turns out to be considerably more complicated than the proof given by Perron [21, page 111] of the corresponding result for C-fractions.

In Section 3 we make a start towards meeting requirement (f) with our δ-fractions, by offering four convergence theorems involving these continued fractions. At the beginning of the section we state, in Theorem A, a version of Poincaré's Theorem on linear homogeneous difference equations, which we use later. Theorem 3.1 is concerned with properties of uniformly convergent δ-fractions. Theorem 3.2 essentially states that the δ-fraction (1.1) with $b_0 = 1$ converges to an analytic function in a neighborhood of $z = 0$, if the sequence $\{d_n\}$ is bounded. It follows from Theorem 3.3 that the δ-fraction (1.1) converges to a meromorphic function in the unit disk $|z| < 1$, if $\lim_{n\to\infty} d_n = 0$. In Section 3 we also introduce the concept of a (p,q) <u>limit periodic δ-fraction</u>. Theorem 3.4 is a very useful convergence theorem for limit periodic δ-fractions of type (1.1). We have obtained other convergence theorems, which will be published elsewhere, for the (2,1) and (1,2) cases. The various techniques used in the proofs of these convergence theorems can be used to obtain further results of this type. For other recent results involving limit periodic continued fractions, we refer the reader to the work of Thron and Waadeland [27, 28, 29] and to the work of Gill [6, 7].

In Section 4, we give the regular δ-fraction expansions for $\tan z$, $\tanh z$ Dawson's integral function, $(1+z)^{\nu}$, $(1+z^2)^{\nu}$, $\operatorname{Arctan} z$, $\operatorname{Arctanh} z$, $\exp(z)$, and $\exp(z^2)$. We have obtained the δ-fraction expansion for many other classical analytic functions, some of which will be presented in another paper. In our opinion, regular δ-fractions do indeed meet requirement (e). We use a _method of successive extensions_ along with frequent equivalence transformations to derive many of these expansions. We display several techniques for establishing the validity of these representations. A powerful new technique, based on Poincaré's Theorem, allows us, in many cases, to establish convergence of a δ-fraction over the whole complex plane (except for poles and cuts) to the analytic function to which it corresponds. Our examples also indicate that (p,q) limit periodic behavior is quite common in the expansions of classical functions.

Finally, we assert that the regular δ-fractions meet the remaining requirement (g), not yet considered. However, the only justification for this assertion that we give, is to point out the following: The class of regular δ-fractions contains the class of regular C-fractions, and the latter have known connections with the Padé table (see [9, page 190]).

We refer the reader to Chapter 2 in the book by Jones and Thron [9] for the basic definitions, formulas, and properties of continued fractions that are employed in this paper. Other valuable reference books on the subject of continued fractions are those by Perron [21] and Wall [33].

2. _Correspondence_. We follow the work of Jones and Thron [10, or 9 (Section 5.1)] on sequences of meromorphic functions corresponding to a formal Laurent series in introducing the basic definitions and notation for this section. We call

$$L = c_m z^m + c_{m+1} z^{m+1} + c_{m+2} z^{m+2} + \cdots \quad ; \quad c_m \neq 0 \quad , \quad m \geqq 0 \quad ,$$

where the c_n are complex numbers, a _formal power series_ (fps). $L = 0$ is also called a (fps). We define a function λ on the family of all such power series L as follows:

$$\lambda(L) = \infty \quad \text{if} \quad L = 0 \quad ; \quad \lambda(L) = m \quad \text{if} \quad L \neq 0 \quad .$$

If $f(z)$ is a function analytic at the origin (i.e., analytic in an open disk containing $z = 0$), then its Taylor series expansion about $z = 0$ will be denoted by $L(f)$. A sequence $\{R_n(z)\}$ of functions, where each $R_n(z)$ is analytic at the origin, will be said to _correspond to a_ (fps) L _at_ $z = 0$ if

$$\lim_{n\to\infty} \lambda(L - L(R_n)) = \infty.$$

If $\{R_n(z)\}$ corresponds to a (fps) L then the <u>order of correspondence</u> ν_n of $R_n(z)$ is defined by

$$\nu_n = \lambda(L - L(R_n)) \quad .$$

Thus if $\{R_n(z)\}$ corresponds to L, then L and $L(R_n)$ agree term-by-term up to and including the term $z^{\nu_n - 1}$. Finally, a continued fraction

$$b_0(z) + \overset{\infty}{\underset{n=1}{K}} \left(\frac{a_n(z)}{b_n(z)}\right)$$

is said to correspond to a (fps) L if its sequence of approximants corresponds to L.

<u>Theorem</u> 2.1. <u>For every formal power series</u>

$$L_0 = 1 + c_1 z + c_2 z^2 + \cdots$$

<u>there exists a uniquely determined regular</u> δ-<u>fraction</u>

$$K = 1 - \delta_0 z + \frac{d_1 z}{1-\delta_1 z} + \frac{d_2 z}{1-\delta_2 z} + \cdots$$

<u>such that</u> K <u>corresponds to</u> L_0.

<u>Proof</u>. We define a sequence $\{L_n\}$ of power series each with constant term 1 as follows. If $L_0 \equiv 1$, choose $L_1 \equiv 1$ and define the δ-fraction K by $K = 1$. If $L_0 \not\equiv 1$ and $c_1 \neq 0$, choose $\delta_0 = 0$ and $d_1 = c_1$. If $L_0 \not\equiv 1$ and $c_1 = 0$ choose $\delta_1 = 1$ and $d_1 = 1$. Then

$$L_0 = 1 - \delta_0 z + d_1 z L_0^* ,$$

where

$$L_0^* = 1 + c_1^* z + c_2^* z^2 + \cdots ,$$

and we define

$$L_1 = \frac{1}{L_0^*} = 1 + c_{1,1} z + c_{1,2} z^2 + c_{1,3} z^3 + \cdots .$$

If $L_0, L_1, \ldots, L_n$ are defined and have constant term 1, then we define L_{n+1} in the following manner: If $L_n \equiv 1$, choose $L_{n+1} \equiv 1$. If $L_n \not\equiv 1$ and

$$L_n = 1 + c_{n,1} z + c_{n,2} z^2 + \cdots ,$$

choose $\delta_n = 0$ and $d_{n+1} = c_{n,1}$ if $c_{n,1} \neq 0$. Otherwise, if $c_{n,1} = 0$, define

$$\delta_n = 1 \quad \text{and} \quad d_{n+1} = 1 \quad .$$

Then, if $L_n \not\equiv 1$, we have

$$L_n = 1 - \delta_n z + d_{n+1} z L_n^* ,$$

where

$$L_n^* = 1 + c_{n,1}^* z + c_{n,2}^* z^2 + \cdots .$$

In this case, we define

$$L_{n+1} = \frac{1}{L_n^*} = 1 + c_{n+1,1}z + c_{n+1,2}z^2 + \cdots .$$

Hence, it follows by induction that L_n is defined for all $n \geqq 0$. It is easy to see that if $L_n \equiv 1$ for some n then $L_k \equiv 1$ for all $k \geqq n$. If this should happen, let m be the least value of n such that $L_n \equiv 1$. Then we have

$$L_k = L(1-\delta_k z) + \frac{L(d_{k+1}z)}{L_{k+1}} , \quad 0 \leqq k \leqq m-1 ,$$

where $d_{k+1} = 1$ if $\delta_k = 1$.

If we set $a_k(z) = d_k z + 0$ and $b_k(z) = 1 - \delta_k z$, then it follows from [10, Theorem 2] that

$$L_0 = L\left(1 - \delta_0 z + \frac{d_1 z}{1-\delta_1 z} + \cdots + \frac{d_{m-1}z}{1-\delta_{m-1}z} + \frac{d_m z}{1}\right)$$

and K corresponds to L_0 if

$$K = 1 - \delta_0 z + \frac{d_1 z}{1-\delta_1 z} + \cdots + \frac{d_{m-1}z}{1-\delta_{m-1}z} + \frac{d_m z}{1} .$$

In the remaining case, where $L_n \not\equiv 1$ for all n, we have shown that there exist sequences $\{\delta_k\}$ and $\{d_k\}$, where $d_k \neq 0$, $\delta_k = 0$ or 1, and $d_{k+1} = 1$ if $\delta_k = 1$ such that

$$L_k = L(1-\delta_k z) + \frac{L(d_{k+1}z)}{L_{k+1}} , \quad k \geq 0 .$$

Again, if we set $a_k(z) = d_k z$ and $b_k(z) = 1 - \delta_k z$, it follows from [9, Theorem 2] that K corresponds to L_0, where

$$K = 1 - \delta_0 z + \frac{d_1 z}{1-\delta_1 z} + \frac{d_2 z}{1-\delta_2 z} + \cdots .$$

With this our proof is complete.

Theorem 2.2. *Each finite δ-fraction*

$$K_n = 1 - \delta_0 z + \frac{d_1 z}{1-\delta_1 z} + \cdots + \frac{d_{n-1}z}{1-\delta_{n-1}z} + \frac{d_n z}{1}$$

and each infinite δ-fraction

$$K_\infty = 1 - \delta_0 z + \frac{d_1 z}{1-\delta_1 z} + \frac{d_2 z}{1-\delta_2 z} + \cdots$$

correspond to a uniquely determined power series

$$L_0 = 1 + c_1 z + c_2 z^2 + \cdots .$$

The order of correspondence ν_k *of* K_∞ *is* $k+1$ *and* ν_k *for* K_n *is* $k+1$ *if* $0 \leqq k < n$ *and* ∞ *if* $k \geqq n$.

Proof. Let the sequences $\{R_n(z)\}$ and $\{S_n(z)\}$ of functions, analytic at the origin, be defined by

$$R_n(z) = \frac{A_n(z)}{B_n(z)} , \quad n \geq 0 ; \quad S_k(z) = \frac{A_k^*(z)}{B_k^*(z)} , \quad 0 \leqq k \leqq n ; \quad S_k(z) \equiv \frac{A_n^*(z)}{B_n^*(z)} , \quad k > n$$

where $A_n(z)/B_n(z)$ is the n-th approximant of K_∞ and $A_k^*(z)/B_k^*(z)$ is the k-th approximant of K_n. Since

$$R_{n+1}(z) - R_n(z) = \frac{(-1)^n d_1 d_2 \cdots d_{n+1} z^{n+1}}{B_n(z) B_{n+1}(z)}$$

and $B_n(0) \equiv 1$, it follows that

$$\nu_n = \lambda(L(R_{n+1}) - L(R_n)) = n + 1 \to \infty$$

monotonically as $n \to \infty$. Also

$$S_{k+1}(z) - S_k(k) = \frac{(-1)^k d_1 d_2 \cdots d_{k+1} z^{k+1}}{B_k^*(z) B_{k+1}^*(z)} \quad \text{if } k < n \text{ and } 0 \text{ if } k \geq n \ .$$

Hence, in this case,

$$\nu_k = \lambda(L(S_{k+1}) - L(S_k)) = k + 1 \quad \text{if } k < n \text{ and } \infty \text{ if } k \geq n \ ,$$

and again $\lim_{k\to\infty} \nu_k = \infty$. Thus we have met the requirements of [10, Theorem 1] and our theorem follows immediately from this result.

Theorem 2.3. (A) *Two regular infinite δ-fractions*

$$K = 1 - \delta_0 z + \mathop{K}_{k=1}^{\infty} (d_k z/(1-\delta_k z))$$

and

$$K^* = 1 - \delta_0^* z + \mathop{K}_{k=1}^{\infty} (d_k^* z/(1-\delta_k^* z))$$

correspond to the same power series

$$L_0 = 1 + c_1 z + c_2 z^2 + \cdots$$

if and only if

$$\delta_k = \delta_k^* , \quad k \geq 0 \quad \text{and} \quad d_k = d_k^* , \quad k \geq 1 \ .$$

(B) *A regular infinite δ-fraction K^* and a regular finite δ-fraction*

$$K_n = 1 - \delta_0 z + \frac{d_1 z}{1-\delta_1 z} + \cdots + \frac{d_{n-1} z}{1-\delta_{n-1} z} + \frac{d_n z}{1}$$

correspond to the same power series L_0 if and only if

$$\delta_k = \delta_k^* , \quad 0 \leq k \leq n-1 \ ; \quad \delta_k^* = 1 , \quad k \geq n$$

$$d_k = d_k^* , \quad 1 \leq k \leq n \ ; \quad d_k^* = 1 \ , \quad k \geq n+1$$

(C) *A regular finite δ-fraction K_n and a regular finite δ-fraction*

$$K_m^* = 1 - \delta_0^* z + \frac{d_1^* z}{1-\delta_1^* z} + \cdots + \frac{d_{m-1}^* z}{1-\delta_{m-1}^* z} + \frac{d_m^* z}{1}$$

correspond to the same power series L_0 if and only if

$$n = m \ ; \quad \delta_k = \delta_k^* , \quad 0 \leq k \leq n-1 \ ; \quad d_k = d_k^* , \quad 1 \leq k \leq n \ .$$

Proof. The techniques used to prove this theorem are similar to the ones employed by Perron [2, p. 108] and Thron [26, p. 208] in their proofs of analogous theorems for C-fractions and T-fractions, respectively. Therefore, to allow space for other items in this article, we omit a detailed proof.

Theorem 2.4. A power series

$$L_0 = 1 + c_1 z + c_2 z^2 + \cdots$$

is the Taylor series about the origin of a rational function

$$R(z) = \frac{1 + a_1 z + \cdots + a_n z^n}{1 + b_1 z + \cdots + b_m z^m}$$

if and only if there exists a finite regular δ-fraction

$$K = 1 - \delta_0 z + \frac{d_1 z}{1-\delta_1 z} + \cdots + \frac{d_{n-1} z}{1-\delta_{n-1} z} + \frac{d_n z}{1}$$

such that K correponds to L_0 .

Proof. Let the degree of any polynomial $P(z)$ be denoted by $\hat{P}$. We prove the above theorem by mathematical induction, where our n-th induction statement is that all rational functions of the form $\frac{P(z)}{Q(z)}$, where $P(0) = Q(0) = 1$ and $\max(\hat{P},\hat{Q}) = n$, have a δ-fraction expansion of the type asserted in the theorem. If $n = 0$, then $P(z)/Q(z) \equiv 1$ and we have only to choose $K = 1$. If $n = 1$, then we have only to consider the three types of rational functions $R_1(z) = 1 + az$ $R_2(z) = 1/(1+bz)$, or $R_3(z) = (1+az)/(1+bz)$, where $a \neq 0$, $b \neq 0$. The desired δ-fraction expansions for R_1 and R_2 are clearly

$$1 + \frac{az}{1} \quad \text{and} \quad 1 - \frac{bz}{1} + \frac{bz}{1} ,$$

respectively. If $a = b$ in R_3 , choose $K = 1$; otherwise, the desired expansion for R_3 is easily seen to be

$$1 + \frac{(a-b)z}{1} + \frac{bz}{1} .$$

Now assume that our theorem is true for all rational functions $R(z) = P(z)/Q(z)$ satisfying $P(0) = Q(0) = 1$ and $\max(\hat{P},\hat{Q}) = k$ for some $k, 0 \leq k \leq n$. Let $R_0(z) = P_0(z)/Q_0(z)$ be an arbitrary rational function satisfying $P_0(0) = Q_0(0) = 1$ and $\max(\hat{P}_0,\hat{Q}_0) = n + 1$. The proof is completed by showing that either $R_0(z) \equiv 1$, in which case we choose $K = 1$, or $R_0(z) \not\equiv 1$ can be expressed in one of the following six forms in a neighborhood of $z = 0$

(a) $$R_0(z) = 1 + \frac{d_1 z}{R_1(z)}$$

(b) $$R_0(z) = 1 + \frac{d_1 z}{1} + \frac{d_2 z}{R_1(z)}$$

(c) $$R_0(z) = (1-z) + \underbrace{\frac{z}{1-z} + \cdots + \frac{z}{1-z}}_{\alpha_1 - 2 \text{ terms}} + \frac{z}{1} + \frac{d_1 z}{R_1(z)}$$

(d) $$R_0(z) = 1 + \frac{d_1 z}{1-z} + \underbrace{\frac{z}{1-z} + \cdots + \frac{z}{1-z}}_{\alpha_2 - 2 \text{ terms}} + \frac{z}{1} + \frac{d_2 z}{R_1(z)}$$

(e) $R_0(z) = (1-z) + \underbrace{\frac{z}{1-z} + \cdots + \frac{z}{1-z}}_{\alpha_1 - 2 \text{ terms}} + \frac{z}{1} + \frac{d_1 z}{1} + \frac{d_2 z}{R_1(z)}$

(f) $R_0(z) = (1-z) + \underbrace{\frac{z}{1-z} + \cdots + \frac{z}{1-z}}_{\alpha_1 - 2 \text{ terms}} + \frac{z}{1} + \frac{d_1 z}{1-z}$

$$+ \underbrace{\frac{z}{1-z} + \cdots + \frac{z}{1-z}}_{\alpha_2 - 2 \text{ terms}} + \frac{z}{1} + \frac{d_2 z}{R_1(z)} ,$$

where $d_1 \neq 0$, $d_2 \neq 0$ are complex numbers, $\alpha_1 \geq 2$, $\alpha_2 \geq 2$ are positive integers, and $R_1(z) = P_1(z)/Q_1(z)$, where $P_1(0) = Q_1(0) = 1$ and $\max(\hat{P}_1, \hat{Q}_1) \leq n$

We omit the rather lengthy case analyses involved in the remainder of the proof, since a full proof will be published elsewhere. However, we mention here, that to arrive at the above cases, we found it convenient to make repeated use of the following observation.

If

$$F(z) = 1 + \frac{az^{\alpha}}{H(z)} ,$$

where $a \neq 0$, α is an integer ≥ 2, and $H(z)$ is a rational function satisfying $H(0) = 1$, then in a neighborhood of $z = 0$

$$F(z) = (1-z) + \frac{z}{1} - \frac{az^{\alpha-1}}{az^{\alpha-1} + H(z)} .$$

<u>Theorem</u> 2.5: <u>A regular infinite δ-fraction</u>

$$K = 1 - \delta_0 z + \mathop{K}_{n=1}^{\infty} (d_n z/(1-\delta_n z))$$

<u>corresponds to the Taylor series expansion about</u> $z = 0$ <u>of a rational function</u>

$$R(z) = \frac{1 + a_1 z + \cdots + a_n z^n}{1 + b_1 z + \cdots + b_m z^m}$$

<u>if and only if there exists an integer</u> $N \geqq 0$ <u>such that</u> $\delta_n = 1$ <u>if</u> $n \geqq N$ <u>and</u> $d_n = 1$ <u>if</u> $n \geqq N+1$.

<u>Proof</u>. By Theorem 2.4 there exists a finite δ-fraction

$$K_n = 1 - \delta_0 z + \frac{d_1 z}{1-\delta_1 z} + \cdots + \frac{d_{n-1} z}{1-\delta_{n-1} z} + \frac{d_n z}{1}$$

such that K_n corresponds to $L(R(z))$. Then by part (B) of Theorem 2.3, the infinite δ-fraction

$$K = 1 - \delta_0 z + \frac{d_1 z}{1-\delta_1 z} + \cdots + \frac{d_{n-1} z}{1-\delta_{n-1} z} + \frac{d_n z}{1-z} + \frac{z}{1-z} + \frac{z}{1-z} + \cdots$$

also corresponds to $L(R(z))$. By part (A) of the same Theorem 2.3 any other infinite δ-fraction corresponding to $L(R(z))$ must be identical to K so we have our desired result.

3. Convergence. We shall later use the following theorem, a version of which was first given by Poincaré [25] in 1885.

Theorem A (Poincaré): For $n = 1,2,3,\ldots$, let D_n be a nontrivial solution of the homogeneous linear difference equation

(3.1) $$D_n = b_n D_{n-1} + a_n D_{n-2} ,$$

where

$$\lim_{n\to\infty} a_n = a , \quad \lim_{n\to\infty} b_n = b .$$

Let the roots x_1 and x_2 of the characteristic equation

(3.2) $$x^2 - bx - a = 0$$

satisfy

(3.3) $$|x_1| > |x_2| .$$

Then

$$\lim_{n\to\infty} \frac{D_n}{D_{n-1}} = x_k$$

where x_k is one of the roots of (3.2).

To fix the ideas of his method of proof Poincaré sketched the proof of a similar result for third order linear difference equations. Perron [22, 23, 24] gave extensions of Poincaré's Theorem and proofs for the general n-th order case through three separate papers, the first two of which appeared in 1909 and the last in 1921. Treatments in English, of the Theorems of Poincaré and Perron on finite differences may be found in the books of Gel'fond [5] and Milne-Thomson [19] on the calculus of finite differences.

The following theorem is an adaptation to δ-fractions of some convergence results given by Jones and Thron [10] dealing with sequences of meromorphic functions.

Theorem 3.1. Let the infinite δ-fraction

$$K = 1 - \delta_0 z + \frac{d_1 z}{1-\delta_1 z} + \frac{d_2 z}{1-\delta_2 z} + \cdots$$

correspond to the power series

$$L_0 = 1 + c_1 z + c_2 z^2 + \cdots .$$

Let D be a domain containing a neighborhood of the origin. Then,

(A) K converges uniformly on every compact subset of D if and only if its sequences of approximants $\{A_n(z)/B_n(z)\}$ is uniformly bounded on every compact subset of D.

(B) If K converges uniformly on every compact subset of D, then $f(z) = \lim_{n\to\infty} A_n(z)/B_n(z)$ is analytic in D and L_0 is the Taylor series expansion of $f(z)$ about $z = 0$.

(C) Two infinite subsequences of approximants of K which converge uniformly on every compact subset of D converge to the same analytic function in D .

Proof. Part (A) and (B) follow immediately from [10, Theorem 4'] . Part (C) may be established as follows: Let $\{F_{n_k}(z)\}$ and $\{F_{m_k}(z)\}$ be two subsequences of approximants of K which converge uniformly on every compact subset of D . Since each of these subsequences corresonds to L_0 , it follows from [10, Theorem 4'] that $\lim_{k\to\infty} f_{n_k}(z) = F_n(z)$ and $\lim_{k\to\infty} f_{m_k}(z) = F_m(z)$, where $F_n(z)$ and $F_m(z)$ are analytic in D and have the property that L_0 is the Taylor expansion of each about the origin. Since D is a domain, it follows from the Identity Theorem for analytic functions that $F_n(z) = F_m(z)$ for all z in D .

The next theorem shows that a δ-fraction converges uniformly in a neighborhood of the origin to an analytic function if $\{\delta_n\}$ is an arbitrary sequence of zeros and ones, provided only that the sequence $\{d_n\}$ is bounded.

Theorem 3.2: If the coefficients d_n of the infinite δ-fraction

$$K = 1 - \delta_0 z + \underset{n=1}{\overset{\infty}{K}} (d_n z/(1-\delta_n z))$$

satisfy the inequality

$$0 < |d_n| \leq M ,$$

then K converges uniformly in the disk $|z| \leq (\sqrt{1+m} + \sqrt{M})^{-2}$ to a function f(z) which is analytic in the interior of this disk.

Proof. With the aid of an equivalence transformation, K can be written in the form

$$K = 1 - \delta_0 z + \underset{n=1}{\overset{\infty}{K}} \left(\frac{d_n E_n(z)}{1}\right) ,$$

where for each n , $E_n(z)$ is one of the three functions

$$z , z/(1-z) , z/(1-z)^2 ,$$

provided $z \neq 1$. If $|z| \leq r$, $0 < r < 1$, then

$$|z| \leq \frac{|z|}{1-|z|} \leq \frac{|z|}{(1-|z|)^2} \leq \frac{r}{(1-r)^2} .$$

From this we also derive

$$\left|\frac{z}{1-z}\right| \leq \frac{r}{(1-r)^2} \quad \text{and} \quad \left|\frac{z}{(1-z)^2}\right| \leq \frac{r}{(1-r)^2} ,$$

since $|1-z| \geq 1 - |z|$.

We impose the further restriction on r that it satisfy the equation

$$\frac{r}{(1-r)^2} = \frac{1}{4M} .$$

We solve this equation for the value of r in the interval $0 < r < 1$ to obtain

$$r = (\sqrt{1+M} + \sqrt{M})^{-2} .$$

By the Convergence Neighborhood Theorem in Jones and Thron's book [9, page 108], Satz 2.25 with $p_2 \equiv 2$ in Perron's book [21, page 64], or by Theorem 10.1 in

Wall's book [33, page 42], K converges uniformly if $|d_n E_n(z)| \leq 1/4$. Since the latter inequality is satisfied whenever $|d_n| \leq M$ and $|z| \leq (\sqrt{1+M} + \sqrt{M})^{-2}$, the uniform convergence is established. It is clear, therefore, that K converges uniformly on every compact subset of $|z| < (\sqrt{1+M} + \sqrt{M})^{-2}$. It follows from part (B) of Theorem 3.1 that f(z) is analytic in this disk, and our proof is complete.

By further restricting the sequence $\{d_n\}$ in Theorem 3.2 we can say the following:

<u>Theorem</u> 3.3: <u>Let</u> $K = 1 - \delta_0 z + \underset{n=1}{\overset{\infty}{K}} (d_n z/(1-\delta_n z))$ <u>be a</u> δ-<u>fraction such that</u>

$$\lim_{n\to\infty} d_n = 0 .$$

<u>Then</u> K <u>converges to a function</u> f(z) <u>which is both meromorphic in the open unit disk</u> $D(0,1) = \{z: |z| < 1\}$ <u>and analytic at</u> $z = 0$. <u>The convergence is uniform on every compact subset of</u> D(0,1) <u>which contains no poles of</u> f(z).

<u>Proof</u>. By an equivalence transformation K can be put into the form

$$K = 1 - \delta_0 z + \frac{d_1 z/(1-\delta_1 z)}{1} + \frac{\underset{n=2}{\overset{\infty}{K}} (E_n(z)/1)}{1} ,$$

where

$$E_n(z) = \frac{d_n z}{(1-\delta_{n-1} z)(1-\delta_n z)} , \quad n \geq 2 .$$

Clearly, each $E_n(z)$ is analytic if $|z| < 1$. It is also easy to see that for each M satisfying $0 < M < 1$ there exists an n_M such that $|E_n(z)| \leq 1/4$ for all $n \geq n_M$ and $|z| \leq M$. Thus, by the Convergence Neighborhood Theorem referenced in the proof of Theorem 3.2, a tail

$$\underset{n=m}{\overset{\infty}{K}} (E_n(z)/1)$$

of K converges uniformly to a function F(z) on $|z| \leq M$ if $m \geq n_M$. By Theorem 3.1, F(z) is analytic in the disk $|z| < M$ and $F(0) = 0$. The remainder of the proof will not be given here since, after making the above observations, it is very similar to the proof suggested by Jones and Thron for their Theorem 7.23 [9, p. 275].

The remaining theorem in this section deals with the convergence of δ-fractions having certain convergence criteria imposed on the sequences $\{d_n\}$ and $\{\delta_n\}$. We shall say that a δ-fraction

$$b_0 - \delta_0 z + \underset{n=1}{\overset{\infty}{K}} (d_n z/(1-\delta_n z))$$

is (p,q) <u>limit periodic</u> if there exist positive integers p and q such that

$$\lim_{\nu\to\infty} d_{p\nu+k} = D_k , \quad k = 0,1,\ldots,p-1$$

and

$$\lim_{\nu\to\infty} \delta_{q\nu+m} = \Delta_m , \quad m = 0,1,\ldots,q-1 ,$$

where each Δ_m is 0 or 1 and the D_k are numbers in the extended complex plane. Thus, in this setting, limit periodic regular C-fractions are in the class of (1,1) limit periodic δ-fractions. In the next section we shall see that the regular δ-fraction expansions for certain analytic functions can serve as examples of (p,q) limit periodic δ-fractins of types (1,1), (2,1), (4,1), and (8,1). We have obtained general convergence results, which will appear elsewhere, for the (1,2) and (2,1) cases. Before we state Theorem 3.4, it will be convenient for us to introduce the symbol $R[\alpha]$ for the ray from α to ∞ in the direction of α defined by

(3.4) $$R[\alpha] = \{\alpha t : t \geqq 1\} , \quad \alpha \neq 0 , \quad \alpha \in \mathbb{C} .$$

Theorem 3.4. Let $K = 1 - \delta_0 z + \mathop{K}_{n=1}^{\infty} (d_n z/(1-\delta_n z))$ be a δ-fraction satisfying

$$\lim_{n\to\infty} d_n = d \quad ; \quad \lim_{n\to\infty} \delta_n = \delta ,$$

where d is a complex constant and δ is either 0 or 1.

(A) If $d = \delta = 0$, then K converges to a function $f(z)$ which is both meromorphic in the complex plane $\mathbb{C}$ and analytic at $z = 0$. The convergence is uniform on every compact subset of $\mathbb{C}$ which contains no poles at $f(z)$.

(B) If $d = 0$ and $\delta = 1$, then the conclusions are the same as in (A) with $\mathbb{C}$ replaced by the punctured plane $\mathbb{C} - \{1\}$.

(C) If $d \neq 0$ and $\delta = 0$ then the conclusions are the same as in (A) with $\mathbb{C}$ replaced by the cut plane $\mathbb{C} - R[p-1/(4d)]$.

(D) If $d \neq 0$ and $\delta = 1$, then the conclusions are the same as in (A) with $\mathbb{C}$ replaced by any domain D such that $0 \in D$ and $D \subset \mathbb{C} - E_d$, where

$$E_d = \{z ; [2 - z - 1/z]/(4d) \in [0,1]\} .$$

If d is real, then E_d is a subset of the set made up of the real line and the unit circle. If $d = 1$, in particular, then E_d is the unit circle

Proof. Let

$$E = \{z : 1 - \delta z = 0\} \cup \{z : \frac{4dz}{(1-\delta z)^2} \in [-\infty,-1]\}$$

and let D be any domain (open connected set) in $\mathbb{C}$ satisfying

$$0 \in D \text{ and } D \cap E = \emptyset .$$

It is sufficient to prove the above theorem for the interior T^0 of T, where T is an arbitrary connected compact set such that $T \subset D$ and $0 \in T^0$. Since D contains no points of E, it follows that the roots $x_1(z)$ and $x_2(z)$ of the quadratic equation

$$x^2 - (1 - \delta z)x - dz = 0$$

have unequal moduli if $z \in D$. Thus, since $T \subset D$ and T is compact, there exist constants θ, C_1, and C_2 $(0 < \theta < 1 , C_1 > 0 , \text{ and } C_2 > 0)$ such that

$$\left|\frac{x_2(z)}{x_1(z)}\right| \leq \theta$$

and

$$C_1 \leq \left|x_1(z)\right| \leq C_2$$

for all $z \in T$. Hence, we have met the hypotheses of Satz 2.42 of Perron [21, page 93], so there exists an integer ν_0 such that the continued fraction

$$K_\nu = 1 - \delta_\nu z + \mathop{K}_{m=\nu+1}^{\infty} (d_m z/(1-\delta_m z))$$

converges uniformly on T if $\nu \geqq \nu_0$. If $P_n(z)$ and $Q_n(z)$ denote the numerator and denominator, respectively, of the n-th approximant of K_ν, then from the determinant formula we have

$$\frac{P_{n+1}(z)}{Q_{n+1}(z)} - \frac{P_n(z)}{Q_n(z)} = \frac{(-1)^n z^{n+1} \prod_{k=1}^{n+1} d_{\nu+k}}{Q_n(z) Q_{n+1}(z)}, \quad n \geqq 0 .$$

Therefore,

$$\lambda(L(P_{n+1}/Q_{n+1}) - L(P_n/Q_n)) = n + 1$$

and hence by Theorem 2.2, or by Theorem 1 of Jones and Thron [10, page 4], there exists a power series

$$L_\nu = 1 + C_1 z + C_2 z^2 + \cdots$$

to which K_ν corresponds at $z = 0$, the order of correspondence being $n + 1$. Each approximant of K_ν is a rational function analytic at the origin since $Q_n(0) \equiv 1$ for all $n \geqq 0$.

Also, since K_ν converges uniformly on T, it converges uniformly on every compact subset of T^0 to a function $F_\nu(z)$. By Theorem 4' of [10, page 15], $F_\nu(z)$ is analytic in T^0 and L_ν is its Taylor series expansion about $z = 0$. Now let A_k and B_k denote the numerator and denominator, respectively, of the k-th approximant of the original continued fraction K. Then for $n \geqq 0$,

$$\frac{A_{\nu+n}}{B_{\nu+n}} = \frac{A_{\nu-1}(P_n/Q_n) + zd_\nu A_{\nu-2}}{B_{\nu-1}(P_n/Q_n) + zd_\nu B_{\nu-2}} .$$

For $z \in T^0$ let

$$f(z) = \lim_{n\to\infty} \frac{A_{\nu+n}}{B_{\nu+n}} = \frac{A_{\nu-1}F_\nu(z) + zd_\nu A_{\nu-2}}{B_{\nu-1}F_\nu(z) + zd_\nu B_{\nu-2}} .$$

Since the numerator and denominator functions of the last expression for $f(z)$ have no common zeros and since the denominator is not identically zero in T^0 (it is equal to 1 at $z = 0$), it follows that $f(z)$ is meromorphic in T^0. Using the facts that $\{P_n(z)/Q_n(z)\}$ converges uniformly to $F_\nu(z)$ on compact subsets of T^0, $0 \in T^0$, and

$$B_{\nu-1}(z)F_\nu(z) + zd_\nu B_{\nu-2}(z)$$

does not vanish at $z = 0$, it is not difficult to verify that $\{A_k(z)/B_k(z)\}$ converges uniformly to $f(z)$ on pole free compact subsets of T^0. The fact that $f(z)$ is analytic at the origin follows from Theorem 3.2. After choosing d, δ,

D and E appropriately for each of the four cases in the statement of the theorem, our proof is complete. Part (A) also follows from Theorem 7.23 of Jones and Thron [9, page 275] dealing with general T-fractions.

4. Expansions. We shall make repeated use in this section of the extension theorem for continued fractions (Satz 1.7) in Perron's book [21, page 16]. Hereafter, we shall refer to this result as Theorem B. We shall also make heavy use of equivalence transformations of continued fractions. For thorough discussions of equivalence transformations, the reader is referred to the book of Jones and Thron [9, Section 2.3] and to Perron's book [21, §2].

We are now ready to give, through a series of five Theorems, regular δ-fraction expansions for a variety of classical analytic functions as well as considerable information about the regions of validity of these expansions.

Theorem 4.1: *The regular δ-fraction expansions of* tan z *and* tanh z *are given by*

$$\tan z = \frac{z}{1-z} + \frac{z}{1} + \frac{z/3}{1} - \frac{z/3}{1} - \frac{z/5}{1} + \frac{z/5}{1} + \frac{z/7}{1} - \frac{z/7}{1} - \frac{z/9}{1} + \cdots$$

and

$$\tanh z = \frac{z}{1-z} + \frac{z}{1} - \frac{z/3}{1} + \frac{z/3}{1} - \frac{z/5}{1} + \frac{z/5}{1} - \frac{z/7}{1} + \frac{z/7}{1} - \frac{z/9}{1} + \cdots$$

These expansions are valid everywhere in the complex plane. The δ-fractions above are (1,1) limit periodic.

Proof. According to [12, page 122], the following representation for tan z is valid everywhere in the complex plane except at points which are poles of the function

$$(4.1)\qquad \tan z = \frac{z}{1} - \frac{z^2}{3} - \frac{z^2}{5} - \frac{z^2}{7} - \cdots - \frac{z^2}{2n+1} - \cdots .$$

By letting the ρ_n in Theorem B take on the values of z and $-z$ it is easily seen that the continued fraction (4.1) can be extended to the continued fraction

$$(4.2)\qquad \frac{z}{1-z} + \frac{z}{1} + \frac{z}{3} - \frac{z}{1} - \frac{z}{5} + \frac{z}{1} + \frac{z}{7} - \frac{z}{1} - \frac{z}{9} + \frac{z}{1} + \cdots ,$$

which by an equivalence transformation can be put into the form

$$(4.3)\qquad \frac{z}{1-z} + \frac{z}{1} + \frac{z/3}{1} - \frac{z/3}{1} - \frac{z/5}{1} + \frac{z/5}{1} + \frac{z/7}{1} - \frac{z/7}{1} - \frac{z/9}{1} + \cdots .$$

The continued fraction (4.3) is a (1,1) limit periodic δ-fraction satisfying

$$\lim_{n\to\infty} d_n = 0 \quad \text{and} \quad \lim_{n\to\infty} \delta_n = 0 .$$

Also, its 2n-th approximant is the n-th approximant of (4.1). Hence, it follows from Theorem 3.4 that (4.3) converges to tan z everywhere in $\mathbb{C}$ except for poles.

From [12, page 123] we obtain

$$(4.4)\qquad \tanh z = \frac{z}{1} + \frac{z^2}{3} + \frac{z^2}{5} + \cdots + \frac{z^2}{2n+1} + \cdots ,$$

valid everywhere in $\mathbb{C}$ except at the poles of tanh z. By extension, (4.4) becomes

(4.5) $$\frac{z}{1-z}+\frac{z}{1}-\frac{z}{3}+\frac{z}{1}-\frac{z}{5}+\frac{z}{1}-\frac{z}{7}+\frac{z}{1}-\cdots,$$

which is equivalent to

(4.6) $$\frac{z}{1-z}+\frac{z}{1}-\frac{z/3}{1}+\frac{z/3}{1}-\frac{z/5}{1}+\frac{z/5}{1}-\frac{z/7}{1}+\frac{z/7}{1}-\cdots.$$

The even approximants of (4.6) are the approximants of (4.4), so again by Theorem 3.4, (4.6) converges to $\tanh z$ everywhere this function is defined.

Theorem 4.2: If $F(z)$ is Dawson's integral function, where,

$$F(z)=e^{-z^2}\int_0^z e^{t^2}dt,$$

then the regular δ-fraction expansion of F, valid everywhere in $\mathbb{C}$, is given by

(4.7) $$F(z)=\frac{z}{1-z}+\frac{z}{1}+\frac{d_3 z}{1}+\frac{d_4 z}{1}+\frac{d_5 z}{1}+\cdots,$$

where for $n \geq 1$

(4.8)
$$d_{4n-1}=-d_{4n}=\frac{(-1)^n n\binom{2n}{n}}{(4n-1)4^{n-1}}$$

$$d_{4n+2}=-d_{4n+1}=\frac{(-1)^n 4^n}{(4n+1)\binom{2n}{n}},$$

and

(4.9) $$|d_{4n-1}|\sim\frac{4\sqrt{n/\pi}}{4n-1}\quad;\quad|d_{4n+2}|\sim\frac{\sqrt{n\pi}}{4n+1}.$$

The δ-fraction (4.7) is (1,1) limit periodic.

Proof. According to McCabe [16], the function $F(z)$ can be represented by the continued fraction

(4.10) $$F(z)=\frac{z}{1}+\frac{2z^2}{3}-\frac{4z^2}{5}+\frac{6z^2}{7}-\frac{8z^2}{9}+-\cdots,$$

and the expansion is valid everywhere in the complex plane. We mention here, also, that Dijkstra [1] has given a certain continued fraction expansion for a generalization of Dawson's integral function. By an equivalence transformation, the continued fraction (4.10) can be put into the form

(4.11) $$F(z)=\frac{z}{1}+\frac{\frac{2z^2}{3}}{1}-\frac{\frac{4z^2}{3\times5}}{1}+\frac{\frac{6z^2}{5\times7}}{1}-\frac{\frac{8z^2}{7\times9}}{1}+-\cdots.$$

By extending (4.11) we obtain the δ-fraction

(4.12) $$\frac{z}{1-z}+\frac{z}{1}-\frac{\frac{2z}{3}}{1}+\frac{\frac{2z}{3}}{1}+\frac{\frac{4}{2}\left(\frac{z}{5}\right)}{1}-\frac{\frac{4}{2}\left(\frac{z}{5}\right)}{1}$$
$$+\frac{\frac{2\times6}{4}\left(\frac{z}{7}\right)}{1}-\frac{\frac{2\times6}{4}\left(\frac{z}{7}\right)}{1}-\frac{\frac{4\times8}{2\times6}\left(\frac{z}{9}\right)}{1}+\frac{\frac{4\times8}{2\times6}\left(\frac{z}{9}\right)}{1}$$
$$-\frac{\frac{2\times6\times10}{4\times8}\left(\frac{z}{11}\right)}{1}+\frac{\frac{2\times6\times10}{4\times8}\left(\frac{z}{11}\right)}{1}+\frac{\frac{4\times8\times12}{2\times6\times10}\left(\frac{z}{13}\right)}{1}-\cdots.$$

It can be shown that (4.12) is the same as the continued fraction (4.7), where the d_n are given by formulas (4.8). It follows from the asymptotic formualas (4.9) (which were determined with the aid of Stirling's formula) that

$$\lim_{n\to\infty} d_n = 0 \ .$$

Also, the 2n-th $(n \geq 0)$ approximant of the continued fraction (4.7) is the n-th approximant of (4.11). Thus it follows from Theorem 3.4 that (4.7) converges to $F(z)$ at all points $z \in \mathbb{C}$, and our proof is complete.

The expansion (4.13) in our next Theorem is easily derived from a well known continued fraction originally due to Lagrange (see [12, page 102] or [21, page 152]). Thus, in our proof of Theorem 4.3, we shall concentrate on justifying the new δ-fraction expansions (4.15) and (4.16) over the region indicated. In the proof of Theorem 4.3 we employ for the first time, a new technique, based on Poincaré's Theorem, for establishing the convergence behavior of (4.15). Murphy [20] has given various continued fraction expansions for $(1+z^2)^{-1/2}$. We deal with the function $(1+z^2)^{\nu}$ for all ν satisfying $0 < \nu < 1$.

Theorem 4.3: (A) *Suppose* ν *is any real number satisfying* $0 < \nu < 1$. *Then the regular* δ*-fraction expansion of* $(1+z)^{\nu}$, *valid for all* $z \in \mathbb{C} - R[-1]$, *is given by*

$$(4.13) \qquad (1+z)^{\nu} = 1 + \frac{\nu z}{1} + \frac{\frac{(1-\nu)z}{2}}{1} + \frac{\frac{(1+\nu)}{6}}{1} + \frac{\frac{(2-\nu)z}{6}}{1} + \frac{\frac{(2+\nu)z}{10}}{1} + \frac{\frac{(3-\nu)z}{10}}{1} + \cdots + \frac{\frac{(n+\nu)z}{4n+2}}{1} + \frac{\frac{(n+1-\nu)z}{4n+2}}{1} + \cdots \ .$$

In particular,

$$(4.14) \qquad \sqrt{1+z} = 1 + \frac{z/2}{1} + \frac{z/4}{1} + \frac{z/4}{1} + \frac{z/4}{1} + \cdots \ .$$

(B) *The regular* δ*-fraction expansion of* $(1+z^2)^{\nu}$ $(0 < \nu < 1)$, *valid for all* $z \in \mathbb{C} - \{R[-i] \cup R[i]\}$, *is given by*

$$(4.15) \qquad (1+z^2)^{\nu} = (1-z) + \frac{z}{1} - \frac{c_1 z}{1} + \frac{c_1 z}{1} - \frac{c_2 z}{1} + \frac{c_2 z}{1} - \frac{c_3 z}{1} + \frac{c_3 z}{1} - + \cdots ,$$

where, if $B(x,y)$ *denotes the Beta function and* $n \geq 1$,

$$c_{2n-1} = \left(\frac{n}{2n-1}\right)\frac{B(n+\nu,1-\nu)}{B(n-\nu,\ \nu)} \to \begin{cases} 0 & \text{if } 0 < \nu < 1/2 \\ \infty & \text{if } 1/2 < \nu < 1 \end{cases} \quad \text{as } n \to \infty$$

$$c_{2n} = \left(\frac{1}{2}\right)\frac{B(n+1-\nu,\nu)}{B(n+\nu,1-\nu)} \to \begin{cases} \infty & \text{if } 0 < \nu < 1/2 \\ 0 & \text{if } 1/2 < \nu < 1 \end{cases} \quad \text{as } n \to \infty$$

In particular,

$$(4.16) \qquad \sqrt{1+z^2} = (1-z) + \frac{z}{1} - \frac{z/2}{1} + \frac{z/2}{1} - \frac{z/2}{1} + \frac{z/2}{1} - + \cdots$$

The δ*-fraction* (4.15) *is* (4,1) *limit periodic if* $\nu \neq 1/2$ *and* (2,1) *limit periodic if* $\nu = 1/2$.

Proof. After substituting z^2 for z in the expansion (4.13) we obtain

$$(4.17) \qquad (1+z^2)^{\nu} = 1 + \frac{\nu z^2}{1} + \frac{\frac{(1-\nu)z^2}{2}}{1} + \frac{\frac{(1+\nu)z^2}{6}}{1} + \frac{\frac{(2-\nu)z^2}{6}}{1} + \frac{\frac{(2+\nu)z^2}{10}}{1} + \frac{\frac{(3-\nu)z^2}{10}}{1} + \cdots + \frac{\frac{(n+\nu)z^2}{4n+2}}{1} + \frac{\frac{(n+1-\nu)z^2}{4n+2}}{1} + \cdots \ .$$

By an equivalence transformation, this continued fraction can be put into the form

(4.18) $$1+\frac{z^2}{b_1}+\frac{z^2}{b_2}+\frac{z^2}{b_3}+\cdots,$$

where $b_1 = 1/\nu$ and

$$b_{2n+1}=\frac{(2n+1)(1-\nu)\cdots(n-\nu)}{\nu(1+\nu)\cdots(n+\nu)}\ ;\quad b_{2n}=\frac{2\nu(1+\nu)\cdots(n-1+\nu)}{(1-\nu)(2-\nu)\cdots(n-\nu)}\ ,\quad (n \geqq 1)\ .$$

With the aid of the functional relations

$$\Gamma(x+1) = x\Gamma(x) \quad\text{and}\quad B(x,y)=\frac{\Gamma(x)\Gamma(y)}{\Gamma(x+y)}\ ,$$

where $\Gamma(x)$ is the gamma function, the following formulas for the b_n can be obtained

(4.19) $$b_{2n-1}=\left(\frac{2n-1}{n}\right)\frac{B(n-\nu,\nu)}{B(n+\nu,1-\nu)}\ ;\quad b_{2n}=\frac{2B(n+\nu,1-\nu)}{B(n+1-\nu,\nu)}\quad (n \geqq 1)\ .$$

Using Theorem B with $\rho_n \equiv z$ $(n \geqq 0)$, we extend the continued fraction (4.18) to obtain

(4.20) $$(1-z)+\frac{z}{1}-\frac{z}{b_1}+\frac{z}{1}-\frac{z}{b_2}+\frac{z}{1}-\frac{z}{b_3}+\frac{z}{1}-\cdots,$$

which, after an equivalence transformation and after setting $c_n = 1/b_n$ $(n \geqq 1)$, becomes the continued fraction (4.15). Let $A_n(z)(B_n(z))$ denote the numerator (denominator) of the n-th approximant of (4.17), $f_n(z) = A_n(z)/B(z)$, $h_n(z) = B_n(z)/B_{n-1}(z)$, and let $g_n(z)$ denote the n-th approximant of (4.15). Then, for all $n \leqq 0$,

$$g_{2n+1}(z) = f_n(z) \quad\text{and}\quad g_{2n}(z)=\frac{A_n(z)-zc_nA_{n-1}(z)}{B_n(z)-zc_nB_{n-1}(z)}\ .$$

Hence,

(4.21) $$\left|g_{4n}(z)-f_{2n}(z)\right| = \frac{|zc_{2n}|\,|f_{2n}(z)-f_{2n-1}(z)|}{|h_{2n+1}(z)-zc_{2n}|}$$

and

(4.22) $$\left|g_{4n+2}(z)-f_{2n+1}(z)\right| = \frac{|zc_{2n+1}|\,|f_{2n+1}(z)-f_{2n}(z)|}{|h_{2n+1}(z)-zc_{2n+1}|}\ .$$

The continued fraction (4.17) is limit periodic because its partial numerators converge to $z^2/4$, and the roots $x_i(z)$ $(i = 1,2)$ of the associated quadratic equation

$$x^2 - x - z^2/4 = 0$$

have unequal moduli if $z \in \mathbb{C} - \{R[-i] \cup R[i]\}$. Therefore, it follows from Theorem A that $\{h_n(z)\}$ must converge to $x_1(z)$ or $x_2(z)$ if z is in this region. We have that $c_n \equiv 1/2$ if $\nu = 1/2$, and with the aid of Stirling's formula it can be verified that

$$\lim_{n\to\infty} c_{2n-1} = 0\ (\infty) \quad\text{if}\quad 0 < \nu < 1/2 \quad (1/2 < \nu < 1)$$

and

$$\lim_{n\to\infty} c_{2n} = \infty\ (0) \quad\text{if}\quad 0 < \nu < 1/2 \quad (1/2 < \nu < 1)\ .$$

We are now in a position to establish the convergence behavior of (4.15) asserted in part (B) . By convention, the continued fraction (4.15) converges to 1 if $z = 0$, so let us now assume that $z \neq 0$ and $z \in \mathbb{C} - \{R[-i] \cup R[i]\}$. Then, using the convergence properties of $\{c_n\}$ and $\{h_n(z)\}$ established above, it follows that the right sides of formulas (4.21) and (4.22) converge to 0 as $n \to \infty$. Hence, since $\lim_{n\to\infty} f_n(z) = (1+z^2)^{\nu}$, it follows that

$$\lim_{n\to\infty} g_{4n}(z) = \lim_{n\to\infty} g_{4n+2}(z) = \lim_{n\to\infty} g_{2n+1}(z) = (1+z^2)^{\nu}$$

and our proof is complete.

<u>Theorem</u> 4.4: (A) <u>The regular δ-fraction expansion of</u> Arctan z <u>is given by</u>

(4.23) $$\operatorname{Arctan} z = \frac{z}{1-z} + \frac{z}{1} + \frac{d_3 z}{1} + \frac{d_4 z}{1} + \frac{d_5 z}{1} + \cdots ,$$

<u>where</u> (<u>for</u> $n \quad 1$),

(4.24)
$$d_{4n} = -d_{4n-1} = \frac{n^2\left[\binom{2n}{n}\right]^2}{(4n-1)4^{2n-1}} \quad \frac{4n}{\pi(4n-1)} \to \frac{1}{\pi} \quad \text{as } n \to \infty$$

$$d_{4n+2} = -d_{4n+1} = \frac{4^{2n}}{(4n+1)\left[\binom{2n}{n}\right]^2} \quad \frac{\pi n}{4n+1} \to \frac{\pi}{4} \quad \text{as } n \to \infty .$$

<u>This expansion is valid for all</u> $z \in \mathbb{C} - \{R[-i] \cup R[i]\}$ <u>except possibly at</u> $z = (\pi/4 - 1/\pi)^{-1}$.

(B) <u>The regular δ-fraction expansion of</u> Arctanh z <u>is given by</u>

(4.25) $$\operatorname{Arctanh} z = \frac{z}{1-z} + \frac{z}{1} + \frac{c_3 z}{1} + \frac{c_4 z}{1} + \frac{c_5 z}{1} + \cdots ,$$

<u>where</u> (<u>for</u> $n \quad 1$)

$$c_{4n-1} = -c_{4n} = d_{4n} \; ; \quad c_{4n+2} = -c_{4n+1} = d_{4n+2}$$

<u>with</u> d_{4n} <u>and</u> d_{4n+2} <u>defined by formulas</u> (4.24). <u>This expansion is valid for all</u> $z \in \mathbb{C} - \{R[-1] \cup R[1]\}$ <u>except possibly at</u> $z = (\pi/4 + 1/\pi)^{-1}$. <u>The δ-fractions</u> (4.23) <u>and</u> (4.25) <u>are</u> (4,1) <u>limit periodic</u>.

<u>Proof</u>. The following known representation of Arctan z is taken from [9, page 202] :

(4.26) $$\operatorname{Arctan} z = \frac{z}{1} + \frac{1^2z^2}{3} + \frac{2^2z^2}{5} + \frac{3^2z^2}{7} + \frac{4^2z^2}{9} + \cdots ,$$

valid if $z \in \mathbb{C} - \{R[-i] \cup R[i]\}$. The continued fraction (4.26) is equivalent to

(4.27) $$\operatorname{Arctan} z = \frac{z}{1} + \frac{\frac{1^2z^2}{1\times 3}}{1} + \frac{\frac{2^2z^2}{3\times 5}}{1} + \frac{\frac{3^2z^2}{5\times 7}}{1} + \frac{\frac{4^2z^2}{7\times 9}}{1} + \cdots$$

The continued fraction (4.23) is derived as an extension of (4.27) as follows: Let $a_n(z)$, $n = 1,2,\ldots$, denote the n-th partial numerator of the continued fraction (4.27) . Then the continued fraction (4.23) is identical to

$$\frac{a_1(z)}{1-\rho_1(z)} + \frac{\rho_1(z)}{1} - \frac{a_2(z)/\rho_1(z)}{1-\rho_2(z)+a_2(z)/\rho_1(z)} + \frac{\rho_2(z)}{1} - \frac{a_3(z)/\rho_2(z)}{1-\rho_3(z)+a_3(z)/\rho_2(z)} + \cdots ,$$

where $\{\rho_n(z)\}$ is defined by

$$\rho_1(z) = z \quad ; \quad \rho_{n+1}(z) = a_{n+1}(z)/\rho_n(z) \quad , \quad n \geq 1 \quad .$$

The formulas (4.24) for the coefficients $\{d_n\}$ of (4.23) and their asymptotic behavior can be derived from the above process with a modest amount of algebraic manipulation and Stirling's formula. The continued fraction (4.23) certainly converges to Arctan z when $z = 0$, so from now on in our convergence investigation we assume $z \neq 0$. Let $f_n(z) = A_n(z)/B_n(z)$ denote the n-th approximant of (4.27), $g_n(z)$ the same for (4.23), and $h_n(z) = B_n(z)/B_{n-1}(z)$. Then $g_{2n}(z) = f_n(z)$, $n = 0,1,2,\ldots$,

$$(4.28) \qquad \left|g_{4n-1}(z) - f_{2n}(z)\right| = \frac{\left|d_{4n}z\right|\left|f_{2n}(z)-f_{2n-1}(z)\right|}{\left|h_{2n}(z) - d_{4n}z\right|} \quad , \quad n \geq 1 \quad ,$$

and

$$(4.29) \qquad \left|g_{4n+1}(z) - f_{2n+1}(z)\right| = \frac{\left|d_{4n+2}z\right|\left|f_{2n+1}(z)-f_{2n}(z)\right|}{\left|h_{2n+1}(z) - d_{4n+2}z\right|} \quad , \quad n \geq 1 \quad .$$

The partial numerators $a_n(z)$ of (4.27) converge to $z^2/4$ as $n \to \infty$. Therefore, by Theorem A, $\{h_n(z)\}$ converges to one of the two roots of

$$x^2 - x - z^2/4 = 0$$

if $z \in \mathbb{C} - \{R[-i] \cup R[i]\}$. These roots are

$$\frac{1 \pm \sqrt{1+z^2}}{2} \quad ,$$

where $\sqrt{}$ denotes the square root with positive real part. In particular, if z is real, it is easy to see that $\lim_{n\to\infty} h_n(z) = (1 + \sqrt{1+z^2})/2$ (since the other choice is a negative number). The sequences $\{d_{4n}z\}$ and $\{d_{4n+2}z\}$ converge to z/π and $\pi z/4$, respectively. We now see from (4.28) and (4.29) that the sequences $\{g_{4n-1}(z)\}$ and $\{g_{4n+1}(z)\}$ will converge to Arctan z in any region $\{f_n(z)\}$ converges to this function, provided $h_{2n}(z) \not\to z/\pi$ and $h_{2n+1}(z) \not\to \pi z/4$ as $n \to \infty$. Thus we investigate the roots of the equations

$$\frac{1 \pm \sqrt{1+z^2}}{2} = \frac{\pi z}{4} \quad \text{and} \quad \frac{1 \pm \sqrt{1+z^2}}{2} = \frac{z}{\pi} \quad .$$

The only possible candidates for roots of the first set of equations are

$$z = 0 \quad \text{and} \quad z = (\pi/4 - 1/\pi)^{-1} \quad 2.140922923 \quad .$$

We have already disposed of the case $z = 0$. For the second value of z above we obtain

$$\lim_{n\to\infty} h_n((\pi/4 - 1/\pi)^{-1}) = \frac{1 + \sqrt{1+(\pi/4-1/\pi)^{-2}}}{2} = \frac{\pi(\pi/4-1/\pi)^{-1}}{4} \quad ;$$

so for $z = (\pi/4-1/\pi)^{-1}$ it follows that

$$\lim_{n\to\infty}(h_{2n}(z) - d_{4n+2}z) = 0 \quad .$$

Therefore, unfortunately, our methods do not allow us to decide the convergence behavior of (4.23) for this value of z. The only possible roots of the second set of equations above having z/π on the right side are $z = 0$ and

$z = -(\pi/4-1/\pi)^{-1}$. But $\{h_{2n+1}(z) - d_{4n+2}z\}$ converges to a nonzero limit at $z = -(\pi/4-1/\pi)^{-1}$, so that the continued fraction (4.23) converges to Arctan z at this point. Thus

$$\lim_{n\to\infty} g_n(z) = \lim_{n\to\infty} f_n(z) = \text{Arctan } z$$

over the region indicated in part (A) of our theorem.

The proof of part (B) will not be given since it parallels the proof of part (A). We give only the continued fraction representation of Arctanh z from which the δ-fraction (4.25) can be derived by an extension. Making use of the fact that Arctanh z = -i Arctanh(iz) and employing (4.27) , it follows that

$$(4.30) \qquad \text{Arctanh } z = \frac{z}{1} \, \frac{\frac{1^2z^2}{1\times 3}}{1} - \frac{\frac{2^2z^2}{3\times 5}}{1} - \frac{\frac{3^2z^2}{5\times 7}}{1} - \frac{\frac{4^2z^2}{7\times 9}}{1} - \cdots ,$$

valid if $z \in \mathbb{C} - \{R[-1] \cup R[1]\}$.

In our next theorem we give a new continued fraction representation for the function $\exp(z^2)$ that turns out to be a regular δ-fraction with all of its partial denominators equal to 1 . The proof that the corresponding δ-fraction for $\exp(z^2)$ converges to $\exp(z^2)$ everywhere demands a new twist not employed in our previous examples.

__Theorem__ 4.5: (A) __The regular δ-fraction expansion of__ exp(z) , __valid for all__ $z \in \mathbb{C}$, __is given by__

$$(4.31) \qquad \exp(z) = 1 + \frac{z}{1} - \frac{z/2}{1} + \frac{z/6}{1} - \frac{z/6}{1} + \frac{z/10}{1} - \frac{z/10}{1} + \frac{z/14}{1} - \frac{z/14}{1} + \cdots$$

(B) __The regular δ-fraction expansion of__ $\exp(z^2)$, __valid for all__ $z \in \mathbb{C}$ __is given by__

$$(4.32) \qquad \exp(z^2) = (1-z) + \frac{z}{1} - \frac{z/1}{1} + \frac{z/1}{1} + \frac{z/2}{1} - \frac{z/2}{1} + \frac{z/3}{1} - \frac{z/3}{1}$$

$$- \frac{z/2}{1} + \frac{z/2}{1} - \frac{z/5}{1} + \frac{z/5}{1} + \frac{z/2}{1} - \frac{z/2}{1} + \frac{z/7}{1} - \frac{z/7}{1}$$

$$- \frac{z/2}{1} + \frac{z/2}{1} - \frac{z/9}{1} + \frac{z/9}{1} + \frac{z/2}{1} - \frac{z/2}{1} + \cdots$$

$$= (1-z) + \mathop{K}_{n=1}^{\infty} (d_n z/1) ,$$

__where for__ $n = 1,2,\ldots$,

$$(4.33) \qquad \begin{array}{lll} d_1 = 1 & d_{8n-4} = \frac{1}{2} & d_{8n-1} = -\frac{1}{4n-1} \\ d_{8n-6} = -\frac{1}{4n-3} & d_{8n-3} = -\frac{1}{2} & d_{8n} = -\frac{1}{2} \\ d_{8n-5} = \frac{1}{4n-3} & d_{8n-2} = \frac{1}{4n-1} & d_{8n+1} = \frac{1}{2} . \end{array}$$

__The δ-fraction__ (4.32) __is__ (8,1) __limit periodic__.

__Proof.__ It is well known that the following representation for exp(z) is valid for all $z \in \mathbb{C}$:

$$(4.34) \qquad \exp(z) = 1 + \frac{z}{1} - \frac{z}{2} + \frac{z}{3} - \frac{z}{2} + \frac{z}{5} - \cdots - \frac{z}{2} + \frac{z}{2n+1} - \cdots .$$

Two sources giving this expansion are [9, page 207] and [12, page 113] . The representation (4.31) is easily derived from (4.34) by an equivalence transformation, and it may be found in [21, page 124] . By replacing z by z^2 in (4.34) and using another equivalence transformation we have for all $z \in \mathbb{C}$ that

(4.35) $$\exp(z^2) = 1 + \frac{z^2}{1} + \frac{z^2}{(-2)} + \frac{z^2}{(-3)} + \frac{z^2}{2} + \frac{z^2}{5} + \frac{z^2}{(-2)} + \frac{z^2}{(-7)} + \cdots ,$$
$$= 1 + \mathop{K}_{n=1}^{\infty} (z^2/b_n) ,$$

where

(4.36) $$b_{2n} = (-1)^n 2 \quad \text{and} \quad b_{2n-1} = (-1)^{n-1}(2n-1) , \quad n \geqq 1 .$$

We extend the continued fraction (4.35) (using Theorem B with $\rho_n \equiv z$) to obtain the continued fraction

(4.37) $$(1-z) + \mathop{K}_{n=1}^{\infty} ((-1)^{n-1} z/c_n) ,$$

where

$$c_{2n-1} \equiv 1 \quad \text{and} \quad c_{2n} \equiv b .$$

The regular δ-fraction

(4.38) $$(1-z) + \frac{z}{1} - \frac{z/b_1}{1} + \frac{z/b_1}{1} - \frac{z/b_2}{1} + \frac{z/b_2}{1} - \frac{z/b_3}{1} + \frac{z/b_3}{1} - \cdots ,$$

where the b_n are given by (4.36) , is equivalent to (4.37) . It is easily verified that the continued fractions (4.38) and (4.32) are identical. Let $g_n(z)$ denote the n-th approximant of (4.32) and let $A_n(z)$ $(B_n(z))$ denote the numerator (denominator) of the n-th approximant of

(4.39) $$\exp(z^2) = 1 + \frac{z^2}{1} - \frac{z^2/2}{1} + \frac{z^2/6}{1} - \frac{z^2/6}{1} + \frac{z^2/10}{1} - \frac{z^2/10}{1} + \cdots$$

derived from (4.31) by replacing z by z^2 . Then (if $f_n(z) = A_n(z)/B_n(z)$ and $h_n(z) = B_n(z)/B_{n-1}(z)$),

$$g_{2n+1}(z) = f_n(z) , \quad (n \geqq 0) ; \text{ and } \quad g_{2n}(z) = \frac{A_n(z) - (z/b_n)A_{n-1}(z)}{B_n(z) - (z/b_n)B_{n-1}(z)} , \quad (n \geqq 1) .$$

Hence,

(4.40) $$|g_{8n}(z) - f_{4n}(z)| = \frac{|z/2|\,|f_{4n}(z) - f_{4n-1}(z)|}{|h_{4n}(z) - z/2|}$$

(4.41) $$|g_{8n+2}(z) - f_{4n+1}(z)| = \frac{|z/(4n+1)|\,|f_{4n+1}(z) - f_{4n}(z)|}{|h_{4n+1}(z) - z/(4n+1)|}$$

(4.42) $$|g_{8n+4}(z) - f_{4n+2}(z)| = \frac{|z/2|\,|f_{4n+2}(z) - f_{4n+1}(z)|}{|h_{4n+2}(z) + z/2|}$$

(4.43) $$|g_{8n+6}(z) - f_{4n+3}(z)| = \frac{|z/(4n+3)|\,|f_{4n+3}(z) - f_{4n+2}(z)|}{|h_{4n+3}(z) + z/(4n+3)|}$$

The sequence of partial numerators of (4.39) converges to 0 as $n \to \infty$. Therefore, it follows from Theorem A that $\lim_{n\to\infty} h_n(z) = 0$ or 1 . Unfortunately for a given z, we do not know how to determine whether $\{h_n(z)\}$ converges to 0 or 1. Hence, we cannot employ expressions (4.41) and (4.43) above and our technique used in the past to determine the convergence behavior of $\{g_{8n+2}(z)\}$ and $\{g_{8n+6}(z)\}$. Therefore, we try a different approach here and investigate first the convergence of a tail of (4.32).

Let

$$K_n = \mathop{K}_{m=1}^{\infty}\left(\frac{zd_{8n-4+m}}{1}\right),$$

where the coefficients of z are given by (4.33). Then the odd part K_n^* of K_n is given by

$$K_n^* = (-z/2) + \frac{\frac{z^2}{2(4n-1)}}{1} - \frac{\frac{z^2}{2(4n-1)}}{1} + \frac{\frac{z^2}{2(4n+1)}}{1} - \frac{\frac{z^2}{2(4n+1)}}{1}$$

$$+ \frac{\frac{z^2}{2(4n+3)}}{1} - \frac{\frac{z^2}{2(4n+3)}}{1} + \frac{\frac{z^2}{2(4n+5)}}{1} - \frac{\frac{z^2}{2(4n+5)}}{1} + \cdots .$$

Suppose $|z| \leq M$ (> 0) and choose n in K_n^* large enough such that

$$\frac{M^2}{2(4n-1)} \leq \frac{1}{4}.$$

It follows from the Convergence Neighborhood Theorem (see [9, page 108] and Theorem 3.1 that K_n^* converges uniformly on the set $|z| \leq M$ to a function $F_n(z)$ that is analytic on $|z| < M$. Now let G_m denote the m-th approximant of K_n, and let $N_m(z)(D_m(z))$ denote the numerator (denominator) of the m-th approximant of K_n^*. Then

$$G_{2m+1}(z) = \frac{N_m(z)}{D_m(z)} \quad \text{and} \quad G_{2m}(z) = \frac{N_m(z) - zd_{8n-3+2m}N_{m-1}(z)}{D_m(z) - zd_{8n-3+2m}D_{m-1}(z)},$$

with the aid of which we obtain

$$(4.44) \quad |G_{2m}(z) - N_m(z)/D_m(z)| = \frac{|zd_{8n-3+2m}|\,|N_m(z)/D_m(z) - N_{m-1}(z)/D_{m-1}(z)|}{|(D_m(z)/D_{m-1}(z)) - zd_{8n-3+2m}|}.$$

If we let $H_m(z) = D_m(z)/D_{m-1}(z)$, then $H_1(z) \equiv 1$ and it can be verified by induction that

$$|H_{2m+1}(z) - 1| \leq \frac{4n-1}{2[4n+2m-1]} \quad \text{and} \quad |H_{2m+2}(z) - 1| \leq \frac{4n-1}{2[4n+2m-1]},$$

provided z satisfies $|z| \leq M$. Hence, we can now say that $\lim_{m\to\infty} H_m(z) = 1$. By splitting $\{G_{2m}(z)\}$ into four subsequences as we did for $\{g_{2m}(z)\}$ and by using (4.44), it can be seen without too much difficulty that $\lim_{m\to\infty} G_{2m}(z) = \lim_{m\to\infty} N_m(z)/D_m(z) = F_n(z)$ if $|z| < M$. Thus given an arbitrary positive number M there is a tail of (4.32) which converges to an analytic function F_n on $|z| < M$. By the Convergence Neighborhood Theorem [9, page 108] applied to (4.32), we have, in particular, that (4.32) converges uniformly on the set $|z| \leq 1/2$. Hence, by arguments similar to those used in the proof of

Theorem 3.4, the δ-fraction (4.32) must converge to $\exp(z^2)$ in a neighborhood of the origin and, therefore, to $\exp(z^2)$ throughout $|z| < M$. Since M was arbitrary, our proof is complete.

References

1. D. Dijkstra, A continued fraction expansion for a generalization of Dawson's Integral, Math. of Comp. 31 (1977), 503-510.

2. D. M. Drew and J. A. Murphy, Branch points, M-fractions, and rational approximants generated by linear equations, J. Inst. Maths. Applics. 19 (1977), 169-185.

3. V.K. Dzjadyk, On the asymptotics of diagonal Padé approximants of the functions sin z , cos z , sinh z , and cosh z , Math. USSR Sbornik 36 (1980), 231-249.

4. J.S. Frame, The Hankel power sum matrix inverse and the Bernoulli continued fraction, Math. of Comp. 33 (1979), 815-816.

5. A.O. Gel'fond, Calculus of Finite Differences (Authorized English translation of the third Russian edition) Hinustan Publishing Corporation (India), 1971.

6. J. Gill, Enhancing the convergence region of a sequence of bilinear transformations, Math. Scand. 43 (1978), 74-80.

7. ______, The use of attractive fixed points in acclerating the convergence of limit-periodic continued fractions, Proc. Amer. Math. Soc. 47 (1975), 119-126.

8. T.L. Hayden, Continued fraction approximation to functions, Numer. Math. 7 (1965), 292-309.

9. W.B. Jones and W.J. Thron, Continued Fractions: Analytic Theory and Applications, Encyclopedia of Mathematics and Its Applications, Vol. 11, Addison-Wesley, Reading, Massachusetts, 1980.

10. W.B. Jones and W.J. Thron, Sequences of meromorphic functions corresponding to a formal Laurent series, Siam J. Math. Anal. 10 (1979), 1-17.

11. S.S. Khloponin, Approximation of functions by continued fractions, Izvestiya VUZ. Matematika 23 (1979), 37-41.

12. A.N. Khovanskii, The Applications of Continued Fractions and Their Generalizations to Problems in Approximation Theory, English translation by Peter Wynn, Noordhoff, 1963.

13. W. Leighton and W.T. Scott, A general continued fraction expansion, Bull. Amer. Math. Soc., (1939), 596-605.

14. A. Magnus, Certain continued fractions associated with the Padé table, Math. Zeitschr. 78 (1962), 361-374.

15. A. Magnus, Expansion of power series into P-fractions, Math. Zeitschr. 80 (1962), 209-216.

16. J.H. McCabe, A continued fraction expansion, with a truncation error estimate, for Dawson's integral, Math. of Comp. 28 (1974), 811-816.

17. J.H. McCabe, A further correspondence property of M-fractions, Math. of Comp. 32 (1978), 1303-1305.

18. J.H. McCabe and J.A. Murphy, Continued fractions which correspond to power series expansions at two points, J. Inst. Maths. Applics. 17 (1976), 233-247.

19. L.M. Milne-Thomson, The Calculus of Finite Differences, MacMillan and Co., London, 1933, Chapter XVII.

20. J.A. Murphy, Certain rational function approximations to $(1+x^2)^{-1/2}$, J. Inst. Maths. Applics. 7 (1971), 138-150.

21. O. Perron, Die Lehre von den Kettenbrüchen, Band II, Teubner, Stuttgart, 1957.

22. ________, Über die Poincarésche lineare Differenzengleichung, J. Reine Angew. Math. 137 (1909), 6-64.

23. ________, Über einen Satz des Herrn Poincaré, J. Reine Angew. Math. 136 (1909), 17-37.

24. ________, Über Summengleichungen und Poincarésche Differenzengleichungen, Math. Ann. 84 (1921), 1-15.

25. H. Poincaré, Sur les Equations Linéaires aux Différentielles ordinaires et aux Différences finies, Amer. J. Math., VIII (1885), 203-258.

26. W.J. Thron, Some properties of continued fractions $1 + d_0 z + K(z/(1+d_n(z))$, Bull. Amer. Math. Soc. 54 (1948), 206-218.

27. W.J. Thron and H. Waadeland, Accelerating convergence of limit periodic continued fractions $K(a_n/1)$, Numer. Math. 34 (1980), 155-170.

28. ________________________, Analytic continuation of fuctions defined by means of continued fractions, Math. Scand. 47 (1980), 72-90.

29. ________________________, Convergence questions for limit periodic continued fractions, Rocky Mountain J. Math. (1981).

30. H.B. Van Vleck, On the convergence of algebraic continued fractions, whose coefficients have limiting values, Trans. Amer. Math. Soc. 5 (1904), 253-262.

31. H. Waadeland, General T-fractions corresponding to functins satisfying certain boundedness conditions, J. Approx. Theory 26 (1979), 317-328.

32. H. Waadeland, Limit periodic general T-fractions and holomorphic functions, J. Approx. Theory 27 (1979), 329-345.

33. H.S. Wall, Analytic Theory of Continued Fractions, Van Nostrand, New York, 1948.

Department of Mathematics
University of Missouri
Columbia, Missouri 65201
U.S.A.

ON THE STRUCTURE OF THE TWO-POINT PADÉ TABLE

Arne Magnus

1. Introduction. The two-point Padé table is a double-entry table of rational functions related to two power series

$$L = c_1 z + c_2 z^2 + c_3 z^3 + \cdots \quad \text{and} \quad L^* = c_0^* + c_{-1}^* z^{-1} + c_{-2}^* z^{-2} + \cdots \quad , \quad c_0^* \neq 0$$

as follows. The two-point Padé approximant at the location (m,n) is the rational function $P_{m,n}/Q_{m,n}$ where the degrees of the polynomials $P_{m,n}$ and $Q_{m,n}$ are $\leq m$ and $\leq n$, respectively, and these two polynomials are selected so that as many of the initial coefficients as possible in the power series $Q_{m,n}L - P_{m,n}$ and $Q_{m,n}L^* - P_{m,n}$ are equal to zero.

We develop the basic properties of the approximants $P_{m,n}/Q_{m,n}$ in the table including existence and uniqueness. We find the actual degrees of $P_{m,n}$ and $Q_{m,n}$ and show that each approximant is equal to at least one adjacent approximant. Finally, we determine the shape and location of larger blocks of equal approximants.

In Section 2 we give the definition of the ordinary one-point Padé table and state without proof some of its basic properties as a background for the two-point table.

In Section 3 we give our definition of the two-point table and prove some of its properties while in Section 4 we study the block structure of equal approximants.

In 1948 Thron [12] introduced the T-fractions $t = 1 + d_0 z + K(\frac{z}{1+d_n z})$ and proved that if $0 < m \leq d_n \leq M < \infty$, $\forall n$ then t converges to two different functions, one holomorphic near 0 , the other holomorphic near ∞ . In 1957 Perron [10] generalized these fractions to $T = \gamma_0 + \delta_0 z + K(\frac{z}{\gamma_n + \delta_n z})$, $\gamma_n \neq 0$, $n \geq 1$ and proved convergence to two power series, L near 0 and L^* near ∞ . The approximants of the general T-fraction T corresponding to L and L^* , [4], are the two-point Padé approximants $P_{m,m}/Q_{m,m}$ of L and L^* down the diagonal $m = n$. This connects the two-point Padé table with the general T-fractions and the closely related M-fractions of McCabe and Murphy [9] .

Baker, Rushbrooke and Gilbert [2] , 1964, appear to be the first to use the term two-point approximant. McCabe [8] , 1975, introduces a two-point table via the M-fractions.

2. On the one-point Padé table. For a comprehensive treatement of the ordinary Padé table see, for example, the literature references [4] , [10] , [14] .

Given a power series with complex coefficients

$$L = c_0 + c_1 z + c_2 z^2 + \cdots$$

and an ordered pair of non-negative integers (m,n) , we seek polynomials

$$P_{m,n} = p_0 + p_1 z + \cdots + p_m z^m \quad \text{and} \quad Q_{m,n} = q_0 + q_1 z + \cdots + q_n z^n$$

so that either

(1.1a) $$L - P_{m,n}/Q_{m,n} = O(z^{m+n+1+j}) \quad , \quad j = \text{maximum}$$

or

(1.1b) $$Q_{m,n} L - P_{m,n} = O(z^{m+n+1}) \quad ,$$

where $O(z^d)$ denotes some power series whose leading term is of degree d or larger.

It is easy to show that the two formulations (1.1a) and (1.1b) are equivalent in the sense that both determine the same unique rational function $P_{m,n}/Q_{m,n}$, called the (m,n) Padé approximant of L .

In recent years, Baker [1] has adopted an alternate definition, namely (1.1a) with the additional restriction that

$$j = \text{maximum} \geq 0 \quad .$$

As a consequence, $P_{m,n}/Q_{m,n}$ may not exist for certain pairs (m,n) but when it does exist it equals the classical Padé approximant.

We point out that among authors who use the classical definitions (1.1a) or (1.1b) there is very little agreement on notation, notably in the use of m and n and the placement of the m and n axes in the Padé table. To some, m denotes the degree of $P_{m,n}$; to others, it denotes the degree of $Q_{m,n}$. To some, the m-axis points to the right, to others it points downward.

The Padé table is drawn in Figure 1 and the approximants $P_{m,n}/Q_{m,n}$ occupies the square with vertices (m,n) , $(m+1\ ,n)$, $(m+1\ ,n+1)$, $(m,n+1)$.

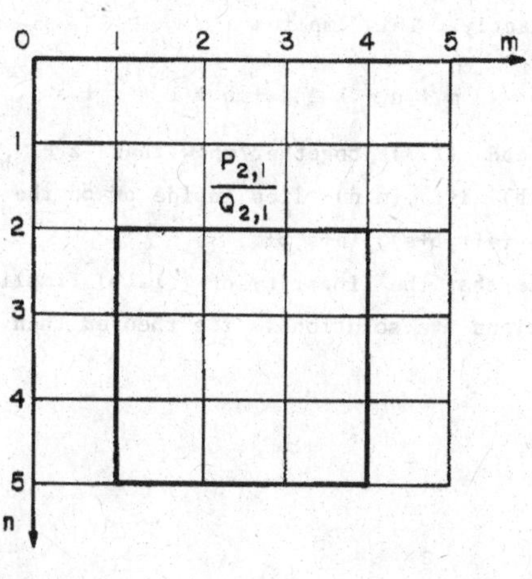

Figure 1. The square block where $m = 1,2,3$ and $n = 2,3,4$ has been enclosed by a heavier outline.

It is customary to require that $c_0 \neq 0$ in which case the following theorem is well known [10].

Theorem 1. All Padé approximants that are equal to a given Padé approximant occupy the squares in a square block. If the upper left-hand corner of such a block has coordinates (a,b), then the corresponding approximant

$$P_{a,b}/Q_{a,b} = (p_0 + p_1 z + \cdots + p_a z^a)/(q_0 + q_1 z + \cdots + q_b z^b)$$

satisfies $p_0 p_a q_0 q_b \neq 0$ and $Q_{a,b}L - P_{a,b} = O(z^{a+b+1+r})$ exactly, where the square block has dimensions $r+1$; that is, $P_{m,n}/Q_{m,n} = P_{a,b}/Q_{a,b}$ if $m = a + i$, $n = b + j$, $i, j = 0,1,\ldots,r$.

Although $P_{m,n}/Q_{m,n}$ is unique, $P_{m,n}$ and $Q_{m,n}$ are in general not unique, but determined as follows.

Theorem 2. In the notation of Theorem 1, if s is fixed, $0 \leqq s \leqq r$ then all approximants where $m = a + s$, $n = b + s$, $b + s + 1$, ..., $b + r$ or $m = a + s$, $a + s + 1$, ..., $a + r$, $n = b + s$ have the form $P_{m,n} = P_{a,b}S_s$ and $Q_{m,n} = Q_{a,b}S_s$, where S_s is an appropriate polynomial of degree $\leqq s$. If $m + n \leqq a + b + r$ then S_s may be chosen arbitrary, while if $m + n = a + b + r + t$, $0 \leqq t \leqq r$ then S_s contains the factor z^t, $S_s = z^t S_{s-t}$, where S_{s-t} is an arbitrary polynomial of degree $\leqq s - t$.

For the proof, we refer to Figure 2 below. We first determine for what (m,n) the polynomials $z^s P_{a,b}$ and $z^s Q_{a,b}$ satisfy (1.1b). Since the exact degrees of $z^s P_{a,b}$ and $z^s Q_{a,b}$ are $a + s$ and $b + s$, respectively, we must have

$$a + s \leqq m \text{ and } b + s \leqq n . \tag{1.2}$$

From $Q_{a,b}L - P_{a,b} = O(z^{a+b+1+r})$, exactly, follows $(z^s Q_{a,b})L - (z^s P_{a,b})L = O(z^{a+b+1+r+s})$, exactly. This implies

$$m + n + 1 \leqq a + b + 1 + r + s . \tag{1.3}$$

Inequalities (1.2) and (1.3) together show that $z^s P_{a,b}$ and $z^s Q_{a,b}$ satisfy (1.1b) iff (m,n) lies inside or on the triangle with vertices $(a+s,b+s)$, $(a+r,b+s)$, $(a+s,b+r)$.

Secondly, we note that the linearity of (1.1b) implies that linear combinations of solutions are solutions. The theorem then easily follows.

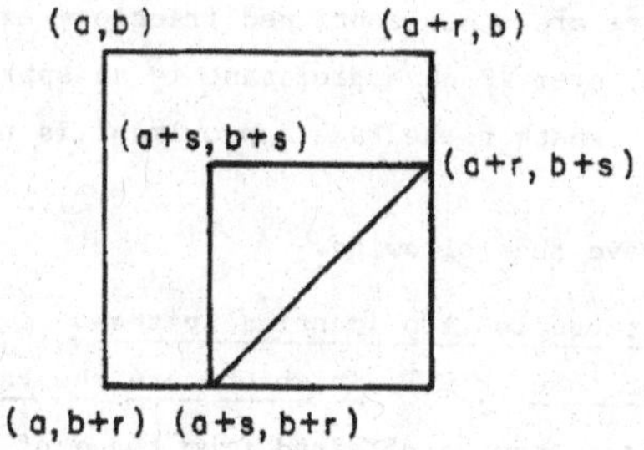

Figure 2

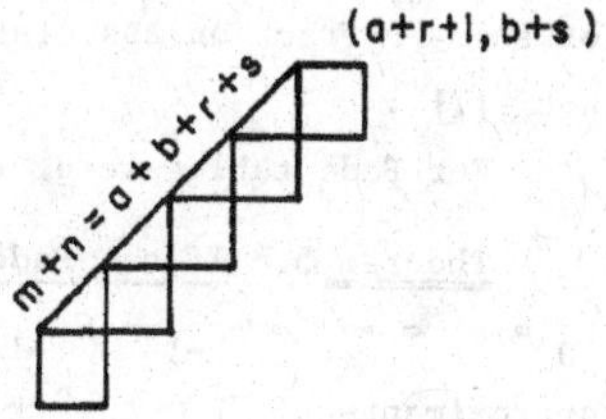

Figure 3

Remark. The line $m+n=a+b+r+s$ in Figure 2 passes through the upper left-hand corner, as shown in Figure 3, of squares in which $z^s P_{a,b}$ and $z^s Q_{a,b}$ are solutions of (1.1b).

The polynomial $P_{a,b}$ and $Q_{a,b}$ may be expressed in terms of the coefficients of L. In particular, we have (see [10]):

Theorem 3. *In the notation of Theorem 1 $Q_{a,b}$ is the product of an arbitrary constant and the determinant*

$$\begin{vmatrix} z^b & z^{b-1} & \cdots & 1 \\ c_{a-b+1} & c_{a-b+2} & \cdots & c_{a+1} \\ \cdots & \cdots & \cdots & \cdots \\ c_a & c_{a+1} & \cdots & c_{a+b} \end{vmatrix} .$$

Some Padé approximants are the convergents or approximants of certain continued fractions, which approximants have this property is related to the concept of normality.

Definition. The Padé approximant $P_{m,n}/Q_{m,n}$ is normal iff $r=0$; that is, $P_{m,n}/Q_{m,n}$ is not equal to any other Padé approximant. The Padé table is *normal* iff every approximant is normal.

Theorem 4. *If the Padé approximants $P_{m,m}/Q_{m,m}$, $m=0,1,2,\ldots$ along the main diagonal of the Padé table are all distinct then L has an associated continued fraction whose m-th approximant is $P_{m,m}/Q_{m,m}$.*

If the approximants $P_{0,0}/Q_{0,0}$, $P_{1,0}/Q_{1,0}$, $P_{1,1}/Q_{1,1}$, $P_{2,1}/Q_{2,1}$, $\ldots$ along a stairstep path in the Padé table are all distinct then L has a regular C-fraction expansion whose consecutive approximants are those Padé approximants. In particular, if the above Padé approximants are normal then the continued fraction expansions in question exist.

The proof is found in [10] . There are other continued fractions expansions related to the Padé table. For example, every Padé approximant is an approximant of some P-fraction associated with L , whether the Padé approximant is normal or not [7] .

For Padé tables were $c_0 = 0$ we have the following.

Theorem 5. If one Padé approximant equals zero identically then $c_0 = c_1 = \cdots = c_{k-1} = 0$, $c_k \neq 0$ for some $k \geq 1$ in which case the Padé approximants of $L = c_k z^k + c_{k+1} z^{k+1} + \cdots$ can be obtained from those of $L' = c_k + c_{k+1} z + \cdots = L/z^k$ as follows. For $0 \leq m \leq k-1$ we have $P_{m,n} \equiv 0$, $Q_{m,n} = z^{\alpha} S_{n-\alpha}$, where $\alpha = \max(0, n-(k-1-m))$ and $S_{n-\alpha}$ is an arbitrary polynomial of degree $\leq n - \alpha$. For $k \leq m$ we have

$$P_{m,n} = z^k P'_{m-k,n} \ , \quad Q_{m,n} = Q'_{m-k,n} \ ,$$

where $P'_{m-k,n}/Q'_{m-k,n}$ is the $(m-k,n)$ Padé approximant of L' . That is, the Padé table is moved k columns to the right and every entry is multiplied by z^k and the first k vacated columns are filled with zeros.

The proof is elementary and therefore omitted.

We note that the block structure of Theorem 1 is carried over except for the first k columns of zeros.

3. The two-point Padé table. Given two power series

$$L = c_1 z + c_2 z^2 + c_3 z^3 + \cdots \quad \text{and} \quad L^* = c_0^* + c_{-1}^* z^{-1} + c_{-2}^* z^{-2} + \cdots \ , \quad c_0^* \neq 0 \tag{3.1}$$

and a pair, (m,n) , of non-negative integers. We seek $P_{m,n} = p_0 + p_1 z + \cdots + p_m z^m$ and $Q_{m,n} = q_0 + q_1 z + \cdots + q_n z^n \not\equiv 0$ so that $P_{m,n}/Q_{m,n}$ approximates L and L^* in some sense as well as possible.

Before we adopt our definition we make a few remarks. Firstly, since $P_{m,n}/Q_{m,n}$ depends on a finite number, $m + n + 1$, of parameters $(p_0, \ldots, p_m, q_0, \ldots, q_n$ up to a factor of proportionality) we expect that the better we approximate one series the less well will we approximate the other. We must therefore commit ourselves to the amount of approximation allotted to each series. In applications, these amounts may depend on the extent to which the two series are known.

Secondly, the two formulations (1.1a) and (1.1b) for the one-point table have their analogs for the two-point table but they are not equivalent. For example, if $m + n + 1$ is even and we require that

$$\begin{aligned} L - P_{m,n}/Q_{m,n} &= O(z^{(m+n+1)/2}) \quad \text{and} \\ L^* - P_{m,n}/Q_{m,n} &= O((\tfrac{1}{z})^{-(m+n+1)/2}) \ , \end{aligned} \tag{3.2}$$

that is, equally good approximation to L and L^* , then this determines a differnt rational function $P_{m,n}/Q_{m,n}$ than does the requirement

$$Q_{m,n}L - P_{m,n} = O(z^{(m+n+1)/2}) \quad \text{and} \quad Q_{m,n}L^* - P_{m,n} = O((\tfrac{1}{z})^{-(m+n+1)/2}) \ .$$

We must therefore make a choice between these two options. The former one of these, expressed in (3.2) is essentially uninteresting since $c^* \neq 0$ implies that $P_{m,n}/Q_{m,n}$ is holomorphic at $z = \infty$ and different from zero there: that is, the actual degrees of $P_{m,n}$ and $Q_{m,n}$ are equal. As a matter of fact, it is easily seen that in this formulation the only distinct approximants are $P_{k,k}/Q_{k,k}$, $k = 0,1,2,\ldots$ and that $P_{m,n}/Q_{m,n} = P_{k,k}/Q_{k,k}$ where $k = \min(m,n)$. The table is therefore essentially a single-entry, not a double-entry table.

We adopt the following definition.

Definition. The two-point Padé approximant $P_{m,n}/Q_{m,n}$ of the two power series L and L^* in (3.1) is determined by

(3.3) $$Q_{m,n}L - P_{m,n} = O(z^r) \quad \text{and}$$

(3.4) $$Q_{m,n}L^* - P_{m,n} = O((\tfrac{1}{z})^{s-\max(m,n)}) \ ,$$

where $O(z^d)$, $(O((\frac{1}{z})^d))$, is some power series in ascending (descending) powers of z whose leading term is of degree d $(-d)$ or higher (lower). We select r and s to be

$$r = \left[\frac{m+n+2}{2}\right] \quad \text{and} \quad s = \left[\frac{m+n+1}{2}\right] \ ,$$

$[x]$ denotes the greatest integer $\leq x$.

We have

$$r + s = m + n + 1 \ ,$$

the number of parameters in $P_{m,n}/Q_{m,n}$.

Jones and Thron [3] also consider the case where $L = c_0 + c_1 z + c_2 z^2 + \cdots$ and $L^* = c^*_\mu z^\mu + c^*_{\mu-1} z^{\mu-1} + \cdots + c^*_0 + \cdots$, $\mu \geq 0$. The two-point Padé approximants obtained by that pair of series are closely related to those of this paper.

If we compare coefficients on both sides of the equations (3.3) and (3.4) we obtain the following systems of linear equations for the coefficients of $P_{m,n}$ and $Q_{m,n}$,

(3.5) $$\sum_{i=0}^{n} q_i c_{k-i} - p_k = 0 \ , \quad k = 0,1,\ldots.r-1$$

(3.6) $$\sum_{i=0}^{n} q_i c^*_{k-i} - p_k = 0 \ , \quad k = \max(m,n)-s+1,\ldots,\max(m,n) \ ,$$

where any coefficient p,q,c,c^* with subscripts outside the ranges given earlier shall be equated to zero.

The cases where $m = 0$ or $n = 0$ will be considered separately. We first consider $m = 0$. Then (3.5) with $k = 0$ implies $p_0 = 0$ or $P_{0,n} \equiv 0$ Similarly, (3.6) gives $q_n = q_{n-1} = \cdots = q_{[n/2]+1} = 0$. The remaining $[n/2]+1$ coefficients of $Q_{0,n}$ can then be determined from the remaining $r - 1 = [n/2]$

equations in (3.5) - $k \neq 0$ - so that $Q_{0,n} \not\equiv 0$. Thus $P_{0,n}/Q_{0,n} \equiv 0$ for all n .

Next we assume $n = 0$. Then $Q_{m,0} = q_0 \neq 0$ and (3.5) implies $p_k = q_0 c_k$ for $k = 0,1,\ldots,r-1 = [m/2]$, while (3.6) shows that $p_k = \sum_{i=0}^{n} q_i c^*_{k-i} = q_0 c^*_k = 0$ for $k = [m/2]+1,\ldots,m$. Thus, $P_{m,0}/Q_{m,0} = \sum_{i=0}^{r-1} c_i z^i$, $m = 0,1,2,\ldots$.

When $m,n \geq 1$ then $\max(m,n) - s + 1 \leq r - 1$ so that the ranges for k in (3.5) and (3.6) overlap. We eliminate the p_k's from the corresponding equations and obtain

$$(r-1) - (\max(m,n) - s) = m + n - \max(m,n)$$

linear equations for the $n + 1$ unknown $q_0,q_1,\ldots,q_n$. Since $m + n - \max(m,n) < n + 1$ there always exists a nontrivial solution $(q_0,q_1,\ldots,q_n)$ which in turn determines $(p_0,p_1,\ldots,p_m)$ uniquely. This proves the following theorem.

Theorem 6. For each pair (m,n) of non-negative integers there exist polynomials $P_{m,n}$ and $Q_{m,n} \not\equiv 0$ of degrees $\leq m$ and $\leq n$, respectively, satisfying (3.3) and (3.4) .

The actual degrees of $P_{m,n}$ and $Q_{m,n}$ are in general less than m and n , respectively, as stated in the next theorem [3] .

Theorem 7. If $n \leq m - s \leq m$ then degree $P_{m,n} \leq m - s$ and degree $Q_{m,n} \leq n$. If $m - s \leq n \leq m$ then degree $P_{m,n} \leq n$ and degree $Q_{m,n} \leq n$. If $n - s \leq m \leq n$ then degree $P_{m,n} \leq m$ and degree $Q_{m,n} \leq m$. If $m \leq n - s \leq n$ then degree $P_{m,n} \leq m$ and degree $Q_{m,n} \leq n - s$. These bounds are best possible.

Proof: If $m \leq n - s \leq n$ then since $P_{m,n}$ contains no term of degree $> m$ and $O((\frac{1}{z})^{s-n})$ contains no term of degree $> n - s$ we see from (3.4) that $Q_{m,n}L^*$ contains no term of degree $> n - s$. Since $c^* \neq 0$ this implies that $Q_{m,n}$ contains no term of degree $> n - s$. As is seen, in Theorem 10 below, the coefficient of the terms of highest and lowest degree in $Q_{m,n}$ depends linearly on some, but not all, of the coefficients in L and L^* . Such coefficients may therefore be chosen so that $Q_{m,n}$ has maximal degree, in this case $n - s$. We also note that the constant term of $Q_{m,n}$ is, in general, different from zero.

The three remaining cases are proven similarly.

We can now show the uniqueness of $P_{m,n}/Q_{m,n}$.

Theorem 8. $P_{m,n}$ and $Q_{m,n}$ may not be unique but $P_{m,n}/Q_{m,n}$ is.

For the proof we first assume $n \leq m$. Assume (3.3) and (3.4) are satisfied by two pairs $(P_{m,n},Q_{m,n})$ and $(\hat{P}_{m,n},\hat{Q}_{m,n})$. Then, since

$$Q_{m,n}\hat{P}_{m,n} - \hat{Q}_{m,n}P_{m,n} = (Q_{m,n}L - P_{m,n})\hat{Q}_{m,n} - (\hat{Q}_{m,n}L - \hat{P}_{m,n})Q_{m,n} ,$$

we see that $Q_{m,n}\hat{P}_{m,n} - \hat{Q}_{m,n}P_{m,n} = O(z^r)$ and similarly that $Q_{m,n}\hat{P}_{m,n} - \hat{Q}_{m,n}P_{m,n} = O((\frac{1}{z})^{s-m-n}) = O((\frac{1}{z})^{-r+1})$, since $m + n + 1 = r + s$. This shows that $Q_{m,n}\hat{P}_{m,n} - \hat{Q}_{m,n}P_{m,n}$ contains no term of degree $\leq r - 1$ nor any term of degree $\geq r$. That is, $Q_{m,n}\hat{P}_{m,n} - \hat{Q}_{m,n}P_{m,n} \equiv 0$ so $P_{m,n}/Q_{m,n}$ is unique.

Next we assume $n - s \leqq m \leqq n$. Then degree $Q_{m,n} \leqq m$ and (3.6) again gives $Q_{m,n}\hat{P}_{m,n} - \hat{Q}_{m,n}P_{m,n} = O((\frac{1}{z})^{s-n-m})$ leading to the same conclusion.

Finally, if $m \leqq n - s \leqq n$ then degree $Q_{m,n} \leqq n - s$ so that degree $(Q_{m,n}\hat{P}_{m,n} - \hat{Q}_{m,n}P_{m,n}) \leqq n - s + m = r - 1$ which, together with $Q_{m,n}\hat{P}_{m,n} - \hat{Q}_{m,n}P_{m,n} = O(z^r)$ again implies the uniqueness of $P_{m,n}/Q_{m,n}$.

In the one-point Padé table all approximants may be different, as is for example the case with $L = \sum_{n=0}^{\infty} z^n/n!$, in contrast to the two-point approximants which always come in pairs or, along the diagonal $m = n$, in triples.

__Theorem__ 9. __For every pair__ L,L^* , __if__ $m + n$ __is even then__

(A) $m \geqq n$ __implies__ $P_{m+1,n}/Q_{m+1,n} = P_{m,n}/Q_{m,n}$ __and__

(B) $m \leqq n$ __implies__ $P_{m,n+1}/Q_{m,n+1} = P_{m,n}/Q_{m,n}$.

__Proof:__ If $m + n$ is even and $m \geqq n$ then r and $s - m$ do not change when m is replaced by $m + 1$ so the required approximations in (3.3) and (3.4) namely $O(z^r)$ and $O((\frac{1}{z})^{s-m})$ are preserved. This, together with the uniqueness of $P_{m+1,n}/Q_{m+1,n}$ proves (A) . The proof of (B) is similar.

We can now introduce the two-point Padé table of L and L^* . As is the case with the one-point Padé table the approximant $P_{m,n}/Q_{m,n}$ is placed in the square with vertices (m,n) , $(m+1,n)$, $(m,n+1)$ and $(m+1,n+1)$. By Theorem 9 equal approximants may be combined in pairs or triples to give the table the structure of a brick wall. The hypotheses of Theorem 7 combined with $m + n =$ even may be rewritten as $3n \leqq m$, $n \leqq m \leqq 3n$, $m \leqq n \leqq 3m$ m and $3m \leqq n$, respectively, dividing the table into four sectors 1, 2, 3 and 4 .

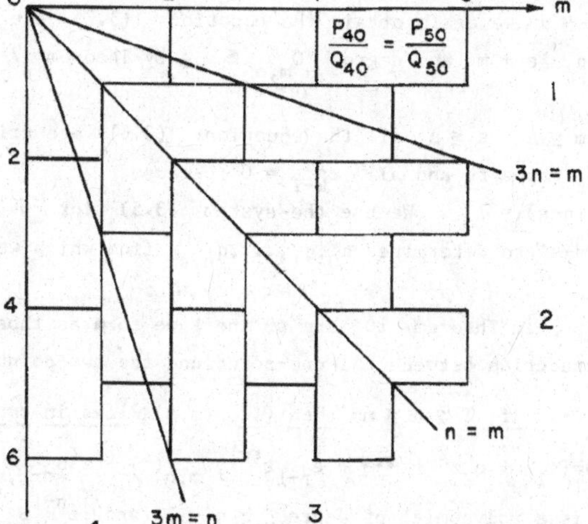

Figure 4

There is an analog of Theorem 3 for the two-point approximants expressing $Q_{m,n}$ in terms of the coefficients of L and L^*.

Theorem 10. Let $\delta_k = c_k^* - c_k$, $k = 0, \pm 1, \pm 2, \ldots$. *If the rank of the coefficient matrix involved is maximal then* $Q_{m,n}$ *is given by*

$$\text{(A)} \qquad Q_{m,n} = \text{constant} \cdot \begin{vmatrix} z^n & \cdots & 1 \\ \delta_{r-2n} & \cdots & \delta_{m-s+1} \\ \cdot & \cdots & \cdot \\ \delta_{m-s} & \cdots & \delta_{m-s+n} \end{vmatrix}, \text{ when } n \leqq m$$

$$\text{(B)} \qquad Q_{m,n} = \text{constant} \cdot \begin{vmatrix} z^m & \cdots & 1 \\ \delta_{r-2m} & \cdots & \delta_{n-s+1} \\ \cdot & \cdots & \cdot \\ \delta_{n-s} & \cdots & \delta_{n-s+m} \end{vmatrix}, \text{ when } n-s \leqq m \leqq n$$

$$\text{(C)} \qquad Q_{m,n} = \text{constant} \cdot \begin{vmatrix} z^{n-s} & \cdots & z^m & \cdots & 1 \\ 0 & \cdots\ 0 & \delta_1 & \cdots & \delta_{m+1} \\ \vdots & \ddots & \vdots & & \vdots \\ 0 & & \vdots & & \vdots \\ \delta_1 & \cdots & \delta_{r-2m} & \cdots & \delta_{n-s+1} \\ \cdot & \cdots & \cdot & \cdots & \cdot \\ \delta_m & \cdots & \delta_{n-s} & \cdots & \delta_{r-1} \end{vmatrix}, \text{ when } 1 \leqq m \leqq n-s \leqq n .$$

Proof: If $n \leqq m$ the overlap in the ranges for k in (3.5) and (3.6) is $k = m-s+1, m-s+2, \ldots, m-s+n$. We eliminate the corresponding p_k from (3.5) and (3.6) and obtain

$$\text{(3.7)} \qquad \sum_{i=0}^{n} q_i \delta_{k-i} = 0 \quad , \quad k = m-s+1, \ldots, m-s+n \quad .$$

One possible solution is the one given in formula (A) of Theorem 10, and since the rank is maximal, equal to n, that is the only solution.

If $n-s \leqq m \leqq n$ we again obtain the equations (3.7) but with $k = n-s+1, \ldots, n-s+m$ and degree $Q_{m,n} \leqq m$, by Theorem 7. This gives formula (B).

Finally, if $m \leqq n-s \leqq n$ all the equations (3.6) are trivial, namely $0 = 0$ since no p_k appears and all $c_{k-i}^* = 0$ because $k - i \geqq (n-s+1) - (n-s) = 1$. We use the system (3.5) for $k = m+1, m+2, \ldots, m+n-s$ to determine $q_0, q_1, \ldots, q_{n-s}$ from which we obtain formula (C).

The determinants in Theorem 10 are of the same form as those in Theorem 3. This suggests a connection between the one-point and the two-point table.

Theorem 11. (A) *If* $1 \leqq n \leqq m$ *then (If* (m,n) *lies in sector* 1 *or* 2)

$$P_{m,n}(z) = (c_1 z + c_2 z^2 + \cdots + c_{r-1} z^{r-1}) Q_{m,n}(z) + z^r S_{n-1,n}(z) \quad ,$$

where $S_{n-1,n}(z)$ *is a polynomial of degree* $\leqq n-1$ *and* $z^{n-1} S_{n-1,n}(\frac{1}{z}) / z^n Q_{m,n}(\frac{1}{z})$

is the one-point Padé approximant at $(n-1,n)$ *of* $\delta_{r-1} + \delta_{r-2}z + \delta_{r-3}z^2 + \cdots$.

(B) *If* $n - s < m \leqq n$ *then (if* (m,n) *lies in sector* 3)

$$P_{m,n}(z) = (c_0^* + c_{-1}^* z + \cdots + c_{r-2m}^* z^{r-2m})Q_{m,n}(z) - z^{r-2m}S_{m-1,m}(z) ,$$

where $S_{m-1,m}(z)$ *is a polynomial of degree* $\leqq m - 1$ *and* $S_{m-1,m}(z)/Q_{m,n}(z)$ *is the one-point Padé approximant at* $(m-1,m)$ *of* $\delta_{r-2m} + \delta_{r-2m+1}z + \delta_{r-2m+2}z^2 + \cdots$

(C) *If* $m \leqq n - s \leqq n$ *then (if* (m,n) *lies in sector* 4) $P_{m,n}(z)/Q_{m,n}(z)$ *is the one-point Padé approximant at* $(m,n-s)$ *of* $L = c_1 z + c_2 z^2 + c_3 z^3 + \cdots$.

Proof: (A) Equation (3.3) implies $Q_{m,n}(c_1 z + \cdots + c_{r-1}z^{r-1}) - P_{m,n} = O(z^r) = -z^r S_{n-1,n} + O(z^{r+n})$ or

$$Q_{m,n}(z)(c_1 z + c_2 z^2 + \cdots + c_{r-1}z^{r-1}) + z^r S_{n-1,n}(z) - P_{m,n}(z) = O(z^{r+n}) .$$

By Theorem 7 , the left-hand side is of degree $\leqq r + n - 1$ and therefore identically equal to 0 . This proves the formula for $P_{m,n}(z)$. Next we insert this expression for $P_{m,n}$ in (3.4) , replace z by $\frac{1}{z}$ and multiply by z^{n+r-1} to obtain

$$(z^n Q_{m,n}(\tfrac{1}{z}))(\delta_{r-1} + \delta_{r-2}z + \delta_{r-3}z^3 + \cdots) - (z^{n-1}S_{n-1,n}(\tfrac{1}{z})) = O(z^{2n}) ,$$

which proves (A) .

(B) If $n - s \leqq m \leqq n$ we replace z by $\frac{1}{z}$ in (3.4) and multiply by z^m and find

$$(z^m Q_{m,n}(\tfrac{1}{z}))(c_0^* + c_{-1}^* z + \cdots + c_{-m+n-s+1}^* z^{m-n+s-1}) - z^m P_{m,n}(\tfrac{1}{z})$$

$$= z^{2m-n+s-1}S_{m-1,m}(\tfrac{1}{z}) + O(z^{2m-n+s}) .$$

The strict inequality $n - s < m$ then implies that $P_{m,n}(z)$ is as stated in part (B) . We insert this in (3.3) , change sign and cancel z^{r-2m} to find

$$Q_{m,n}(z)(\delta_{r-2m} + \delta_{r-2m+1}z + \delta_{r-2m+2}z^2 + \cdots) - S_{m-1,m}(z) = O(z^{2m}) ,$$

which proves part (B) .

Finally, when $m \leqq n - s \leqq n$ then degree $P_{m,n}$ + degree $Q_{m,n} \leqq m + n - s = r - 1$, which combined with (3.3) shows that $P_{m,n}/Q_{m,n}$ is the one-point Padé approximant of L at $(m,n-s)$. This proves part (C) .

In addition to the equal approximants given by the brick wall structure of Theorem 9 there are equal approximants on opposite sides of the diagonal $m = n$ as follows.

Theorem 12. *If* (m,n) *lies in sector* 2 *or* 3 *then* $P_{m,n}/Q_{m,n} = P_{n,m}/Q_{n,m}$

Proof: If m and n are interchanged then r and $s - \max(m,n)$ remain unchanged. The approximations required by $P_{n,m}$ and $Q_{n,m}$ in (3.3) and (3.4) are therefore furnished by $P_{m,n}$ and $Q_{m,n}$. Since, furthermore, Theorem 7 states that degree $P_{m,n} \leqq \min(m,n) \leqq n$ and degree $Q_{m,n} \leqq \min(m,n) \leqq m$ we see by the uniqueness of $P_{n,m}/Q_{n,m}$ that $P_{m,n}/Q_{m,n} = P_{n,m}/Q_{n,m}$.

The remaining approximants, in sectors 1 and 4, excluding those in common with the approximants in sectors 2 and 3 are in general distinct except, of course, for the brick structure.

Theorem 13. If $m+n$ is even and $n < m-s \leqq m$ or $m < n-s \leqq n$ then $P_{m,n}/Q_{m,n}$ is in general not equal to any other approximant except the ones given by Theorem 9.

Proof: As in the proof of Theorem 7 we may select the coefficients of L and L^* so that $P_{m,n}$ and $Q_{m,n}$ have maximal degree and so that the approximations in (3.3) and (3.4) are $O(z^r)$, not $O(z^{r+1})$, and $O((\frac{1}{z})^{s-\max(m,n)})$, not $O((\frac{1}{z})^{s-\max(m,n)+1})$, respectively. If $P_{m,n}/Q_{m,n}$ appears at another location $(\tilde{m},\tilde{n})$ in the table where $\tilde{m}+\tilde{n}$ is even we must have

$$\tilde{r} \leqq r \quad \text{and} \quad \tilde{s} - \max(\tilde{m},\tilde{n}) \leqq s - \max(m,n) .$$

A straightforward manipulation of all the inequalities involved leads to $m = \tilde{m}$ and $n = \tilde{n}$ which proves the theorem.

4. Blocks of equal two-point approximants. Theorems 9 and 12 state which approximants are equal in any two-point Padé table. For special choices of L and L^* there may be additional equal approximants. The case where $P_{m,n} \equiv 0$ is treated separately.

Theorem 14. (A) If $P_{m,n} \equiv 0$ for some (m,n) and $L = c_p z^p + c_{p+1} z^{p+1} + \cdots$ $c_p \neq 0$ then $p \geqq 1 + \min(m,n)$.
(B) If L is as in (A) then $P_{m,n} \equiv 0$ for $m = 0,1,\ldots,p-1$ and all n and also for $n = 0,1,\ldots,p-1$ and $m = 0,1,\ldots,n+2p-1$.

Proof: (A) If $n \leqq m$ and $Q_{m,n} = q_\alpha z^\alpha + \cdots + q_{n-\gamma} z^{n-\gamma}$, where $0 \leqq \alpha \leqq n-\gamma \leqq n$ and $q_\alpha q_{n-\gamma} \neq 0$ then $Q_{m,n} L - 0 = O(z^{\alpha+p})$, exactly and $Q_{m,n} L^* - 0 = O((\frac{1}{z})^{\gamma-n})$, exactly. Thus, $r \leqq \alpha + p$ and $s - m \leqq \gamma - n$. Adding these inequalities we find $r + s - m = n + 1 \leqq p - (n - \gamma - \alpha) \leqq p$. Similarly, if $m \leqq n$ we find $m + 1 \leqq p$. This proves (A).
(B) If $0 \leqq n - m =$ even and $0 \leqq m \leqq p - 1$ we shall construct $Q_{m,n}$ as in (A) so that $P_{m,n} \equiv 0$ and $Q_{m,n}$ satisfy (3.3) and (3.4). We must have $r \leqq \alpha + p$ and $s - n \leqq \gamma - n$ or equivalently

$$\frac{n+m}{2} + 1 - p = \frac{n-m}{2} - (p - 1 - m) \leqq \alpha \leqq n - \gamma \leqq \frac{n-m}{2} .$$

We select $\alpha = \max(0, \frac{n+m}{2} + 1 - p)$, $\varepsilon = \frac{n-m}{2} - \alpha \geqq p - 1 - m \geqq 0$ and let S_ε be any polynomial of degree $\leqq \varepsilon$. Then $Q_{m,n} = z^\alpha S_\varepsilon$ is of degree $\leqq n$ and together with $P_{m,n} \equiv 0$ satisfies (3.3) and (3.4). The case $0 \leqq n - m =$ odd is then trivial because of Theorem 9. Thus, the first p columns contain only zero approximants below the diagonal $m = n$.

If $0 \leqq m - n = \text{even}$ and $0 \leqq n \leqq p - 1$ then $P_{m,n} \equiv 0$ and $Q_{m,n}$ satisfy (3.3) and (3.4) iff $r \leqq \alpha + p$ and $s - m \leqq \gamma - n$ which is equivalent to

$$\frac{m-n}{2} - (p - 1 - n) \leqq \alpha \leqq n - \gamma \leqq \min(n, \frac{m-n}{2}) .$$

In order that we have any solution we must therefore have $\frac{m-n}{2} - (p - 1 - n) \leqq \frac{m-n}{2}$ which is trivial because $p - 1 - n \geqq 0$, and $\frac{m-n}{2} - (p - 1 - n) \leqq n$ which is equivalent to $m \leqq n + 2p - 2$. The case $0 \leqq m - n = \text{odd}$ leads to $m \leqq n + 2p - 1$ This proves (B).

When the approximant $P_{m,n}/Q_{m,n} \not\equiv 0$ we have found it convenient to introduce the so-called minimal solution of (3.3) and (3.4) and to determine a point (a,b) in terms of which all the pairs (m',n') can be found for which $P_{m',n'}/Q_{m',n'} = P_{m,n}/Q_{m,n}$. To this end, we give the following definition.

Definition. The minimal solution at (m,n) is the unique solution of (3.3) and (3.4) for which $Q_{m,n} = q_0 + \cdots + q_\ell z^\ell$, $q_\ell = 1$ and ℓ is minimal.

Theorem 15. The minimal solution exists and $P_{m,n}$ and $Q_{m,n}$ have no common factor except possibly a power of z.

Proof: Suppose $(P_{m,n}, Q_{m,n})$ and $(\tilde{P}_{m,n}, \tilde{Q}_{m,n})$ are both minimal solutions then $(P_{m,n} - \tilde{P}_{m,n}, Q_{m,n} - \tilde{Q}_{m,n})$ is also a solution since (3.3) and (3.4) are linear. But $Q_{m,n} - \tilde{Q}_{m,n}$ is of lower degree than ℓ, thus $Q_{m,n} - \tilde{Q}_{m,n} \equiv 0$ which shows that $P_{m,n} \equiv \tilde{P}_{m,n}$ and $Q_{m,n} \equiv \tilde{Q}_{m,n}$.

Suppose $S = az - b$, $ab \neq 0$ is a common factor. Then (3.3) and (3.4) imply

$$(Q_{m,n}/S)L - (P_{m,n}/S) = O(z^r)(-b^{-1})(1 + \frac{a}{b}z + \frac{a^2}{b^2}z^2 + \cdots) = O(z^r) ,$$

$$(Q_{m,n}/S)L^* - (P_{m,n}/S) = O((\tfrac{1}{z})^{s-\max(m,n)})a^{-1}(\tfrac{1}{z})(1 + \frac{b}{a}\frac{1}{z} + \frac{b^2}{a^2}(\tfrac{1}{z})^2 + \cdots)$$

$$= O((\tfrac{1}{z})^{s+1-\max(m,n)}) .$$

Together they show that $(P_{m,n}/S, Q_{m,n}/S)$ is a nontrivial solution where degree $Q_{m,n}/S = \ell - 1$, contradicting the fact that $(P_{m,n}, Q_{m,n})$ is the minimal solution. This proves the theorem.

In order to determine the point (a,b) we first need a point where the minimal solution gives the exact approximation $O(z^r)$ and $O((\frac{1}{z})^{s-\max(m,n)})$.

Theorem 16. If $0 \leqq m - n = \text{even}$, $(P_{m,n}, Q_{m,n})$ is the minimal solution and the approximations in (3.3) and (3.4) are $O(z^{r+u})$ and $O((\frac{1}{z})^{s-m+v})$ exactly then $(z^{u+v}P_{m,n}, z^{u+v}Q_{m,n})$ is the minimal solution at (m',n') where $m' = m + 3u + v$ and $n' = n + u + v$ and the approximations are exactly $O(z^{r'})$ and $O((\frac{1}{z})^{s'-m'})$. A similar statement holds when m and n are interchanged.

<u>Proof:</u> First we note that $n' \leq m'$, $m'-n'$ = even and degree $z^{u+v}P_{m,n} \leq u+v+m \leq m'$, degree $z^{u+v}Q_{m,n} \leq u+v+n = n'$. Then (3.3) and (3.4) imply $(z^{u+v}Q_{m,n})L - (z^{u+v}P_{m,n}) = O(z^{r+2u+v})$ and $(z^{u+v}Q_{m,n})L^* - (z^{u+v}P_{m,n}) = O((\frac{1}{z})^{s-m-u})$. This shows that $(z^{u+v}P_{m,n}, z^{u+v}Q_{m,n})$ is a solution at (m',n') since $r' = [\frac{m'+n'+2}{2}] = r+2u+v$ and $s'-m' = [\frac{m'+n'+1}{2}] - m' = s-m-u$. It also shows that the orders of approximation are exactly $O(z^{r'})$ and $O((\frac{1}{z})^{s'-m'})$. Since $(P_{m,n}, Q_{m,n})$ is the minimal solution we see that $z^{u+v}P_{m,n}$ and $z^{u+v}Q_{m,n}$ cannot have a common factor $S = az-b$, $ab \neq 0$. Also, $(z^{u+v-1}P_{m,n}, z^{u+v-1}Q_{m,n})$ is not a solution at (m',n') since $(z^{u+v-1}Q_{m,n})L - (z^{u+v-1}P_{m,n}) = O(z^{r'-1})$ and not $O(z^{r'})$ as required. This shows that $(z^{u+v}P_{m,n}, z^{u+v}Q_{m,n})$ is the minimal solution at (m',n'). The roles of m and n in this proof may be interchanged. Thus, the theorem is proved.

Without loss of generality we may now assume that $(P_{m,n}, Q_{m,n})$ is the minmal solution, that the approximations in (3.3) and (3.4) are exact and that $P_{m,n} \not\equiv 0$. Consider now the case $0 \leq m-n$ even. We set

$$Q_{m,n} = q_\alpha z^\alpha + \cdots + q_{n-\gamma}z^{n-\gamma}, \quad 0 \leq \alpha \leq n-\gamma \leq n, \quad q_\alpha q_{n-\gamma} \neq 0 .$$

From (3.5) it then easily follows that $p_0 = p_1 = \cdots = p_\alpha = 0$ so that we may write

$$P_{m,n} = p_{\alpha+1}z^{\alpha+1} + \cdots + p_{m-\beta}z^{m-\beta}, \quad \alpha+1 \leq m-\beta \leq m, \quad p_{m-\beta} \neq 0 .$$

We now define a and b by

$$a = m-\alpha-\beta \quad \text{and} \quad b = n-\alpha-\gamma .$$

Since the exact degrees of $z^{-\alpha}P_{m,n}$ and $z^{-\alpha}Q_{m,n}$ are a and b, respectively, and $z^{-\alpha}Q_{m,n}$ has the value $q_\alpha \neq 0$ at $z = 0$ we see that $P_{m,n}/Q_{m,n}$ cannot be a two-point approximant at any point (m',n') where $m' < a$ or $n' < b$.

We investigate for what points (m',n') the pair of polynomials $(z^{-\alpha+\delta}P_{m,n}, z^{-\alpha+\delta}Q_{m,n})$ is a solution of (3.3) and (3.4). The integer δ must be non-negative in order that $z^{-\alpha+\delta}Q_{m,n}$ be a polynomial. We set

$$m' = a+\delta+k \quad \text{and} \quad n' = b+\delta+\ell, \quad 0 \leq k,\ell$$

since degree $z^{-\alpha+\delta}P_{m,n} = a+\delta$ and degree $z^{-\alpha+\delta}Q_{m,n} = b+\delta$. Then $r' = [\frac{m'+n'+2}{2}] = [\frac{m-\alpha-\beta+\delta+k+n-\alpha-\gamma+\delta+\ell+2}{2}] = r-\alpha+\delta+[\frac{-\beta-\gamma+k+\ell}{2}]$ and (3.3) implies $(z^{-\alpha+\delta}Q_{m,n})L - (z^{-\alpha+\delta}P_{m,n}) = O(z^{r-\alpha+\delta})$. From $r' \leq r-\alpha+\delta$ follows $[\frac{-\beta-\gamma+k+\ell}{2}] \leq 0 \Leftrightarrow k+\ell \leq \beta+\gamma+1$. Thus, (m',n') lies in a triangle with vertices $(a+\delta, b+\delta)$, $(a+\delta+\beta+\gamma+1, b+\delta)$ and $(a+\delta, b+\delta+\beta+\gamma+1)$, see Figure 5.

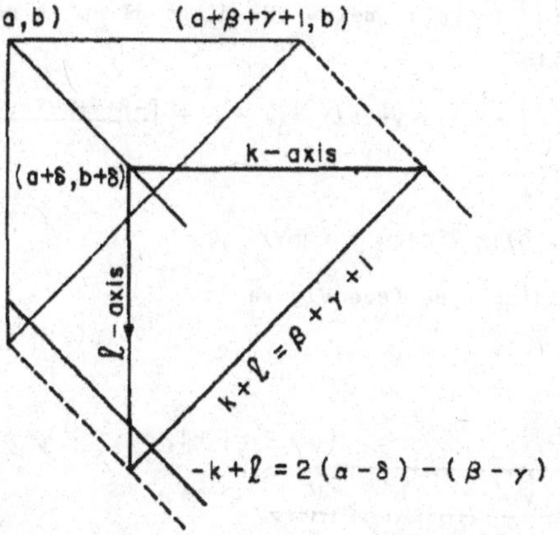

Figure 5

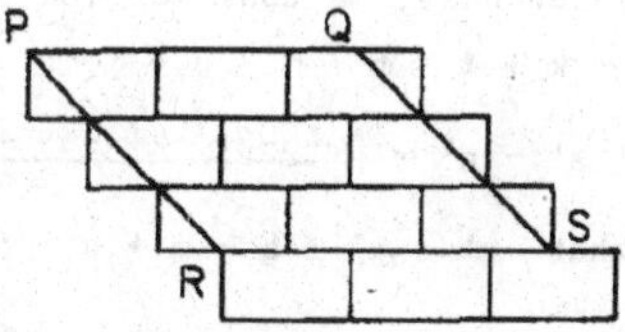

Figure 6

We first restrict our attention to the portion of these triangles which lies above the diagonal $n' = m'$, that is

$$-k + \ell \leqq m - n - \beta + \gamma \ .$$

From (3.4) we get $(z^{-\alpha+\delta}Q_{m,n})L^* - (z^{-\alpha+\delta}P_{m,n}) = O((\frac{1}{z})^{s-m+\alpha-\delta})$ and $s' - m' = \lceil\frac{m' + n' + 1}{2}\rceil - m' = s-m + \lceil\frac{\beta - \gamma - k + \ell + 1}{2}\rceil$. Thus $s' - m' \leqq s - m + \alpha - \delta \Leftrightarrow \lceil\frac{\beta - \gamma - k + \ell + 1}{2}\rceil \leqq \alpha - \delta \Leftrightarrow$

$$-k + \ell \leqq 2(\alpha - \delta) - (\beta - \gamma) \ . \tag{4.1}$$

In the k,ℓ-coordinate system, see Figure 5, this inequality determines a half plane above the bounding line. As δ increases one unit the triangle and the k,ℓ-coordinate system moves one unit down along the diagonal through (a,b) and the bounding line for the half plane moves up two units cutting off a larger part of the next triangle.

The largest value of δ for which both $k + \ell \leqq \beta + \gamma + 1$ and (4.1) is satisfied is $\delta = \alpha + \gamma$ in which case $k = \beta + \gamma$, $\beta + \gamma + 1$ and $\ell = 0$. For any $\delta, 0 \leqq \delta \leqq \alpha + \gamma$ we find $\ell \leqq \alpha - \delta + \gamma$ and $n' \leqq b + \alpha + \gamma = n$. Thus, if $n' \leqq m'$, $P_{m,n}/Q_{m,n}$ appears at all points (m',n') inside the polygon bounded by the lines $n' = b$, $m' - n' = a - b + \beta + \gamma + 1$, $n' = n$ and $m' - n' = a - b - 2\alpha + \beta - \gamma$; that is, the location of the approximants that are equal to $P_{m,n}/Q_{m,n}$ and lie above the diagonal $m' = n'$ form a parallelogram (see Figure 6). The line $m' - n' = a - b - 2\alpha + \beta - \gamma$ cannot meet the vertical line $m' = a$ above the diagonal $m' = n'$ and below a, that is, on the segment between (a,b) and $(a, b+\beta+\gamma+1)$ since we may then choose $k = 0$ and ℓ such that the corresponding entry (m',n') belongs to a "brick" of Theorem 9 which extends to the left of the line $m' = a$ and that would place an approximant equal to $P_{m,n}/Q_{m,n}$ to the left of that line.

Secondly, we consider the points (m',n') below the diagonal $m' = n'$, that is, $-k + \ell \geqq m - n - \beta + \gamma$. We find

$$s' - n' = \left\lceil \frac{m-\alpha-\beta+\delta+k+n-\alpha-\gamma+\delta+\ell+1}{2} \right\rceil - (n-\alpha-\gamma+\delta+\ell) = s - n + \left\lceil \frac{-\beta+\gamma+k-\ell+1}{2} \right\rceil ,$$

so that $s' - n' \leqq s - m + \alpha - \delta \Leftrightarrow$

$$k - \ell \leqq 2(\alpha-\delta) - 2(m-n) + (\beta-\gamma) ,$$

which is a half plane below the bounding line (see Figure 7).

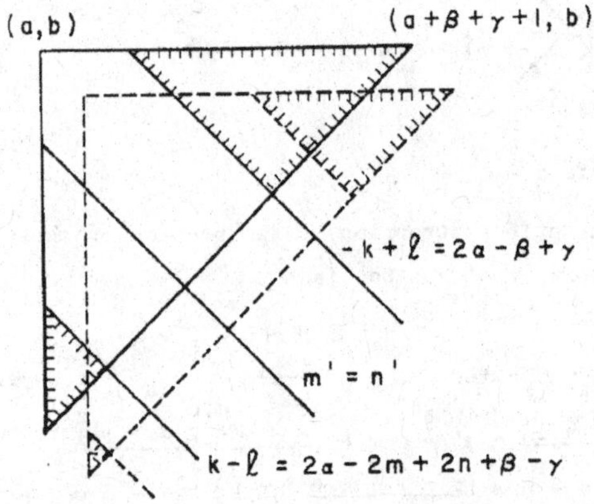

Figure 7

The (m',n') below the diagonal again lie in a parallelogram but of different size than the one above the diagonal unless $a = b$ in which case they lie symmetric about the diagonal $m' = n'$.

Such parallelograms may not appear just anywhere. We refer to Figure 8 where $b < a$. The portion, A, of the triangles which lies between $m' = n'$, $m' = 3n'$ and above $n' = a$; that is, $b \leqq n' \leqq a$ cannot contain any approximant equal to $P_{m,n}/Q_{m,n}$ since the symmetric locations in A' (symmetric with respect to $m' = n'$) lie to the left of the triangles and cannot contain that approximant, contradicting Theorem 12.

The areas B between $m' = n' - 2a$ and $m' = n' + 2a$ and below $n' = a$ cannot contain $P_{m,n}/Q_{m,n}$ either since then there would be such a fraction in A due to the symmetry expressed in Theorem 12 and to the parallelogram shape of the blocks of equal approximants.

The argument is analogous if $a < b$. If $a = b$ then the area A in Figure 8 reduces to a single line segment along the line $n' = a$ and A' lies on, not to the left of, $m' = a$ giving no contradiction. The parallelogram blocks are then placed as in C–C or D–D of Figure 8. Thus, we have the following theorem.

Theorem 17. If $P_{m,n} \neq 0$ then all approximants equal to $P_{m,n}/Q_{m,n}$ appear in one or two parallelogram blocks of contiguous locations. There are three cases.

(A) There is one block of the form PQRS in Figure 8 above $m' = 3n'$ or above $m' = n' + 2a$ or there is one block of the form P'Q'R'S' below $n' = 3m'$ or below $m' = n' - 2a$.

(B) There are two blocks as in (A) one above and one below the diagonal $m' = n'$

(C) There are two blocks, symmetric about $m' = n'$ possibly abutting along $m' = n'$.

As an example, we have computed some of the approximants of the two series

$$L = z + z^2 + \sum_{n=5}^{\infty} z^n \quad \text{and} \quad L^* = 1 \quad .$$

In addition to the equal approximants given in Theorems 9 and 12 there are at least two pairs of blocks of equal approximants outlined by heavy lines in Figure 9 .

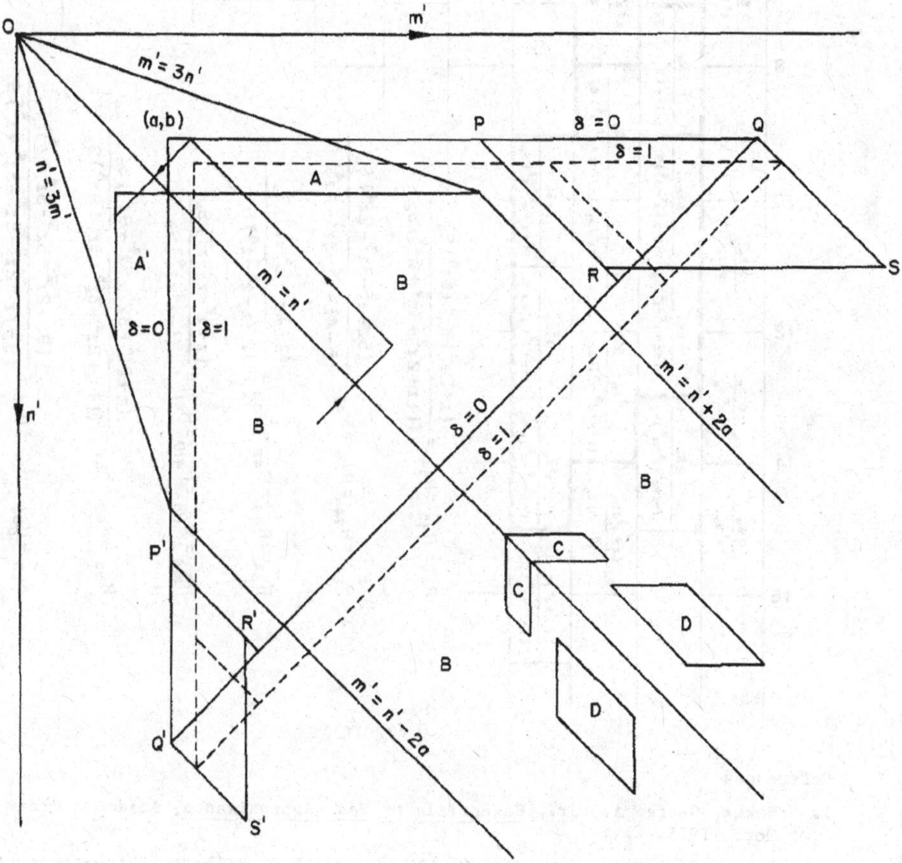

Figure 8

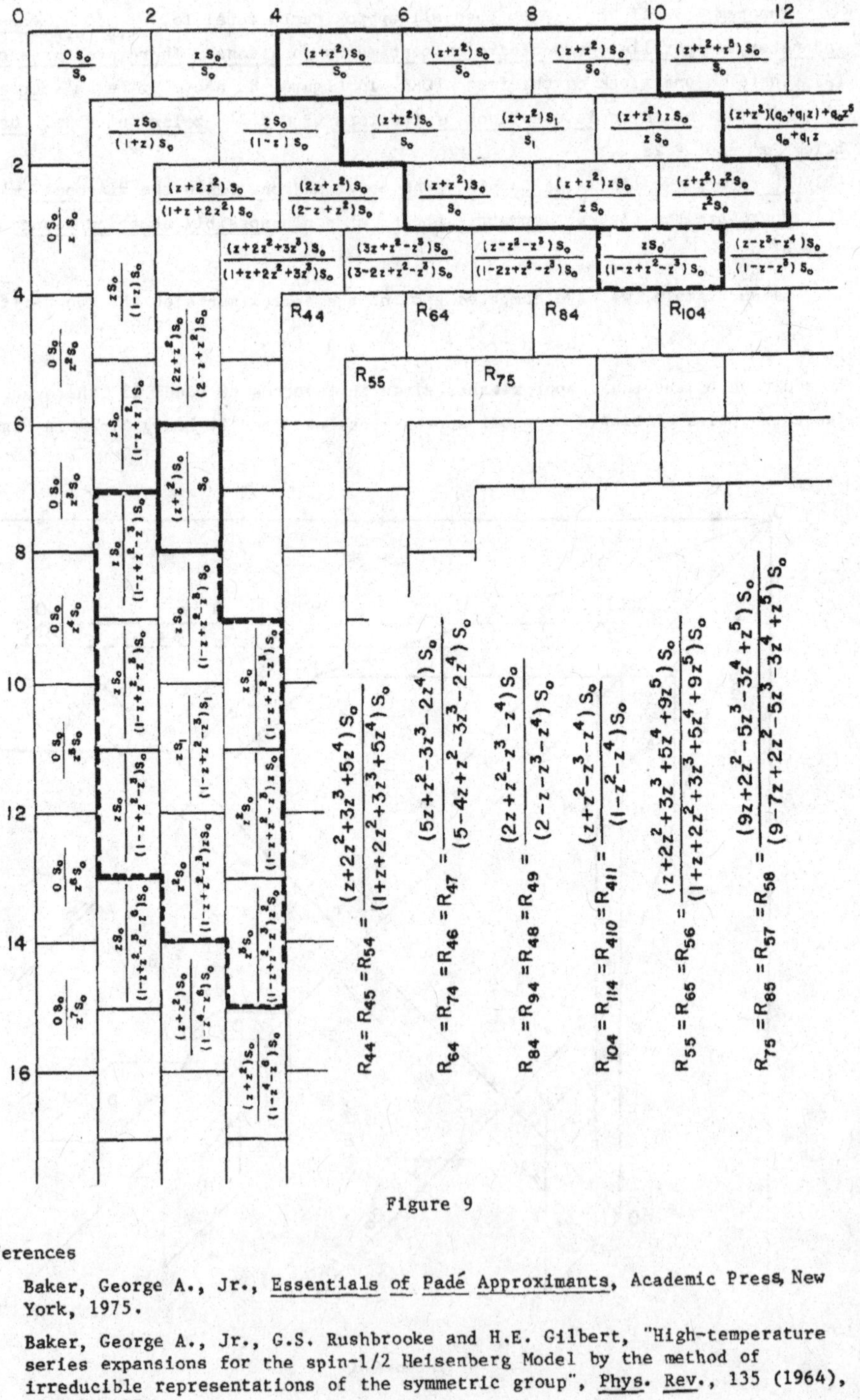

Figure 9

References

1. Baker, George A., Jr., Essentials of Padé Approximants, Academic Press, New York, 1975.

2. Baker, George A., Jr., G.S. Rushbrooke and H.E. Gilbert, "High-temperature series expansions for the spin-1/2 Heisenberg Model by the method of irreducible representations of the symmetric group", Phys. Rev., 135 (1964), A1272-A1277.

3. Jones, William B. and W.J. Thron, "Two-point Padé tables and T-fractions", Bull. Amer. Math. Soc. 83 (1977), 388-390.

4. Jones, William B. and W.J. Thron, Continued Fractions, Analytic Theory and Applications, Encyclopedia of Mathematics and its Applications, 11, Addison-Wesley Pub. Co., Reading, Mass., 1980.

5. Magnus, Arne, "Certain continued fractions associated with the Padé table", Math. Zeit., 78 (1962), 361-374.

6. Magnus, Arne, "On P-expansions of power series", Norske. Vid. Selsk. Skr. (1964), #3, 1-14.

7. Magnus, Arne, "P-fractions and the Padé table", Rocky Mtn. J. Math., 4 (1974), 257-259.

8. McCabe, J.H., "A formal extension of the Padé table to include two-point Padé quotients", J. Inst. Maths. Appl. 15 (1975), 363-372.

9. McCabe, J.H. and J.A. Murphy, "Continued fractions which correspond to power series expansions at two points", J. Inst. Math. Appl., 17 (1976), 233-247.

10. Perron, Oskar, Die Lehre von den Kettenbrüchen, II, B.G. Teubner, Stuttgart, 1957.

11. Avram, Sidi, "Some aspects of two-point Padé approximants", J. Comput. Appl. Math., 6 (1980), 9-17.

12. Thron, W.J., "Some properties of continued fractions $1 + d_0z + K(z/(1+d_nz))$," Bull. Amer. Math. Soc., 54 (1948), 206-218.

13. Thron, W.J., "Two-point Padé table, T-fractions and sequences of Schur", Padé and rational approximation, E.B. Saff and R.S. Varga, eds., Academic Press, New York, (1977), 215-226.

14. Wall, H.S., Analytic Theory of Continued Fractions, Van Nostrand, New York, 1948.

Arne Magnus

Mathematics Department

Colorado State University

Fort Collins, Colorado 80523

A CLASS OF ELEMENT AND VALUE REGIONS FOR CONTINUED FRACTIONS

Marius Overholt

0. Introduction. We shall be dealing with continued fractions

$$\underset{n=1}{\overset{\infty}{K}}\left(\frac{a_n}{1}\right) \quad , \quad a_n \neq 0 \quad . \tag{0.1}$$

See [1] for notation and background. The present paper is based upon my thesis [2] , which originated from a suggestion by W.J. Thron. His idea was to require $|a_n - a| \leq r$, and investigate the location of the approximants

$$f_n = \frac{A_n}{B_n} = \frac{a_1}{1} \underset{+}{} \frac{a_2}{1} + \cdots + \frac{a_n}{1} \tag{0.2}$$

of (0.1) subject to this condition. The expected results were to be used to obtain truncation error estimates for (0.1) . The idea was to some extent motivated by the case where (0.1) is limit-periodic of period 1 , i.e. where $a_n \to a$. In this case given any neighborhood E of a we can find an n such that $a_{n+m} \in E$ for all $m \geq 1$. So we are then seeking information about the location of the approximants of the tail

$$\underset{m=1}{\overset{\infty}{K}}\left(\frac{a_{n+m}}{1}\right) \quad .$$

At least as stated in this form, the question is not tied to closed discs in any essential way. Any other choice of neighborhoods is equally appropriate.

As is usual, a set $E \subset C$ such that $a_n \in E$, $n \geq 1$, shall be called a simple element region for (0.1) . Given a set $E \subset \mathbb{C}$, one may ask for information about the location of the approximants (0.2) of the continued fractions of the form (0.1) having E as a simple element region. The following concept is then essential: A closed set V, $\emptyset \neq V \subset \hat{\mathbb{C}}$, is a value region corresponding to the simple element region E if

$$\frac{E}{1+V} \subset V \quad , \quad E \subset V \quad . \tag{0.3}$$

It is easily seen that if $a_n \in E$, $n \geq 1$, then necessarily $f_n \in V$. See [1] for details, and for a thorough discussion of element and value regions generally. Note that since V is closed we have $f = \lim_{n\to\infty} f_n \in V$ if (0.1) converges. Generally speaking it is very difficult to start with an E and then find a satisfactory V (A set V is satisfactory if it is not too large; $V = \mathbb{C}$ is always a possible choice, but is not satisfactory). In the theory of continued fractions one usually starts by prescribing V and then determines an E belonging to V . For a properly chosen V this approach is often very efficient. Sometimes it is not too difficult to determine explicitly the best possible (i.e. maximal) E corresponding to V . The approach has, however, the drawback that one to a certain extent loses control over E .

Given a simple element region E there exists a best possible V corresponding to E. Putting

$$U_{n+1}(E) = \frac{E}{1+U_n(E)} \quad , \quad U_1(E) = E \quad ,$$

we obtain recursively a sequence $\{U_n(E)\}_{n=1}^{\infty}$ of sets. For each $n \geq 1$ one clearly has

$$\{f_n \in \hat{\mathbb{C}} \mid a_m \in E \ , \ m \geq 1\} = U_n(E) \quad ,$$

and hence

$$U(E) = \overline{\bigcup_{n=1}^{\infty} U_n(E)}$$

is the required best possible V. But even for very simple E the explicit determination of the $U_n(E)$ is usually out of question.

An instructive (but trivial) example, where the approach outlined above actually can be carried out, is the following: Let $0 \leq r \leq R < \infty$, and put $E = [r,R]$, i.e. E is a segment of the non-negative real axis. Clearly $U_n(E) = [r_n,R_n]$ where

$$r_{n+1} = \frac{r}{1+R_n} \quad , \quad R_{n+1} = \frac{R}{1+r_n} \quad ,$$

and $r_1 = r$, $R_1 = R$. So we have

$$U(E) = \overline{\bigcup_{n=1}^{\infty} [r_n,R_n]} \quad .$$

Consequently $U(E) \subset \left[\frac{r}{1+R},R\right]$, and in some cases the inclusion may be replaced by equality.

1. __A useful class of element regions.__ The approach outlined above can not be carried out except in trivial cases. We may, however, modify this approach by trying to determine a sequence of sets $\{U_n^*(E)\}$ satisfying the condition

(1.1) $$U_{n+1}^*(E) \supset \frac{E}{1+U_n^*(E)} \quad , \quad U_1^*(E) = E \quad .$$

It is easily seen that

(1.2) $$U_n(E) \subset U_n^*(E) \quad ,$$

and hence

(1.3) $$U(E) \subset U^*(E) = \overline{\bigcup_{n=1}^{\infty} U_n^*(E)} \quad .$$

To make this approach work, we shall want all the sets $U_n^*(E)$ to be, roughly speaking, sets of the "same type". This is because we form $U_{n+1}^*(E)$ from $U_n^*(E)$ by (1.1). It would clearly simplify matters if we could use the same procedure regardless of the value of n. This would give the process of forming the $U_n^*(E)$ the character of a relatively simple iteration process.

The choice $E = \{z \in \mathbb{C} \mid |z-a| \leq r\}$ is not suitable for the approach outlined above. We shall therefore leave the original choice of closed discs as simple element regions in favor of another class of sets. These are natural generalizations of the segments $[r,R]$ of the positive real axis. We shall use sets of the form

$$E = \{z \in \mathbb{C} \,|\, |\arg z| \leq \Gamma\ ,\ r \leq |z| \leq R\}\ ,\quad 0 \leq \Gamma \leq \pi\ ,$$

as simple element regions. They will be denoted by $S(r,R,-\Gamma,\Gamma)$ and called <u>symmetric segments</u>. They are special cases of general <u>segments</u> $S(r,R,\gamma,\Gamma)$, $\gamma \leq \Gamma$. (The notation is self-explanatory.) The general (unsymmetrical) segments will not be considered to any great extent in this paper. See [2] for details of these.

If S_1 and S_2 are segments, then so are S_1S_2 and S_1^{-1}. One has

(1.4) $$S(r_1,R_1,\gamma_1,\Gamma_1)\cdot S(r_2,R_2,\gamma_2,\Gamma_2) = S(r_1r_2,R_1R_2,\gamma_1+\gamma_2,\Gamma_1+\Gamma_2)$$

(1.5) $$S^{-1}(r_1,R_1,\gamma_1,\Gamma_1) = S(R_1^{-1},r_1^{-1},-\Gamma_1,-\gamma_1)\ .$$

The approximants (0.2) of (0.1) are formed by repeated use of the three operations $z \to 1+z$, $z \to z^{-1}$, $z \to wz$. Segments are preserved under the last two of these. If S is a segment, then $1+S$ is generally not a segment. We can, however, avoid this difficulty by enclosing $1+S$ in a segment. The simplest case is when S is a symmetric segment contained in the half-plane $\mathrm{Re}\, z \geq 0$. We have:

<u>Observation</u> 1. <u>Let</u> S <u>be a symmetric segment contained in the half-plane</u> $\mathrm{Re}\, z \geq 0$. <u>Then</u>

(1.6) $$1+S \subset S(|1+re^{i\Gamma}|,\ 1+R,-\arg(1+Re^{i\Gamma}),\arg(1+Re^{i\Gamma}))\ .$$

A proper verification is trivial, and shall be replaced by an illustration from which the truth of Observation 1 follows immediately.

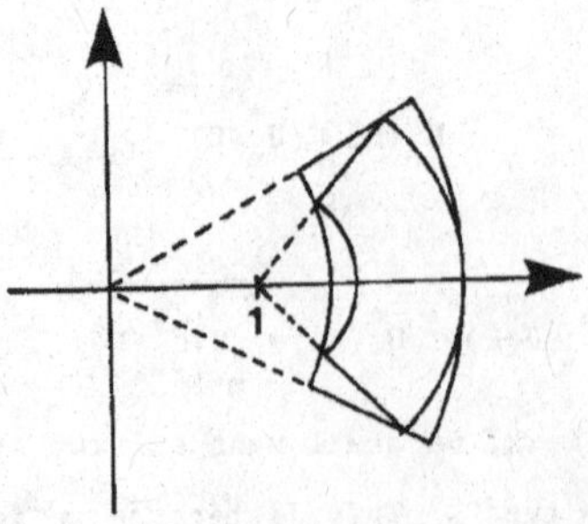

Fig. 1

We furthermore have:

Observation 2. Let S_1 and S_2 be symmetric segments, and let S_2 be contained in $\mathrm{Re}\, z \geq 0$. If

$$r_3 \leq \frac{r_1}{1+R_2} \quad , \quad R_3 \geq \frac{R_1}{\left|1+r_2 e^{i\Gamma_2}\right|} \quad ,$$

$$\Gamma_3 \geq \Gamma_1 + \arg(1+R_2 e^{i\Gamma_2}) \quad ,$$

then

$$S_1/(1+S_2) \subset S(r_3, R_3, -\Gamma_3, \Gamma_3) \quad .$$

Proof. Apply Observation 1 to $1 + S_2$, then use (1.4) and (1.5) .

The above observation makes it possible to determine $U^*_{n+1}(E)$ from $U^*_n(E)$ when E is a symmetrical segment, and all the $U^*_n(E)$ are symmetrical segments contained in $\mathrm{Re}\, z \geq 0$. We have the

Proposition 3. Let S be a symmetrical segment. Put

$$r_{n+1} = \frac{r}{1+R_n} \quad , \quad r_1 = r \quad , \quad R_{n+1} = \frac{R}{\left|1+r_n e^{i\Gamma_n}\right|} \quad , \quad R_1 = R \quad ,$$

$$\Gamma_{n+1} = \Gamma + \arg(1 + R_n e^{i\Gamma_n}) \quad , \quad \Gamma_1 = \Gamma \quad .$$

If $\Gamma_k \leq \frac{\pi}{2}$ for $1 \leq k \leq n$, then

$$U_n(S) \subset S_n \quad .$$

In particular

$$U(S) \subset S(\inf_{n\geq 1} r_n \ , \ \sup_{n\geq 1} R_n \ , \ -\sup_{n\geq 1} \Gamma_n \ , \ \sup_{n\geq 1} \Gamma_n)$$

if $\Gamma_n \leq \frac{\pi}{2}$ for all $n \geq 1$.

The above proposition is an immediate consequence of Observation 2 . It gives a result of the type (1.2) for symmetric segments S that are such that S_n is contained in $\mathrm{Re}\, z \geq 0$ for all $n \geq 1$. A necessary, though by no means sufficient, condition for this is that S itself be contained in $\mathrm{Re}\, z \geq 0$. The second part of the proposition above is not, however, of the type (1.3) , but gives the smallest segment S' , such that $U^*(S) \subset S'$. If $\Gamma_n \leq \frac{\pi}{2}$ for all $n \geq 1$, then it is not difficult to see that

$$\sup_{n\geq 1} R_n = R_1 = R \quad , \quad \inf_{n\geq 1} r_n = r_2 = \frac{r}{1+R} \quad ,$$

in Proposition 3 , so that the first difficulty is to establish conditions to insure $\Gamma_n \leq \frac{\pi}{2}$ for all $n \geq 1$. The second one is to determine $\sup_{n\geq 1} \Gamma_n$. As the Γ_n are given by a coupled set of nonlinear recurrence relations, only partial answers can be expected. We can, however, use Observation 2 to produce a corresponding value region for certain symmetrical segments S . We have:

Proposition 4. Let $0 \leq \Gamma' \leq \frac{\pi}{2}$ and

$$r' = \frac{r}{1+R} \ , \quad R' = R \ , \quad \Gamma' = \Gamma + \arg(1+Re^{i\Gamma'}) \ .$$

Then $S(r',R',-\Gamma',\Gamma')$ is a value region corresponding to $S(r,R,-\Gamma,\Gamma)$.

Clearly $S \subset S'$. Further Observation 2 establishes $S/(1+S') \subset S'$, so the definition of a value region is satisfied.

We need a criterion for $\Gamma' \leq \frac{\pi}{2}$. As $\phi - \arg(1+Re^{i\phi})$ is a strictly increasing function of ϕ in the interval $0 \leq \phi \leq \frac{\pi}{2}$, see Fig. 2 , one easily deduces that a necessary and sufficient condition for the existence of a Γ' with $0 \leq \Gamma' \leq \frac{\pi}{2}$ and

$$\Gamma' = \Gamma + \arg(1+Re^{i\Gamma'})$$

is that

$$\Gamma \leq \frac{\pi}{2} - \arctan R \ ,$$

which is equivalent to the condition

$$R \leq \cot \Gamma \ .$$

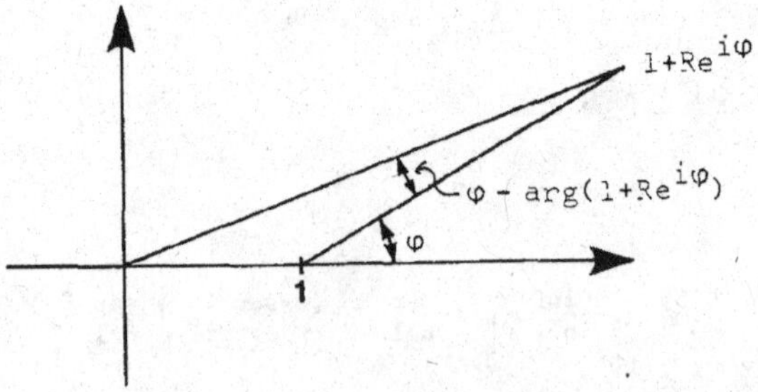

Fig. 2

Proposition 4 makes it possible, for certain segments S , to find a value region S' corresponding to S , or for certain segments S' to find a simple element region S belonging to S' . These results are stated as Theorems 5 and 6 below.

Theorem 5. Let $0 \leq \Gamma \leq \frac{\pi}{2}$ and $0 \leq r \leq R \leq \cot\Gamma$. Let

$$r \leq |a_n| \leq R \ , \quad |\arg a_n| \leq \Gamma \ ,$$

for all $n \geq 1$. Then

$$\frac{r}{1+R} \leq \left| \mathop{K}_{m=1}^{n} \left(\frac{a_n}{1}\right) \right| \leq R \ ,$$

and

$$\left|\arg \mathop{K}_{m=1}^{n} \left(\frac{a_m}{1}\right)\right| \leq \arctan\left(\frac{1+R\sqrt{1+(1-R^2)\tan^2\Gamma}}{1-R^2 \tan^2\Gamma} \tan\Gamma\right)$$

for all $n \geq 1$.

The results for the radii were already established in Proposition 4 ; the result for the angle is obtained by straightforward computation of Γ' from

$$\Gamma' - \arg(1+Re^{i\Gamma'}) = \Gamma \quad .$$

For details, see [2] .

Theorem 6. Let $0 \leq \Gamma' \leq \frac{\pi}{2}$ and $0 \leq r' \leq \frac{R'}{1+R'}$. Let

$$r'(1+R') \leq |a_n| \leq R' \quad ,$$

$$|\arg a_n| \leq \Gamma' - \arctan \frac{R'\sin\Gamma'}{1+R' \cos \Gamma'} \quad ,$$

for all $n \geq 1$. Then

$$r' \leq \left| \mathop{K}_{m=1}^{n} \left(\frac{a_m}{1}\right)\right| \leq R' \quad ,$$

and

$$\left|\arg \mathop{K}_{m=1}^{n} \left(\frac{a_m}{1}\right)\right| \leq \Gamma' \quad ,$$

for all $n \geq 1$.

This is an almost immediate consequence of Proposition 4 . For details see [2].

The results above are in general weaker than what Proposition 3 could give, if it were possible to exploit this result effectively. One may, however, get a certain improvement by combining Proposition 3 and Proposition 4 .

We may use Proposition 3 to determine the S_n up to some n_0 , taking care to check that $\Gamma_n \leq \frac{\pi}{2}$, $1 \leq n \leq n_0$, and then use Proposition 4 to determine an S' , $\Gamma' \leq \frac{\pi}{2}$ such that

$$\frac{S}{1+S'} \subset S' \quad ,$$

$$S_{n_0} \subset S' \quad .$$

Then we have

$$U(S) \subseteq S' \cup \bigcup_{n=1}^{n_0} S$$

Such a result may under certain circumstances be appreciably better than Theorem 5 . See [2] for details.

2. More complicated cases. The case where S is symmetrical and the angle of S is so small that $\Gamma \leq \frac{\pi}{2}$ (Proposition 3) or $\Gamma' \leq \frac{\pi}{2}$ (Proposition 4) is rather special. It is, however, the case that is "nearest to" $S = [r,R]$, and actually degenerates to this case if $\Gamma = 0$. When $S = [r,R]$ we can, essentially, get the best possible value region explicitly. As we move further away from this case,

by increasing the angle and rotating the segment out into the plane (unsymmetrical segment), we get more complicated discussions and less good results. This is because, generally speaking, the smallest segment containing $1+S$ approximates $1+S$ less and less well as S gets further away from the positive real axis. See Fig. 3 and Fig. 4.

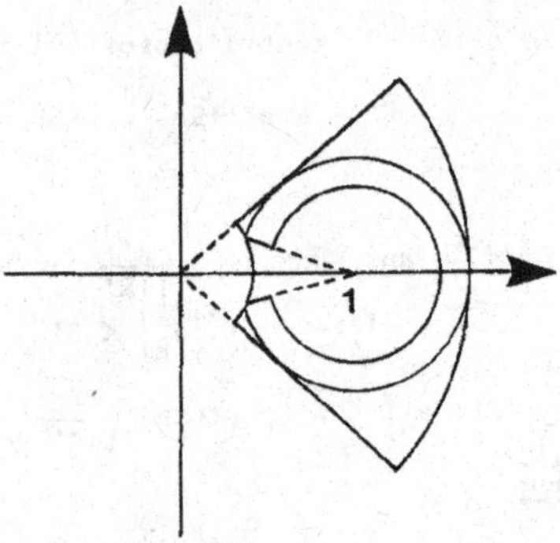

Fig. 3

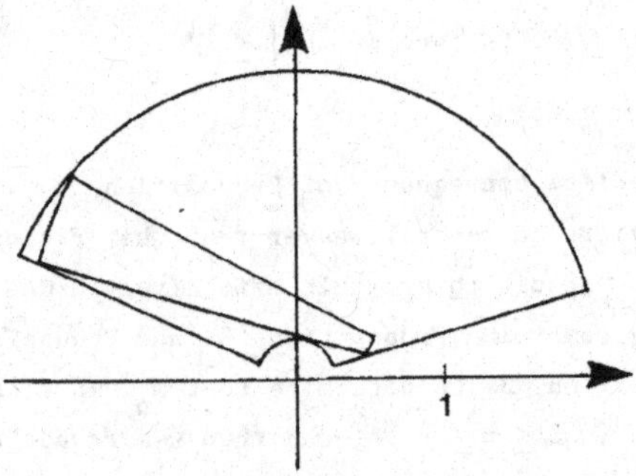

Fig. 4

In Fig. 3 the segment is symmetric, but its angular measure is large. In Fig. 4 the segment has a small angular measure, but it is unsymmetric. The figures illustrate the fact that if the position or shape of S is awkward, then a lot of information is lost in replacing the $1+S$ by the smallest segment containing it.

Analogues of Observation 2 can be given for symmetrical segments with large angular measure, and for general segments. The latter case includes the first as a special case, and the first includes Observation 2 as a special case. See [2] for details. These analogues, especially for general segments, are quite complicated, involving the distinguishing of different cases. They are not as easy to work with as Observation 2, and the results obtained are not so simple. However, useful results may still be obtained, especially if one is willing to start with S' and then determine a simple element region S belonging to S'. As an example we have the following:

Theorem 7. Let $-\frac{\pi}{2} \le \gamma \le 0 \le \Gamma \le \frac{\pi}{2}$ and $0 \le r' \le \frac{R'}{1+R'}$,

$$r'(1+R') \le |a_n| \le R' ,$$

$$\gamma' + \arctan \frac{R'\sin\Gamma'}{1+R'\cos\Gamma'} \le \arg a_n \le \Gamma' + \arctan \frac{R'\sin\gamma'}{1+R'\cos\gamma'} ,$$

for all $n \ge 1$. Then

$$r' \le \left| \overset{n}{\underset{m=1}{K}} \left(\frac{a_m}{1}\right) \right| \le R' , \quad \gamma' \le \arg \overset{n}{\underset{m=1}{K}} \left(\frac{a_m}{1}\right) \le \Gamma' ,$$

for all $n \ge 1$.

Theorem 7 is the simplest theorem for the case when S' is not symmetric. For $\gamma' = -\Gamma'$ we get back Theorem 6 .

3. Another approach. Suppose that we know that V is a value region corresponding to the simple element region E' . Then obviously V is also a value region corresponding to any simple element region E for which $E \subset E'$. So given a simple element region E we can try to enclose it in another simple element region for which we know a corresponding value region. The value region corresponding to E obtained in this way will often be a poor one. The result may, however, immediately be improved by repeated application of the following simple

Proposition 8. Let V be a value region corresponding to the simple element region E . Then

$$\overline{E} \cup \overline{\frac{E}{1+V}} \tag{3.1}$$

is a value region corresponding to E , and if $0 \in V$, then

$$\overline{\frac{E}{1+V}} \tag{3.2}$$

is a value region corresponding to E .

The correctness of the proposition above is immediate by the definition of value region. In [2] the above approach has been used as an alternative to the approach of determining a sequence $\{U_n^*(E)\}$. A natural idea is to use the well-known value result for parabolic simple element regions of the form

$$P(\alpha,p) = \{z \in \mathbb{C} \,|\, |z| - \mathrm{Re}(ze^{-2i\alpha}) \le 2p(\cos\alpha - p)\} , \tag{3.3}$$

where $-\frac{\pi}{2} < \alpha < \frac{\pi}{2}$ and $0 \le p \le \frac{\cos\alpha}{2}$. These element regions are connected with the important parabola theorem, the first version of which was proved by Scott and Wall [3] . For details of the parabola theorem, see [1] . Put

$$H(\alpha,p) = \{z \in \mathbb{C} \mid \mathrm{Re}(ze^{-i\alpha}) \ge -p\} . \tag{3.4}$$

Then the following is true:

Proposition 9. $H(\alpha,p)$ is a value region corresponding to the simple element region $P(\alpha,p)$.

For a proof, see [1] .

Given a segment S we can now try to determine all α,p such that

(3.5) $$S \subset P(\alpha,p) \quad .$$

This will give, by Proposition 9, a family of value regions $H(\alpha,p)$, all corresponding to S. Their intersection is trivially also a value region corresponding to the simple element region S. In this connection we notice that $P(\alpha,p_1) \subset P(\alpha,p_2)$ and $H(\alpha,p_1) \subset H(\alpha,p_2)$ if $p_1 < p_2$. It is then natural first to determine whether there exists an α such that

$$S(r,R,\gamma,\Gamma) \subset P\left(\alpha, \frac{\cos\alpha}{2}\right) \quad .$$

The following way of handling this problem was suggested by W.J. Thron and seems to be the simplest. Put $a_n = c_n^2$ in (0.1), i.e. use the transformation $z = w^2$. The condition (3.6) in the z-plane is then equivalent to the condition

(3.7) $$S(r^{\frac{1}{2}},R^{\frac{1}{2}},\frac{\gamma}{2},\frac{\Gamma}{2}) \subset \{w \in \mathbb{C} \,|\, |\mathrm{Im}(we^{-i\alpha})| \leq \frac{\cos\alpha}{2}\}$$

in the w-plane. The set to the right of the inclusion sign is a closed strip in the w-plane; see Fig. 5. Note that the points $w = \pm\frac{i}{2}$ lie on the boundary of the strip. The result is the following

<u>Proposition</u> 10. <u>Let</u> r, R, γ, Γ <u>be given</u>. <u>There exists an</u> α <u>such that</u> (3.6) <u>is satisfied if and only if</u>

(3.8)
$$R \leq \frac{1}{4} \quad , \quad \text{if} \quad (-\infty,0) \cap S \neq \emptyset \quad ,$$
$$R \leq \frac{1}{4}\,\frac{\cos^2(\frac{\Gamma+\gamma}{2})}{\sin^2(\frac{\Gamma-\gamma}{2})} \quad , \quad \text{if} \quad (-\infty,0) \cap S = \emptyset \quad .$$

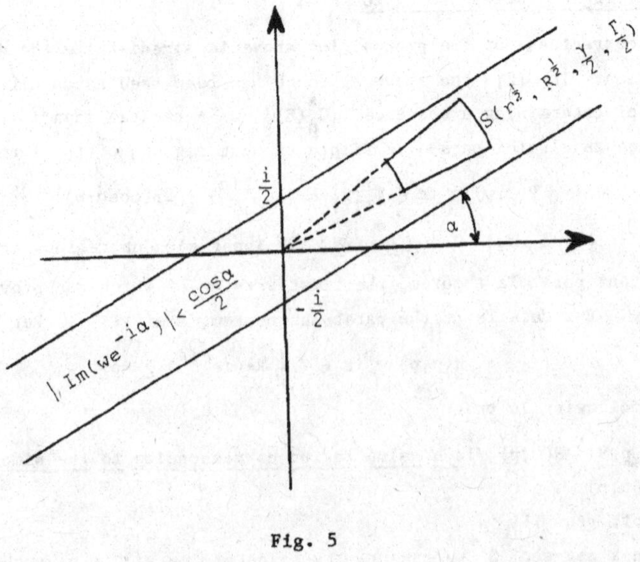

Fig. 5

The proof is based upon simple geometric considerations. For details, see [2] .

The above result furnishes a criterion for a segment to be contained in some parabola $P(\alpha,\frac{\cos\alpha}{2})$. One can also, in the same manner, determine for which α (3.6) is true. For $R > \frac{1}{4}$ the answer is that (3.6) holds for all α with $\alpha_1 \leq \alpha \leq \alpha_2$, where

$$\alpha_1 = \arctan\left(\tan\frac{\Gamma}{2} - \frac{1}{2}R^{-\frac{1}{2}}\sec\frac{\Gamma}{2}\right) ,$$

$$\alpha_2 = \arctan\left(\tan\frac{\Gamma}{2} + \frac{1}{2}R^{-\frac{1}{2}}\sec\frac{\Gamma}{2}\right) .$$

For details, see [2] . It is not difficult to determine the best, i.e. smallest, p for each α . In principle the application of Propositions 9 and then 8 is now straightforward. Unfortunately to obtain good results Proposition 8 may have to be used repeatedly, especially if the segment has a small angular measure. But the results grow very complicated already if Proposition 8 is used twice. For the segment $S(0,\cot\Gamma,-\Gamma,\Gamma)$ with $0 < \Gamma < \frac{\pi}{2}$ Theorem 5 gives the value region $S(0,\cot\Gamma,-\frac{\pi}{2},\frac{\pi}{2})$. What is of interest here are of course the angles, since the result is trivial for the radii. But even a very simple application of the approach outlined in this section will give a result better than this if only Γ is chosen small enough. Since $|z| \leq \cot(\arg z) \Rightarrow |\mathrm{Im}\, z| \leq 1$ it is clear that at least for symmetrical segments the approach using the parabola value result in this case is superior to the approach in Section 2 , except for quite narrow segments.

4. <u>Truncation error estimation</u>. This section will illustrate an application of Theorem 5 in Section 2 . We shall consider (0.1) with

$$a_n \in S(r,R,-\Gamma,\Gamma) \tag{4.1}$$

where $R \leq \cot\Gamma$. We want an estimate for

$$|f - f_n| , \tag{4.2}$$

the truncation error of (0.1) . Here $f = \lim f_n$. The question thus presupposes the convergence of (0.1) . But this convergence will follow from the argument, and will in fact be geometric.

We shall apply a variant of a formula used by Hayden [4] and Roach [5] in truncation error investigations. For references to the literature on truncation error estimates, see [1] . Let B_n be the n-th denominator of (0.1) . Then it is well known that

$$\frac{B_n}{B_{n-1}} = 1 + \frac{a_n}{1} + \cdots + \frac{a_2}{1} , \quad n \geq 1 . \tag{4.3}$$

This is easily proved by induction on n . See [1, p. 71] . Now let A_n denote the n-th numerator of (0.1) , and $S_n(w)$ the n-th linear fractional transformation used in defining (0.1) . See [1, p. 20] for details. Then it is well known that

$$(4.4)\qquad S_n(w) = \frac{A_n + A_{n-1}w}{B_n + B_{n-1}w} .$$

This gives

$$(4.5)\qquad S_n(z) - S_n(w) = \frac{A_n + A_{n-1}z}{B_n + B_{n-1}z} - \frac{A_n + A_{n-1}w}{B_n + B_{n-1}w}$$

$$= \frac{(A_{n-1}B_n - A_nB_{n-1})(z-w)}{(B_n + zB_{n-1})(B_n + wB_{n-1})}$$

$$= \frac{\left(\frac{A_{n-1}}{B_{n-1}} - \frac{A_n}{B_n}\right)(z-w)}{\left(\frac{B_n}{B_{n-1}} + z\right)\left(1 + w\frac{B_{n-1}}{B_n}\right)} .$$

Since $f_{n+1} = S_n(a_{n+1})$, $f_n = S_n(0)$ straightforward computation now gives

$$(4.6)\qquad f_{n+1} - f_n = - \frac{1}{1 + \cfrac{1}{\cfrac{a_{n+1}}{1} + \cdots + \cfrac{a_2}{1}}} (f_n - f_{n-1}) , \quad n \geq 1 .$$

Here we have $f_0 = 0$. This formula gives

$$(4.7)\qquad f_{n+1} - f_n = (-1)^n a_1 \frac{1}{1 + \cfrac{1}{\cfrac{a_{n+1}}{1} + \cdots + \cfrac{a_2}{1}}} \cdots \frac{1}{1 + \cfrac{1}{a_2}} .$$

Hence

$$(4.8)\qquad f = \sum_{n=0}^{\infty} (f_{n+1} - f_n)$$

$$= a_1 \sum_{n=0}^{\infty} (-1)^n \frac{1}{1 + \cfrac{1}{\cfrac{a_{n+1}}{1} + \cdots + \cfrac{a_2}{1}}} \cdots \frac{1}{1 + \cfrac{1}{a_2}} .$$

Now suppose that E is a simple element region for (0.1) , and that V is a value region corresponding to E . If

$$(4.9)\qquad V \subset \{z \in \mathbb{C} \,\big|\, \left|z - \left(\frac{1}{2\varepsilon} - \frac{\frac{3}{2} - \varepsilon}{2 - \varepsilon}\right)\right| \leq \frac{1}{2\varepsilon} + \frac{\frac{3}{2} - \varepsilon}{2 - \varepsilon} - 1\}$$

for some ε , $0 < \varepsilon < 1$, then

$$(4.10)\qquad \frac{1}{1 + \frac{1}{V}} \subset \{z \in \mathbb{C} \,\big|\, |z| \leq 1 - \varepsilon\} .$$

This implies that

$$(4.11)\qquad \left| \frac{1}{1 + \cfrac{1}{\cfrac{a_{n+1}}{1} + \cdots + \cfrac{a_2}{1}}} \right| \leq 1 - \varepsilon ,$$

and hence

(4.12) $$\left|f - f_n\right| \leq \sum_{m=n}^{\infty} \left|f_{m+1} - f_m\right| \leq \left|a_1\right| \sum_{m=n}^{\infty} (1-\varepsilon)^m$$

$$= \frac{\left|a_1\right|}{\varepsilon} (1-\varepsilon)^n \quad .$$

Thus we have geometric convergence. We get in particular the following:

Theorem 11. Let S be symmetric, with $R \leq \cot \Gamma$. Let Γ' be given by Theorem 5 . Then if $a_n \in S$, (0.1) converges and

(4.13) $$\left|f - f_n\right| \leq \frac{R}{1 - \frac{R}{\left|1+Re^{i\Gamma'}\right|}} \left(\frac{R}{\left|1+Re^{i\Gamma'}\right|}\right)^n \quad .$$

References

1. W.B. Jones and W.J. Thron, Continued fractions, Analytic Theory and Applications, ENCYCLOPEDIA OF MATHEMATICS AND ITS APPLICATIONS No. 11 Addison Wesley, Reading, Mass., 1980.

2. Marius Overholt, Om visse simple elementområder for kjedebröker. Thesis (in Norwegian) for the degree cand. real., Universitetet i Trondheim, Matematisk Institutt, NLHT 1981.

3. W.T. Scott and H.S. Wall, A convergence theorem for continued fractions, Trans. Amer. Math. Soc. 47 (1940), 155-172.

4. T.L. Hayden, Continued fraction approximation to functions, Numer. Math. 7 (1965), 292-309.

5. F.A. Roach, Boundedness of value regions and convergence of continued fractions, Proc. Amer. Math. Soc. 62 (1977), 299-304.

Marius Overholt

Matematikk

Rogaland Distriktshögskole

4000 Stavanger

Norway

PARAMETERIZATIONS AND FACTORIZATIONS OF ELEMENT REGIONS FOR CONTINUED FRACTIONS $K(a_n/1)$

Walter M. Reid

1. Introduction. The study of convergence of continued fractions

(1.1a) $$K(a_n/1) = \frac{a_1}{1} + \frac{a_2}{1} + \frac{a_3}{1} + \cdots$$

involves a search for sequences $\{E_n\}$ of convergence regions, that is, regions $E_n \subseteq \mathbb{C}$ for which the condition

(1.1b) $$0 \neq a_n \in E_n \quad , \quad n = 1,2,3,\ldots$$

implies the convergence of the continued fractions to a finite value in $\mathbb{C}$. A fruitful approach in this search, introduced by Lane [7], begins with a second sequence $\{V_n\}$ of closed circular regions in $\hat{\mathbb{C}}$, each containing zero, called value regions, a sequence then of closed disks, closed complements of disks, or closed half-planes. From the sequence of value regions a corresponding sequence of element regions, candidates for convergence regions, is defined as the sequence of sets of those $w \in \mathbb{C}$ which determine for the linear fractional transformations (l.f.t's) which generate the continued fraction a specified mapping property between the value regions, that is,

(1.2) $$E_n = \{w : s(w,V_n) \subseteq V_{n-1}\}$$

where $s(w,v) = \frac{w}{1+v}$ for $v \in V_n$, as in (DN1). This approach has been employed by Jones and Thron [4] and Jones and Snell [2,3] to obtain sequences of element regions and by Jones and Thron [5] to find families of twin-element regins (E_1,E_2), a special and very important instance of a sequence resulting from setting $E_1 = E_{2n-1}$ and $E_2 = E_{2n}$ for twin-value regions $V_0 = V_{2n}$ and $V_1 = V_{2n-1}$.

By defining element regions through (1.2), constraints are placed on the approximants, f_n, of the continued fractions generated by (1.1). Indeed, by (DN4), the nth approximant satisfies

(1.3) $$f_n = S_n(0) \in S_n(V_n)$$

since $0 \in V_n$. Hence from (1.2) and from (DN2b) with $m = 0$,

(1.4) $$S_n(V_n) = S_{n-1}(s(w,V_n)) \subseteq S_{n-1}(V_{n-1}) \subseteq V_0 \quad , \quad n = 1,2,3,\ldots \; .$$

If V_0 is a closed disk, then since each S_n is an l.f.t. and each V_n is a closed circular region, it follows that $\{S_n(V_n)\}$ is a nested sequence of closed disks. Consequently (1.3) implies that

(1.5) $$|f_m - f_n| \leq 2R \quad , \quad m \geq n \; ,$$

where R_n denotes the radius of the disk $S_n(V_n)$. Upper bounds on R_n will provide truncation error estimates for $K(a_n/1)$. The nestedness of $\{S_n(V_n)\}$ implies that $\lim R_n = R \geq 0$. If R can be shown to equal 0 , then the continued fraction converges and the element regions are established as convergence regions. In case $K(a_n/1)$ is known to converge, tight bounds on R_n may reveal the rate of convergence. In addition, the issue of uniform convergence may be considered.

Thron [11] simplified the study of R_n to consideration of a function of two variables, a_n and the independent variable h_{n-1} defined in (DN10) . Use of (1.5) in addressing the issues of convergence depends upon having parameterizations of or estimates for a_n and h_{n-1} ; therein lies an inherent weakness of this method. In many cases E_n as defined in (1.2) is determined by implicit conditions on w rather than explicit parameterizations. On the other hand estimates for the h_{n-1} are gained only by rather involved arguments using (DN10) . Determination of parameterizations for $w \in E_n$, then, has received considerable attention in this analysis.

In the special case with disk and complement of disk as twin-value regions, Thron [11] showed directly, using (1.5) , that for $r > 1$ the twin-element regions

(1.6) $$E_1 = \{w : |w| \geq 2(r-\cos(\arg w))\} \quad , \quad E_2 = \{w : 0 < |w| \leq r\}$$

are twin-convergence regions. In this case the parameterizations for the a_n are apparent. He also proved that convergence is not uniform over the entire convergence regions, but that it is uniform if a_{2n} is restricted to suitable subsets, called contractions, of E_2 .

In cases where each pair of value regions in a sequence includes a half-plane, the parameterizations of a_n are transparent. Indeed, when the value regions are all half-planes or half-planes alternating with either disks or complements of disks, Jones and Snell [2,3] and Jones and Thron [5, Theorem 3.2] identified the element regions as having parabolas or alternating ellipses and branches of hyperbolas, respectively, as boundaries. In [2] , the parameterization of a_n was applied to show that $R = 0$ thus proving the element regions to be convergence regions.

Except in certain cases, however, when the value regions are strictly disks and complements of disks, the element region parameterizatins were not clear. Consequently, as in Jones and Thron [4,5 (Theorems 5.1 and 5.4)] , the element regions' descriptions remained as implicit conditions on $w \in E_n$.

One of those exceptional cases, however, was studied by Lange and Thron [9] and later more generally by Lange [8] , a case of twin element regions (E_1,E_2) which factored; that is, for $c_n^2 = a_n$, c_n satisfied

(1.7) $$|c_{2n}+i\Gamma| \leq \rho \quad , \quad |c_{2n}-i\Gamma| \leq \rho$$
$$|c_{2n-1}+i(1+\Gamma)| \geq \rho \quad , \quad |c_{2n-1}-i(1+\Gamma)| \geq \rho$$

for $\Gamma \in \mathbb{C}$ and $|\Gamma| < \rho < |1+\Gamma|$. Here tight estimates for h_{n-1} were derived and used to prove convergence by showing that $R = 0$. In fact since specific values of a_n were not considered here, a stronger convergence result was proved. The convergence proved to be uniform over the entire element regions. On the other hand, in [8] , other issues of convergence were not addressed. The bounds on R display no sensitivity to precise information about specific continued fractions when a_n is known, provide less useful truncation error estimates, and reveal less about the rate of convergence. For these considerations, parameterization of the a_n is needed.

But apart from those exceptional cases, when the value regions are strictly disks and complements of disks, neither parameterizations of the a_n nor tight estimates of the h_{n-1} were known. Granting then the possibility that $\lim R_n = R > 0$, Jones and Thron nevertheless succeeded in proving convergence by a powerful indirect argument for several families of both sequences of element regions and twin-element regions, the latter subsuming the earlier result (1.6) and (1.7) . They were even able to relax the value region postulate of $0 \in V_n$ in [5] and considered only pre-value regions. These indirect arguments were applied again by Jones and Snell [3] in cases where half-planes appeared among the value regions yielding the element regions mentioned above having conic sections as boundaries. But while these indirect arguments did yield convergence regions, they did not address the issues of uniform convergence, truncation error estimates, and rates of convergence. For these considerations, again, either parameterization of the a_n or factorization of E_n for parameterization of $c_n^2 = a_n$ to estimate h_{n-1} is needed.

This article furnishes in Theorems 1 and 2 the complete parameterizations and factorizations, respectively, of the element regions derived from the general case involving value regions of disks and complements of disks. Consequently parameterizations are offered for elements of the twin-convergence regions of Jones and Thron [5, Theorems 5.1 and 5.4] , which in turn subsume the earlier case (1.6) of Thron. These parameterizations have been applied by Jones and Reid [These Proceedings] in gaining truncation error estimates and uniform convergence results for $K(a_n/1)$ over contractions of the larger family of twin-convergence regions, thus extending the results of Thron [11] . The parameterizations of Theorem 1 also apply to the case of the continued fractions $K(a_n/b_n)$ studied by Jones and Thron [4, Lemma 2.1] . On the other hand, Theorem 2 presents factorizations for the factorable element regions corresponding to value regions, two cases giving the earlier result (1.7) of Lange as a special case, and the third giving a different form. Also necessary and sufficient conditions for factorization of this kind are given. It is hoped that these factorizations will aid in extending the arguments of Lange and Thron to yield the stronger uniform convergence results for larger families of convergence regions.

To simplify the context and notation we limit the setting to two value regions (V_1,V_0) and the single element region E. Let

(1.8a) $$c_0,c_1 \in \mathbb{C} \text{ with } |c_1| \neq |1+c_1| \text{ , and } r_0,r_1 > 0 \text{ .}$$

For $i = 0,1$ let V_i take the form either of a

(1.8b) $$\text{closed disk } (D_i): \quad V_i = \{z : |z-c_i| \leq r_i \text{ , } |c_i| \leq r_i\}$$

or

(1.8c) $$\text{closed complement of disk } (cD_i): \quad V_i = \{z : |z-c_i| \geq r_i \text{ , } r_i \leq |c_i|\} \text{ .}$$

In case either of disk or complement of disk, we will call c_i the <u>center</u> of V_i and r_i the <u>radius</u> of V_i. Then corresponding to (V_1,V_0) is the element region

(1.9) $$E = \{w : s(w,V_1) \subseteq V_0)\} \text{ .}$$

Because of (1.1b) we shall be concerned that E contain nonzero elements.

For sets A and B we will use the following notation:

$\mathrm{Int}(A)$ for the interior of A,

$\partial(A)$ for the boundary of A, and

$B-A$ for the complement of A with respect to B.

2. <u>Parameterizations of Element Regions.</u>

<u>Theorem 1. For r_0,r_1,c_0,c_1 and value regions (V_0,V_1) as in (1.8) taking the specific forms indicated below, E is parameterized as follows:</u>

<u>If</u> $|1+c_1| < r_1 \leq |c_1|$ <u>for</u> $(-1 \notin cD_1,V_0)$ <u>or</u> $|c_1| \leq r_1 < |1+c_1|$ <u>with</u> $|c_0| < r_0$ <u>for</u> $(-1 \notin D_1, 0 \in \mathrm{Int}\, D_0)$ <u>then</u>

(2.1) $$E = \{w = |w|e^{i\theta} : 0 \leq |w| \leq B_1(\theta) \text{ , } \theta \in \Omega\} \text{ .}$$

<u>If</u> $r_1 < |1+c_1|$, $r_1 \leq |c_1|$, $r_0 < |c_0|$, <u>and</u> $\dfrac{r_0}{|c_0|} < \dfrac{r_1}{|1+c_1|}$ <u>for</u> $(-1 \in \mathrm{Int}(cD_1), cD_0)$ <u>then</u>

(2.2) $$E = \{w = |w|e^{i\theta} : B_{-1}(\theta) \leq |w| \leq B_1(\theta), \theta \in \Omega\} \text{ .}$$

<u>If</u> $|c_1| \leq r_1$, $|1+c_1| \leq r_1$, <u>and</u> $r_0 \leq |c_0|$ <u>for</u> $(-1 \in D_1, cD_0)$ <u>then</u>

(2.3) $$E = \{w = |w|e^{i\theta} : B_1(\theta) \leq |w|\} \text{ .}$$

<u>If</u> $|c_1| \leq r_1 < |1+c_1|$ <u>and</u> $r_0 \leq |c_0|$ <u>for</u> $(-1 \notin D_1, cD_0)$ <u>then</u>

(2.4) $$E = \mathbb{C} - \{w = |w|e^{i\theta} : B_1(\theta) < |w| < B_{-1}(\theta) \text{ , } \theta \in \Omega\} \text{ ,}$$

<u>where</u>

(2.5) $$B_{\pm 1}(v) = t(\theta) \pm \sigma_0\sigma_1\sqrt{t^2(\theta)-k} \text{ ,}$$

(2.6a) $$\Omega = \begin{cases} [-\pi,\pi] \ \underline{\text{unless}} \\ c_0(1+c_1) \neq 0 \ , \ r_1 \leq |1+\Gamma_1| \ , \ \underline{\text{and}} \ r_0 \leq |c_0| \ \underline{\text{in which case}} \\ \{\theta : |\theta-\phi| \leq \cos^{-1}\left(\frac{-\sigma_0\sigma_1 r_0 r_1 - \sqrt{k}}{|c_0||1+c_1|}\right)\} \ , \end{cases}$$ (2.6b)

(2.7) $$t(\theta) = -\sigma_0\sigma_1 r_0 r_1 - |c_0||1+c_1|\cos(\theta-\phi) \ ,$$

(2.8) $$\phi = \arg((-c_0)(1+c_1)) \ ,$$

(2.9) $$k = (r_0^2-|c_0|^2)(r_1^2-|1+c_1|^2) \ , \ \underline{\text{and}}$$

(2.10a) for $i = 0,1$

$$\sigma_i = \begin{cases} 1 & \underline{\text{if}} \ V_i \ \underline{\text{is a disk,}} \ |c_i| \leq r_i \\ -1 & \underline{\text{if}} \ V_i \ \underline{\text{is the complement of a disk,}} \ r_i \leq |c_i| \ . \end{cases}$$

Further, let

(2.10b) $$\sigma_2 = \operatorname{sgn}(r_1^2-|1+c_1|^2) \ .$$

Remarks. 1) In relating the influence of the σ's, notice that $\sigma_1 = \sigma_2$ means $-1 \in \operatorname{Int}(V_1)$, $\sigma_2 = 0$ means $-1 \in \partial(V_1)$, and $\sigma_1 = -\sigma_2$ means $-1 \notin V_1$.

2) The conditions $0 \in \operatorname{Int}(D_0)$ for (2.1) and $\frac{r_0}{|c_0|} < \frac{r_1}{|1+c_1|}$ for (2.2) are necessary to insure that $E - \{0\} \neq \emptyset$.

3) In the case of (2.1) where V_0 is a disk and $-1 \notin V_1$, E is a convex set as will be proved following (2.27c) .

4) All element regions in Theorem 1 are symmetric about the axis $\theta = \phi$. When the regions are convex, the extreme values of $\partial(E)$ occur on that axis.

Corollaries. We now relate the parameterizations of Theorem 1 to element regions which are known to be convergence regions.

A. Sequences. For sequences $\{E_n^*\}$ of convergence regions defined by (1.2) for sequences $\{V_n^*\}$ of value regions where each V_n^* has center $\Gamma_n \in \mathbb{C}$ with $|\Gamma_n| \neq |1+\Gamma_n|$ and radius ρ_n , we assign $E_n^* = E$ for $V_1 = V_n^*$, $V_0 = V_{n-1}^*$ with $c_1 = \Gamma_n$, $r_1 = \rho_n$, $c_0 = \Gamma_{n-1}$, $r_0 = \rho_{n-1}$, and $\phi_n = \arg[(-\Gamma_{n-1})(1+\Gamma_n)]$.

1. [Jones and Snell, 3, Theorem 3.1]. (Disks and Complements of Disks). When $|\Gamma_{n-1}| < \rho_{n-1} < |1+\Gamma_{n-1}|$ or when $|1+\Gamma_i| < \rho_i < |\Gamma_i|$ for $i = n, n-1$, then $E_n^* = E$ as in (2.1) with $\Omega_n = [-\pi,\pi]$. On the other hand if $|\Gamma_n| < \rho_n < |1+\Gamma_n|$ while $|1+\Gamma_{n-1}| < \rho_{n-1} < |\Gamma_{n-1}|$, then $E_n^* = E$ as in (2.4) with $\sigma_0\sigma_1 = -1$ and Ω_n as in (2.6b) .

2. [Jones and Thron, 4, Lemma 2.1]. (Disks). Here the continued fraction $K(a_n/b_n)$ is considered with $|\Gamma_n| < \rho_n < |1+\Gamma_n|$. Consequently $E_n^* = E$ as in (2.1) with $\sigma_0\sigma_1 = 1$ and $\Omega_n = [-\pi,\pi]$ upon replacing $1+c_1$ with $b_n + \Gamma_n$ and $1+\bar{c}_1$ with $\bar{b}_n + \bar{\Gamma}_n$. This parameterization, of course, then depends on b_n .

B. Twin-Convergence Regions. For twin-convergence regions (E_1^*, E_2^*) corresponding to the twin-value regions (V_1^*, V_2^*) with centers Γ_1, Γ_2, and radii ρ_1, ρ_2, respectively, where (E_1^*, E_2^*) are defined by

(2.11) $$E_1^* = \{w : s(w, V_1^*) \subseteq V_2^*\} \quad , \quad E_2^* = \{w : s(w, V_2^*) \subseteq V_1^*\} \quad ,$$

we assign

(2.12a) $E_1^* = E$ with $c_1 = \Gamma_1$, $r_1 = \rho_1$, $c_0 = \Gamma_2$, $r_0 = \rho_2$, so that

(2.12b) $\phi_1 = \arg[(-\Gamma_2)(1+\Gamma_1)]$, $k_1 = (\rho_2^2 - |\Gamma_2|^2)(\rho_1^2 - |1+\Gamma_1|^2)$, and

(2.13a) $E_2^* = E$ with $c_1 = \Gamma_2$, $r_1 = \rho_2$, $c_0 = \Gamma_1$, $r_0 = \rho_1$, so that

(2.13b) $\phi_2 = \arg[(-\Gamma_1)(1+\Gamma_2)]$, $k_2 = (\rho_1^2 - |\Gamma_1|^2)(\rho_2^2 - |1+\Gamma_2|^2)$.

The parameterizations apply to regions determined by Jones and Thron [5] . They considered a more general setting of pre-value regions, relaxing the condition that $0 \in V_1^* \cap V_2^*$. The results offered here consider only the convergence regions obtained from value regions.

1. [Jones and Thron, 5, Theorem 5.1] (Disk-Disk). When $|\Gamma_1| < \rho_1 < |1+\Gamma_1|$ and $|\Gamma_2| \leq \rho_2 < |1+\Gamma_2|$, then $\sigma_0\sigma_1 = 1$ and $E_1^* = E$ and $E_2^* = E$ both as in (2.1) with $\Omega_1 = \Omega_2 = [-\pi, \pi]$ under assignments (2.12) and (2.13) , respectively.

2. [Jones and Thron, 5, Theorem 5.4]. (Disk-Complement of Disk). When

(2.14) $$V_1^* = \{z : |z-\Gamma_1| \leq \rho_1\} \text{ with } |\Gamma_1| < \rho_1 \neq |1+\Gamma_1| \quad ,$$

(2.15) $$V_2^* = \{z : |z-\Gamma_2| \geq \rho_2\} \text{ with } |1+\Gamma_2| < \rho_2 \leq |\Gamma_2| ,$$

then, with $\sigma_0\sigma_1 = -1$,

(2.16) $$E_1^* = E \text{ as in } (2.4) \text{ if } |\Gamma_1| < \rho_1 < |1+\Gamma_1| \text{ and}$$

(2.17) $$E_1^* = E \text{ as in } (2.3) \text{ if } |\Gamma_1| < |1+\Gamma_1| < \rho_1$$

while in either case

(2.18) $$E_2^* = E \text{ as in } (2.1) \text{ with } \Omega_2 = [-\pi, \pi] \quad .$$

When Γ_1 and Γ_2 are related to each other through $\Gamma \in \mathbb{C}$ in the manner

(2.19) $$\Gamma_1 = -(1+\Gamma_2) = \Gamma \quad ,$$

then, from (2.12, 13) ,

(2.20) $E_1^* = E$ for $c_1 = \Gamma$, $r_1 = \rho_1$, $c_0 = -(1+\Gamma)$, $r_0 = \rho_2$ so that

(2.21) $\phi_1 = \arg(1+\Gamma)^2 = 2\arg(1+\Gamma)$, $k_1 = (\rho_2^2 - |1+\Gamma|^2)(\rho_1^2 - |1+\Gamma|^2)$,

and

(2.22) $E_2^* = E$ for $c_1 = -(1+\Gamma)$, $r_1 = \rho_2$, $c_0 = \Gamma$, $r_0 = \rho_1$ so that

(2.23) $\qquad \psi_2 = \arg(-\Gamma)^2 = 2\arg\Gamma \quad , \quad k_2 = (\rho_1^2-|\Gamma|^2)(\rho_2^2-|\Gamma|^2) \quad ,$

and we obtain two particularly important special cases of twin-convergence regions for disk-complement of disk value regions.

3. [Lange, 8]. In the case of (2.19) with

$$|\Gamma| < \rho < |1+\Gamma| \tag{2.24}$$

and $\rho_1 = \rho = \rho_2$, then $\sigma_0\sigma_1 = -1$ and from (2.16) , (2.20) , and (2.21) , $E_1^* = E$ as in (2.4) with $k_1 = (\rho^2-|1+\Gamma|^2)$ while from (2.18) , (2.22) , and (2.23) , $E^* = E$ as in (2.1) with $k = (\rho^2-|\Gamma|^2)$. These are the twin convergence regions, with indices interchanged, whose factorizations appear in (1.7) . See Corollary 2 in Section 3 .

4. [Jones and Thron, 5, Corollary 5.7] . In the case of (2.19) with

$$|\Gamma| < |1+\Gamma| < \rho \tag{2.25}$$

and $\rho_1 = \rho$, $\rho_2 = |1+\Gamma|$, then $\sigma_0\sigma_1 = -1$ and $E_1^* = E$ as in (2.3) with (2.20) and (2.21) where $k_1 = 0$, while $E_2^* = E$ as in (2.1) with assignments (2.22) and (2.23) where $k_2 = (\rho^2-|\Gamma|^2)(|1+\Gamma|^2-|\Gamma|^2)$. For an application of these parameterizations in the study of uniform convergence, see Jones and Reid [1] . Actually in [1] the regions E_1 and E_2 are interchanged. These parameterizations reduce to those of Thron [1.6] when $\Gamma = 0$.

<u>Proof of Theorem 1.</u>

The proof divides into two parts. Part I considers the case where $\sigma_2 = 0$ and $-1 \in \partial V_1$. The remainder of the proof, Part II, studies the several cases when $\sigma_2 \neq 0$; that is, when $-1 \notin \partial V_1$.

Part I. $(-1 \in \partial V_1)$. In this part, $r_1 = |1+c_1| > 0$ and $\sigma_2 = 0$. Since $-1 \in \partial V_1$ and $s(w,-1) = \infty$ for $w \neq 0$, V_0 must be the complement of a disk.

a) $(V_1,V_0) = (-1 \in \partial(D_1), cD_0)$. In this case, $-\sigma_0 = \sigma_1 = 1$. It can easily be shown that for $w = |w|e^{i\theta}$,

$$s(w,V_1) = \{z : \mathrm{Re}(ze^{i(\arg(1+c_1)-\theta)}) \geq \frac{|w|}{2|1+c_1|}\} \quad ,$$

a half-plane. Consequently $s(w,V_1) \subseteq V_0 = \{z : |z-c_0| \geq r_0\}$ if and only if

$$\mathrm{Re}[c_0e^{i(\arg(1+c_1)-\theta)}] + r_0 \leq \frac{|w|}{2|1+c_1|}$$

which is equivalent to $|w| \geq 2|1+c_1|(r_0-|c_0|\cos(\theta-\psi)) = 2t(\theta)$ for ψ and $t(\theta)$ as defined by (2.8) and (2.7) , respectively. Since $r_1 = |1+c_1|$, $k = 0$ and since $\sigma_0\sigma_1 = -1$, $2t(\theta) = B_{-1}(\theta)$. Thus $w \in E$ if and only if (2.3) holds.

b) $(V_1,V_0) = (-1 \in \partial(cD_1), cD_0)$. Here $-\sigma_0 = -\sigma_1 = 1$. Again a half-plane appears

$$s(w,V_1) = \{z : \mathrm{Re}(ze^{i(\arg(1+c_1)-\theta)}) \le \frac{|w|}{2|1+c_1|}\} ,$$

so that $s(w,V_1) \subseteq V_0 = \{z : |z-c_0| \ge r_0\}$ if and only if

$$\mathrm{Re}[c_0 e^{i(\arg(1+c_1)-\theta)}] - r_0 \ge \frac{|w|}{2|1+c_1|} ,$$

which is equivalent to

$$0 \le |w| \le 2|1+c_1|[-r_0-|c_0|\cos(\theta-\phi)] = 2t(\theta)$$

which is of the form (2.1) with $\sigma_0\sigma_1 = 1$, $r_1 = |1+c_1|$, and $k = 0$, when $t(\theta) \ge 0$, as is the case when $|\theta-\phi| \le \cos^{-1}(-\frac{r_0}{|c_0|})$, that is, when $\theta \in \Omega$. Ω , in turn, contains nonzero elements if and only if $r_0 < |c_0|$; that is, if and only if, $0 \in \mathrm{Int}(V_0)$.

Part II. In case $r_1 \ne |1+c_1|$ and $-1 \notin \partial(V_1)$, then $\sigma_2 = \pm 1$. The arguments of Jones and Thron [5, Lemma 5.5] yield the results

$$s(w,V_1) = \begin{cases} \{z ; |z-D| \le q\} & \text{if } \sigma_1\sigma_2 = -1 \\ \{z : |z-D| \ge q\} & \text{if } \sigma_1\sigma_2 = 1 \end{cases}$$

where $D = \dfrac{w(1+\bar{c}_1)}{r_1^2-|1+c_1|^2}$ and $q = \dfrac{r_1|w|}{|r_1^2-|1+c_1|^2|}$. Consequently $s(w,V\ )\subseteq V$ if and only if

$$-(q + |D+c_0|) + r_0 \ge 0 \quad \text{when } \sigma_0 = 1 \ , \ \sigma_1\sigma_2 = -1 \ ,$$

$$(q - |D+c_0|) - r_0 \ge 0 \quad \text{when } \sigma_0 = -1 \ , \ \sigma_1\sigma_2 = 1 \ , \text{ and}$$

$$-(q - |D+c_0|) - r_0 \ge 0 \quad \text{when } \sigma_0 = -1 \ , \ \sigma_1\sigma_2 = -1 \ .$$

These cases may be summarized in the single relation

$$\sigma_1\sigma_2(q + \sigma_0|D+c_0|) + \sigma_0 r_0 > 0$$

or equivalently

$$(2.26) \qquad \sigma_1\sigma_2|w|r_1 + \sigma_0 r_0|r_1^2-|1+c_1|^2| + \sigma_0\sigma_1\sigma_2|w(1+\bar{c}_1) + c_0(r_1^2-|1+c_1|^2)| \ge 0 \ .$$

Hence $w \in E$ as in (1.9) if and only if w satisfies (2.26) . The possibilities for the product $\sigma_0\sigma_1\sigma_2$ are either

(2.27a) $\qquad \sigma_0\sigma_1\sigma_2 = -1$ or

(2.27b) $\qquad \sigma_0\sigma_1\sigma_2 = 1$ with $\sigma_0 = \sigma_1\sigma_2 = -1$.

Notice that the case

(2.27c) $\qquad \sigma_0 = \sigma_1\sigma_2 = 1$ is disallowed since if $\sigma_1\sigma_2 = 1$, that is, if $-1 \in V_1$, then $s(w,-1) = \infty$ for $w \neq 0$ and $s(w,V_1) \subset V_0$ imply that V_0 must be the complement of a disk. Hence $\sigma_1\sigma_2 = 1$ implies $\sigma_0 = -1$.

We can verify remark (3) at this point: if V_0 is a disk and $-1 \notin V_1$ that is, if $\sigma_0 = 1$ and $\sigma_1\sigma_2 = -1$, then E is a convex set. Let $w_1, w_2 \in E$ and $w_3 = \alpha w_1 + (1-\alpha)w_2$ for $0 < \alpha < 1$. Let $A = (1+\bar{c}_1)$, $B = c_0(r_1^2 - |1+c_1|^2)$, $C = r_0\left|r_1^2 - |1+c_1|^2\right|$. Then from (2.26) $w_1, w_2, w_3 \in E$ if and only if they each satisfy

$$(2.28) \qquad |w|r_1 + |wA+B| \leq C$$

and E is convex if and only if $w_1, w_2 \in E$ implies $w_3 \in E$. Now

$$\begin{aligned} |w_3|r_1 + |w_3A+B| &= |\alpha w_1 + (1-\alpha)w_2|r_1 + |(\alpha w_1 + (1-\alpha)w_2)A+B| \\ &= |\alpha w_1 + (1-\alpha)w_2|r_1 + |\alpha(w_1A+B) + (1-\alpha)(w_2A+B)| \\ &\leq \alpha(|w_1|r_1 + |w_1A+B|) + (1-\alpha)(|w_2|r_1 + |w_2A+B|) \\ &\leq \alpha C + (1-\alpha)C = C \quad . \end{aligned}$$

The last inequality is justified since $w_1, w_2 \in E$. Hence $w_3 \in E$ and E is convex. Now we continue with the cases (2.27a,b) . If $\sigma_0\sigma_1\sigma_2 = -1$, then (2.26) is equivalent to

$$(2.29) \qquad \sigma_1\sigma_2|w|r_1 + \sigma_0 r_0\left|r_1^2 - |1+c_1|^2\right| \geq \left|w(1+\bar{c}_1) + c_0(r_1^2 - |1+c_1|^2)\right| \quad .$$

Now if the left hand side of (2.29) is nonnegative; that is, if

$$(2.30) \qquad \sigma_1\sigma_2|w| \geq -\sigma_0 \frac{r_0}{r_1}\left|r_1^2 - |1+c_1|^2\right| \quad ,$$

then we may square both sides of (2.29) and multiply by $\sigma_0\sigma_1\sigma_2 = -1$ to obtain equivalently

$$(2.31) \qquad \sigma_0\sigma_1\sigma_2\left|w(1+\bar{c}_1) + c_0(r_1^2 - |1+c_1^2|)\right|^2 \geq \sigma_0\sigma_1\sigma_2(\sigma_1\sigma_2|w|r_1 + \sigma_0 r_0\left|r_1^2 - |1+c_1|^2\right|) \quad .$$

On the other hand, if $\sigma_0\sigma_1\sigma_2 = 1$ with $\sigma_0 = \sigma_1\sigma_2 = -1$, then (2.29) may be written in the form

$$\left|w(1+\bar{c}_1) + c_0(r_1^2 - |1+c_1|^2)\right| \geq r_1|w| + r_0\left|r_1^2 - |1+c_1|^2\right| \quad .$$

Both sides are inherently nonnegative and we may square both sides to get an equivalent inequality. Observe, however, that since $\sigma_0 = \sigma_1\sigma_2 = -1$ we may substitute

$$(r_1|w| + r_0\left|r_1^2 - |1+c_1|^2\right|)^2 = (\sigma_1\sigma_2 r_1|w| + \sigma_0 r_0\left|r_1^2 - |1+c_1|^2\right|)^2$$

and multiply the relation by $\sigma_0\sigma_1\sigma_2 = 1$ to again obtain (2.31) . Thus in cases (2.27a or b), (2.29) is equivalent to (2.31), if $\sigma_0\sigma_1\sigma_2 = 1$, or both (2.30) and (2.31), if $\sigma_0\sigma_1\sigma_2 = -1$. Expand (2.31) and collect like terms in $|w|$ to obtain

(2.32) $$\sigma_0\sigma_1\sigma_2\{|w|^2(r_1^2-|1+c_1|^2) + (r_0^2-|c_0|^2)|r_1^2-|1+c_1|^2|^2$$
$$+ 2[\sigma_0\sigma_1\sigma_2 r_0 r_1|w|\,|r_1^2-|1+c_1|^2|-\mathrm{Re}[w(1+\bar{c}_1)(\bar{c}_0)(r_1^2-|1+c_1|^2)]\} \le 0 \ .$$

Factoring out $(r_1^2-|1+c_1|^2)$, recognizing by (2.10b) that

$$\sigma_2(r_1^2-|1+c_1|^2) = |r_1^2-|1+c_1|^2| \ ,$$

and noting the two forms for

$$\mathrm{Re}[w(1+\bar{c}_1)(\bar{c}_0)] = -\mathrm{Re}[w(1+\bar{c}_0)(-\bar{c}_0)]$$
$$= -|w|\,|c_0|\,|1+c_1|\cos(\theta-\psi) \ ,$$

and dividing throughout by $|r_1^2-|1+c_1|^2|$, we obtain first

(2.33) $$\sigma_0\sigma_1\{|w|^2+2[\sigma_0\sigma_1 r_0 r_1|w|+\mathrm{Re}[w(1+\bar{c}_1)(-\bar{c}_0)]]+(r_0^2-|c_0|^2)(r_1^2-|1+c_1|^2)\} \le 0$$

and second, from (2.7) , (2.8) , and (2.9) ,

(2.34) $$\sigma_0\sigma_1\{|w|^2-2|w|t(\theta)+k\} \le 0 \ .$$

Finally complete the square to find

(2.35) $$\sigma_0\sigma_1(|w|-t(\theta))^2 \le \sigma_0\sigma_1(t^2(\theta)-k) \ .$$

Thus for $\sigma_2 \neq 0$, that is, for $r_1 \neq |1+c_1|$, $w \in E$ if and only if

(2.36a) relation (2.35) holds when $\sigma_0\sigma_1\sigma_2 = 1$ or

(2.36b) both (2.30) and (2.35) hold when $\sigma_0\sigma_1\sigma_2 = -1$.

In anticipation of finding solutions of (2.35) let

$$\Delta_1 = \{\theta : t^2(\theta)-k \ge 0\} \ ,$$
$$\Delta_2 = \{\theta : t(\theta) \ge 0\} \ .$$

If $\sigma_0\sigma_2 = -1$ then $k \le 0$, $|t(\theta)| \le \sqrt{t^2(\theta)-k}$, and consequently

(2.37a) $$t(\theta) - \sqrt{t^2(\theta)-k} \le 0 \le t(\theta) + \sqrt{t^2(\theta)-k}$$

(2.37b) $$\text{for } \theta \in \Delta_1 = [-\pi,\pi] \ .$$

If $\sigma_0\sigma_2 = 1$ then $k \ge 0$, $t(\theta) \ge \sqrt{t^2(\theta)-k}$, and consequently

(2.38a) $$0 \le t(\theta) - \sqrt{t^2(\theta)-k} \le t(\theta) + \sqrt{t^2(\theta)-k}$$

(2.38b) $$\text{for } \theta \in \Delta_1 \cap \Delta_2 \ .$$

We consider separate cases to discover the form of $\Delta_1 \cap \Delta_2$ when $\sigma_0\sigma_2 = 1$. If $c_0(1+c_1) = 0$ then $|t(\theta)| = r_0 r_1 \ge \sqrt{(r_0^2-|c_0|^2)(r_1^2-|1+c_1|^2)} = \sqrt{k}$, and $\Delta_1 = [-\pi,\pi]$. Furthermore $t(\theta) = -\sigma_0\sigma_1 r_0 r_1 \ge 0$ if and only if $\sigma_0\sigma_1 = -1$, in which case $\Delta_2 = [-\pi,\pi]$. Hence if $\sigma_0\sigma_2 = 1$ and $c_0(1+c_1) = 0$ then

(2.39a) $\Delta_2 \neq \emptyset$ only for $\sigma_0\sigma_1 = -1$ and in that case

(2.39b) $$\Delta_1 \cap \Delta_2 = \Delta_1 = \Delta_2 = [-\pi,\pi] \ .$$

On the other hand, if $c_0(1+c_1) \neq 0$, $\sigma_0\sigma_2 = 1$, and $k \geq 0$, then $t^2(\theta)-k \geq 0$ if and only if $|t(\theta)| \geq k$, so that $\theta \in \Delta_1$ if and only if

(2.40a) $$\cos(\theta-\psi) \leq \frac{-\sigma_0\sigma_1 r_0 r_1 - \sqrt{k}}{|c_0||1+c_1|} \quad \text{or}$$

(2.40b) $$\cos(\theta-\psi) \geq \frac{-\sigma_0\sigma_1 r_0 r_1 + \sqrt{k}}{|c_0||1+c_1|} ,$$

while $\theta \in \Delta_2$ if and only if

(2.41) $$\cos(\theta-\psi) \leq \frac{-\sigma_0\sigma_1 r_0 r_1}{|c_0||1+c_1|} .$$

Clearly

(2.42) $$\Delta_1 \cap \Delta_2 = \{\theta : \cos(\theta-\psi) \leq \frac{-\sigma_0\sigma_1 r_0 r_1 - \sqrt{k}}{|c_0||1+c_1|}\} .$$

Now we consider those cases, for $c_0(1+c_1) \neq 0$, when $\Delta_1 \cap \Delta_2 \neq \emptyset$ and further when $\Delta_1 \cap \Delta_2 = [-\pi,\pi]$.

If $\sigma_0\sigma_1 = -1$ then $\frac{r_0 r_1 - \sqrt{k}}{|c_0||1+c_1|} > -1$ is equivalent to $(r_0|1+c_1|+r_1|c_0|)^2 > 0$ which clearly holds. Hence for $\sigma_0\sigma_1 = -1$, $\Delta_1 \cap \Delta_2 \neq \emptyset$. If, in particular, $\sigma_0 = -\sigma_1 = \sigma_2 = 1$ then in fact $\Delta_1 \cap \Delta_2 = [-\pi,\pi]$. This can be seen since

$$\frac{r_0 r_1 - \sqrt{k}}{|c_0||1+c_1|} \geq 1 ,$$

or equivalently

$$r_0 r_1 - |c_0||1+c_1| \geq \sqrt{k} .$$

The left hand side is nonnegative since $|c_0| < r_0$ and $|1+c_1| < r_1 < |c_1|$. Upon squaring both sides and collecting like terms this relation is seen to be equivalent to $(r_0|1+c_1|-r_1|c_0|)^2 > 0$. Thus

(2.43a) $$\text{for } \sigma_0\sigma_1 = -1 , \quad \Delta_1 \cap \Delta_2 \neq \emptyset \quad \text{and}$$

(2.43b) $$\text{for } \sigma_0 = -\sigma_1 = \sigma_2 = 1 , \quad \Delta_1 \cap \Delta_2 = [-\pi,\pi] .$$

On the other hand, if $\sigma_0\sigma_1 = 1$ then $\Delta_1 \cap \Delta_2 \neq \emptyset$ provided that

$$\frac{-r_0 r_1 - \sqrt{k}}{|c_0||1+c_1|} > -1$$

or equivalently

(2.44) $$|c_0||1+c_1| - r_0 r_1 > \sqrt{k} .$$

Since we are now considering the case $\sigma_0\sigma_1 = \sigma_0\sigma_2 = 1$, and by (2.27c) $\sigma_0 = \sigma_1 = \sigma_2 = 1$ is disallowed, we are in fact considering the case $\sigma_0 = \sigma_1 = \sigma_2 = -1$, which implies $r < |c_0|$ and $r_1 < |1+c_1|$. Thus the left hand side of (2.44) is positive and upon squaring one sees that (2.44) is equivalent to

$$(r_0|1+c_1|-r_1|c_0|)^2 > 0$$

which holds if and only if

(2.45) $$\frac{r_1}{|1+c_1|} \neq \frac{r_0}{|c_0|} .$$

Assuming (2.45), then, $\Delta_1 \cap \Delta_2 \neq \emptyset$ for $\sigma_0 = \sigma_1 = \sigma_2 = -1$. We consider one other matter before proceeding. If $\sigma_0\sigma_1\sigma_2 = -1$, then for $\theta \in \Delta_1$,

(2.46a) $\left|-r_1t(\theta)+r_0|r_1^2-|1+c_1|^2|\right| \leq r_1\sqrt{t^2(\theta)-k}$ for $\sigma_0\sigma_1 = -1$, $\sigma_2 = 1$,

(2.46b) $\left|-r_1t(\theta)+r_0|r_1^2-|1+c_1|^2|\right| \geq r_1\sqrt{t^2(\theta)-k}$ for $\sigma_0\sigma_1 = 1$, $\sigma_2 = -1$.

These relations may be seen as follows. Squaring both sides of each, collecting like terms, simplifying with the values of the σ's, and dividing each throughout by $|r_1^2-|1+c_1|^2|$, each is shown to be equivalent to

(2.47) $$r_1^2|c_0|^2 + 2r_0r_1|c_0||1+c_1|\cos(\theta-\psi) + r_0^2|1+c_1|^2 \geq 0 ,$$

which holds since the left hand side is bounded below by the nonnegative quantity $(r_1|c_0|-r_0|1+c_1|)^2$. Finally, let

(2.48) $$\Omega = \{\theta : B_1(\theta) = t(\theta) + \sigma_0\sigma_1\sqrt{t^2(\theta)-k} > 0\} .$$

<u>Case 1</u>. If $\sigma_0\sigma_1 = 1$ and $(V_1,V_0) = (D_1,D_0)$ or (cD_1,cD_0), from (2.35) we know that $w \in E$ only if

$$(|w|-t(\theta))^2 \leq t^2(\theta) - k$$

which for $\theta \in \Delta_1$ has solutions $|w|$ such that

$$\left||w|-t(\theta)\right| < \sqrt{t^2(\theta)-k}$$

or equivalently

(2.49a) $$t(\theta) - \sqrt{t^2(\theta)-k} \leq |w| \leq t(\theta) + \sqrt{t^2(\theta)-k}$$

(2.49b) $$\text{for } \theta \in \Omega = \{\theta : B_1(\theta) = t(\theta) + \sqrt{t^2(\theta)-k} > 0\} .$$

If $\sigma_0\sigma_1 = -1$ then by (2.37), (2.49) reduces to

(2.50a) $$0 \leq |w| \leq t(\theta) + \sqrt{t^2(\theta)-k}$$

(2.50b) $$\text{for } \theta \in \Omega = \Delta_1 = [-\pi,\pi] .$$

a) If $\sigma_0 = \sigma_1 = -1$ with $\sigma_2 = 1$ then $\sigma_0\sigma_1\sigma_2 = 1$ and by (2.36a) $w \in E$ if and only if (2.50) holds. Thus (2.1) is proved for $(V_1,V_0) = (-1 \notin \mathrm{Int}(cD_1),cD_0)$.

b) If $\sigma_0 = \sigma_1 = 1$ with $\sigma_2 = -1$ then $\sigma_0\sigma_1\sigma_2 = -1$ and by (2.36b) $w \in E$ if and only if both (2.50) and (2.30) hold, where (2.30) takes the form

(2.51) $$|w| \leq \frac{r_0}{r_1}(|1+c_1|^2-r_1^2) .$$

Now (2.50) will imply (2.51) for $\theta \in \Omega = [-\pi,\pi]$ if

$$(2.52) \qquad t(\theta) + \sqrt{t^2(\theta)-k} \le \frac{r_0}{r_1}(|1+c_1|^2-r_1^2) \quad ,$$

or equivalently

$$(2.53) \qquad r_1\sqrt{t^2(\theta)-k} \le -r_1 t(\theta) + r_0(|1+c_1|^2-r_1^2) \quad .$$

Since the right hand side of (2.53) satisfies

$$-r_1 t(\theta) + r_0(|1+c_1|^2-r_1^2) = |1+c_1|(r_0|1+c_1|+r_1|c_0|\cos(\theta-\psi))$$
$$\ge |1+c_1|(r_0|1+c_1|-r_1|c_0|) > 0 \quad ,$$

the last inequality following since $\sigma_0 = 1$ implies $r_0 > |c_0|$ and $\sigma_2 = -1$ implies $|1+c_1| > r_1$. Thus (2.53) is equivalent to (2.46b) and hence (2.52) holds. Consequently (2.50a) implies (2.51) . Then $w \in E$ if and only if (2.50) . Thus (2.1) is proved for $(V_1,V_0) = (-1 \notin \mathrm{Int}(D_1),D_0)$.

c) If $\sigma_0 = \sigma_1 = \sigma_2 = -1$ then $\sigma_0\sigma_1\sigma_2 = -1$ and by (2.36b) $w \in E$ if and only if both (2.49) and (2.30) hold, where (2.30) takes the form

$$(2.54) \qquad |w| \ge \frac{r_0}{r_1}(|1+c_1|^2-r_1^2|) \quad .$$

Now since $\sigma_0\sigma_2 = 1$ and $k \ge 0$, from (2.38) , (2.49a) will imply (2.54) for $\theta \in \Omega = \Delta_1 \cap \Delta_2$ if and only if

$$(2.55) \qquad t(\theta) - \sqrt{t^2(\theta)-k} \ge \frac{r_0}{r_1}(|1+c_1|^2-r_1^2)$$

or equivalently

$$(2.56) \qquad r_1 t(\theta) - r_0(|1+c_1|^2-r_1^2) \ge r_1\sqrt{t^2(\theta)-k} \quad .$$

From (2.39) since $\sigma_0\sigma_1 = 1$ we know that $c_0(1+c_1) \ne 0$. Now after simplification it is clear that the left hand side of (2.56) is nonnegative if and only if θ satisfies

$$(2.57) \qquad \cos(\theta-\psi) \le \frac{-r_0|1+c_1|}{r_1|c_0|} \quad .$$

Now the right hand side of (2.57) exceeds -1 if and only if

$$(2.58) \qquad \frac{r_1}{|1+c_1|} > \frac{r_0}{|c_0|} \quad .$$

On the other hand it can be shown that the inequalities

$$(2.59) \qquad \frac{-r_0 r_1-\sqrt{k}}{|c_0||1+c_1|} < \frac{-r_0|1+c_1|}{r_1|c_0|} < \frac{-r_0 r_1+\sqrt{k}}{|c_0||1+c_1|}$$

are each equivalent to

$$(|1+c_1|^2-r_1^2)(r_1^2|c_0|^2-r_0^2|1+c_1|^2) > 0$$

which again holds if and only if (2.58) is true. Assuming (2.58) then, we know that (2.59) holds. Now by (2.59), (2.57) eliminates (2.40b) while (2.40a) implies (2.57). Further (2.57) implies the equivalence of (2.56) and (2.46b). Thus (2.55) holds and assures us that (2.49a) implies (2.54). Thus assuming (2.58), $w \in E$ if and only if (2.49) holds for $\Omega = \Delta_1 \cap \Delta_2$ as in (2.42). Further $\Omega - \{0\} \neq \emptyset$ since (2.58) implies (2.45). Thus (2.2) is proved for $(V_1,V_0) = (-1 \in \mathrm{Int}(cD_1), cD_0)$.

Case 2. If $\sigma_0\sigma_1 = -1$ then $(V_1,V_0) = (cD_1,D_0)$ or (D_1,cD_0) and from (2.35) $w \in E$ only if

(2.60) $$(|w|-t(\theta))^2 \geq t^2(\theta) - k .$$

When $\theta \notin \Delta_1$, that is, when $t^2(\theta) - k < 0$, then any value of $|w|$ is a solution, while for $\theta \in \Delta_1$, (2.60) has solutions $|w|$ which satisfy

$$\big||w|-t(\theta)\big| \geq \sqrt{t^2(\theta)-k} \;;$$

that is, either

(2.61a) $$|w| \leq t(\theta) - \sqrt{t^2(\theta)-k}$$

or

(2.61b) $$|w| \geq t(\theta) + \sqrt{t^2(\theta)-k} .$$

Also notice, for the first two cases below, that when $\sigma_2 = 1$, (2.46a) is equivalent to

(2.62a) $$t(\theta) - \sqrt{t^2(\theta)-k} \leq \frac{r_0}{r_1}(r_1^2-|1+c_1|^2) \leq t(\theta) + \sqrt{t^2(\theta)-k} .$$

(2.62b) $$\text{for } \Omega = \{\theta : B_1(\theta) = t(\theta) - \sqrt{t^2(\theta)-k} > 0\} .$$

a) If $-\sigma_0 = \sigma_1 = \sigma_2 = 1$, the case of $(V_1,V_0) = (-1 \in \mathrm{Int}(D_1), cD_0)$, then $\sigma_0\sigma_1\sigma_2 = -1$ and by (2.36b), $w \in E$ if and only if $|w|$ satisfies both (2.61) and (2.30), which in this case has the form

(2.63) $$|w| \geq \frac{r_0}{r_1}(r_1^2-|1+c_1|^2) .$$

Since $\sigma_0\sigma_2 = -1$, $k < 0$, and by (2.37), (2.61) reduces to (2.61b) for $\theta \in \Delta_1 = [-\pi,\pi]$. By (2.62), (2.61b) implies (2.63). Thus $w \in E$ if and only if $|w|$ satisfies (2.61b) for $\theta \in [-\pi,\pi]$, proving (2.3).

b) If $\sigma_0 = -\sigma_1 = \sigma_2 = 1$, the case $(V_1,V_0) = (-1 \notin \mathrm{Int}(cD_1), D_0)$, then $\sigma_0\sigma_1\sigma_2 = -1$ and again by (2.36b) $w \in E$ if and only if $|w|$ satisfies both (2.61) and (2.30), where, this time, (2.30) stipulates that

(2.64) $$|w| \leq \frac{r_0}{r_1}(r_1^2-|1+c_1|^2) .$$

Since $\sigma_0\sigma_2 = 1$, $k > 0$ and by (2.38a), (2.61) admits both small and large values of $|w|$ for $\theta \in \Delta_1 \cap \Delta_2$. However, in view of (2.62), (2.64) allows only solutions to (2.61a). If $c_0(1+c_1) = 0$, then, from (2.39),

$\theta \in \Delta_1 \cap \Delta_2 = [-\pi,\pi]$. If $c_0(1+c_1) \neq 0$ then for $\theta \in \Omega = \Delta_1 \cap \Delta_2$ as in (2.42). This proves (2.1) in this case.

c) If $-\sigma_0 = \sigma_1 = -\sigma_2 = 1$ then $(V_1,V_0) = (-1 \notin \mathrm{Int}(D_1), cD_0)$, $\sigma_0\sigma_1\sigma_2 = 1$ and by (2.36a) $w \in E$ if and only if $|w|$ satisfies (2.61) for $\theta \in \Omega = \Delta_1 \cap \Delta_2 = [-\pi,\pi]$ if $c_0(1+c_1) = 0$ while if $c_0(1+c_1) \neq 0$ then $w \in E$ if and only if $|w|$ satisfies (2.61) for $\theta \in \Omega = \Delta_1 \cap \Delta_2$ as in (2.42) and for any $|w|$ for $\theta \notin \Delta_1 \cap \Delta_2$. In either case $\mathbb{C} - E$ is more conveniently parameterized as is done in (2.4) .

This completes the proof of Theorem 1 .

3. Factorizations of Element Regions.

Theorem 2. *For* r_0,r_1,c_0,c_1 *and value regions* (V_1,V_0) *as in* (1.8) *taking the specific forms indicated,* E *factors as follows:*

If $|1+c_1| < r_1 \leq |c_1|$, $|c_0| \leq r_0$ *for* $(-1 \notin cD_1,D_0)$, *then*

(3.1) $$E = \{v^2 : |v+ic^*| \leq \sqrt{r_0r_1} , |v-ic^*| \leq \sqrt{r_0r_1}\} .$$

If $|c_1| \leq r_1 < |1+c_1|$, $r_0 \leq |c_0|$ *for* $(-1 \notin D_1,cD_0)$, *then*

(3.2) $$E = \{v^2 : |v+ic^*| \geq \sqrt{r_0r_1} , |v-ic^*| \geq \sqrt{r_0r_1}\} .$$

If $r_1 < |1+c_1|$, $r_1 < |c_1|$, $r_0 \leq |c_0|$ *for* $(-1 \in \mathrm{Int}(cD_1),cD_0)$, *then*

(3.3) $E = E_1 \cup E_2$, *where*

$$E_1 = \{v^2 : |v| \geq \sqrt{|c^*|^2-r_0r_1} , |v+ic^*| \leq \sqrt{r_0r_1}\} ,$$

$$E_2 = \{v^2 : |v| \geq \sqrt{|c^*|^2-r_0r_1} , |v-ic^*| \leq \sqrt{r_0r_1}\} .$$

Furthermore these factorizations occur if and only if

(3.4) $$r_1|c_0| = r_0|1+c_1|$$

and

(3.5) $$(c^*)^2 = (-c_0)(1+c_1) .$$

Corollaries. Here the factorizations of Theorem 2 are related to cases of twin-convergence regions whose setting was outlined in (2.11, 12, 13) .

1. If $|\Gamma_2| < \rho_2 < |1+\Gamma_2|$ and $|1+\Gamma_1| < \rho_1 < |\Gamma_1|$ then $E_1^* = E$ as in (3.1) under the assignment (2.12a) if and only if, from (3.4) and (3.5) , $\rho_2|1+\Gamma_1| = \rho_1|\Gamma_2|$, and $(c_1^*)^2 = (-\Gamma_2)(1+\Gamma_1)$. On the other hand, $E_2^* = E$ as in (3.2) with assignment (2.13a) if and only if, again from (3.4) and (3.5) , $\rho_1|1+\Gamma_2| = \rho_2|\Gamma_1|$ and $(c_2^*) = (-\Gamma_1)(1+\Gamma_2)$.

2. [Lange, 8]. Under the assumptions of Corollary B3 in Section 2 , that is, with assignment (2.19) when $|\Gamma| < \rho < |1+\Gamma|$ and $\rho_1 = \rho = \rho_2$, $E_1^* = E$ as in (3.2) under assignment (2.20) with $c_1^* = \pm(1+\Gamma)$ since $r_1|c_0| = \rho_1|1+\Gamma| = \rho|1+\Gamma| = \rho_2|1+\Gamma| = r_0|1+c_1|$ shows (3.4) to be satisfied. On the other hand, $E_2^* = E$ as in (3.1) under assignment (2.22) with $c_2^* = \pm\Gamma$ since

$r_1|c_0| = \rho_1|\Gamma| = \rho|\Gamma| = \rho_2|1-(1+\Gamma)| = r_0|1+c_1|$ showing again that (3.4) is satisfied. Upon interchanging the indices on E_1^*, E_2^* we obtain the result in (1.7)

<u>Proof of Theorem 2.</u>

From (2.33), in order for $v^2 \in E$ it must satisfy

$$\sigma_0\sigma_1\{|v|^4 - 2|v|^2 r_0 r_1 + |r_0^2-|c_0|^2||r_1^2-|1+c_1|^2| + 2\mathrm{Re}[v^2(-\bar{c}_0)(1+\bar{c}_1)]\} \leq 0 \quad . \tag{3.6}$$

Now for and only for c^* as defined in (3.5), it follows that

$$\mathrm{Re}[v^2(-\bar{c}_0)(1+\bar{c}_1)] = \mathrm{Re}[(v\bar{c}^*)^2] = |vc^*|^2 - 2[\mathrm{Re}(-iv\bar{c}^*)]^2 \tag{3.7}$$

$$= |v|^2|c_0||1+c_1| - 2[\mathrm{Re}(v\overline{(ic^*)})]^2 \quad .$$

Thus the last factor in (3.6) takes the form

$$|v|^4 - 2|v|^2[r_0r_1-|c_0||1+c_1|] + |r_0^2-|c_0|^2||r_1^2-|1+c_1|^2| - 4[\mathrm{Re}(v\overline{(ic^*)})]^2 \quad . \tag{3.8}$$

Further notice that

$$|r_0^2-|c_0|^2||r_1^2-|1+c_1|^2| = (r_0r_1-|c_0||1+c_1|)^2 = (r_0r_1-|c^*|^2)^2 \quad , \tag{3.9}$$

The left hand equality holding if and only if (3.4) holds. Further (3.4) holds only in the case

$$\sigma_0\sigma_2 = 1 \tag{3.10}$$

where σ_0, σ_2 are defined in (2.10). Assuming (3.4), we can then write (3.8) as

$$(|v|^2-(r_0r_1-|c^*|^2))^2 - 4[\mathrm{Re}(v\overline{(ic^*)})]^2 \tag{3.11}$$

which in turn may be factored to give (3.6) the form

$$\sigma_0\sigma_1 Q_{-1}(v)Q_{+1}(v) \leq 0 \tag{3.12}$$

where $Q_{\pm1}(v)$ are defined in two equivalent forms as

$$Q_{\pm1}(v) = |v|^2 \pm 2\mathrm{Re}(v\overline{(ic^*)}) + |c^*|^2 - r_0r_1 \tag{3.13a}$$

$$= |v \pm ic^*|^2 - r_0r_1 \quad . \tag{3.13b}$$

To summarize, then, given (3.4) and (3.5), in order to have $v^2 \in E$ it is necessary that (3.12) hold.

<u>Case I.</u> $(V_1,V_0) = (D_1,cD_0)$ or (cD_1,D_0). In this case $\sigma_0\sigma_1 = -1$. From (3.10) $\sigma_0 = -\sigma_1 = \sigma_2$, which occurs if and only if $-1 \notin V_1$. Further $\sigma_0\sigma_1\sigma_2 = -1$ and by (2.36b) $v^2 \in E$ if and only if both (3.12) and (2.30) hold, where (3.12) takes the form

$$Q_{-1}(v)Q_{+1}(v) \geq 0 \quad . \tag{3.14}$$

Our attention now turns to the specific values of σ_2.

a) If $\sigma_2 = 1$ then $\sigma_0 = -\sigma_1 = \sigma_2 = 1$ and $(V_1,V_0) = (-1 \notin \mathrm{Int}(cD_1),D_0)$, while (2.30) takes the form

(3.15) $$|v|^2 \le \frac{r_0}{r_1}(r_1^2-|1+c_1|^2) = r_0r_1 - |c^*|^2 \quad ,$$

the right hand equality following from (3.4). This relation implies that (3.14) is true if and only if

(3.16) $$Q_{-1}(v) \le 0 \quad \text{and} \quad Q_{-1}(v) \le 0 \;.$$

To this end we will show that if either $Q_{-1}(v) > 0$ or $Q_{+1}(v) > 0$ then (3.14) fails. Indeed, if $Q_{-1}(v) > 0$ then by (3.13a) for $Q_{-1}(v)$ and (3.15), $0 \ge |v|^2 - (r_0r_1-|c^*|^2) > 2\mathrm{Re}[v(\overline{ic^*})]$. Consequently from (3.15) and (3.13a) for $Q_1(v)$ we have

$$\begin{aligned} Q_1(v) &= |v|^2 - (r_0r_1-|c^*|^2) + 2\mathrm{Re}[v(\overline{ic^*})] \\ &< 2(|v|^2-(r_0r_1-|c^*|^2)) \le 0 \end{aligned}$$

so that $Q_1(v) < 0$ and (3.14) fails. A similar argument shows that if $Q_{+1}(v) > 0$ then $Q_{-1}(v) < 0$. Thus (3.14) and (3.15) imply (3.16). Conversely, if (3.16) holds, then clearly (3.14) is satisfied. Further, from (3.13a) for both $Q_{\pm 1}(v)$, it is clear that

$$|v|^2 - (r_0r_1-|c^*|^2) \le -2|\mathrm{Re}(v(\overline{ic^*}))| \le 0$$

which implies (3.15). Thus $v^2 \in E$ if and only if (3.16) holds, or equivalently from (3.13b), if and only if

$$|v \pm ic^*| \le \sqrt{r_0r_1}$$

which proves (3.1).

b) If $\sigma_2 = -1$ then by (3.10), and since $\sigma_0\sigma_1 = -1$, it follows that $-\sigma_0 = \sigma_1 = -\sigma_2 = 1$, the case of $(V_1,V_0) = (-1 \notin \mathrm{Int}(D_1), CD_0)$. Then $\sigma_0\sigma_1\sigma_2 = 1$ and by (2.36a), $v^2 \in E$ if and only if (3.14) is satisfied. In this case $|(-ic^*)-(ic^*)| = |2c^*|^2 = 4|c_0||1+c_1| > 4r_0r_1$. Consequently from (3.13b),

$$\{v\colon |v-ic^*| < \sqrt{r_0r_1}\,,\ |v+ic^*| < \sqrt{r_0r_1}\} = \{v\colon Q_{-1}(v) < 0\,,\ Q_{+1}(v) < 0\} = \emptyset \;.$$

Thus $v^2 \in E$ if and only if both $Q_{-1}(v) \ge 0$ and $Q_{+1}(v) \ge 0$; that is,

$$|v+ic^*| \ge \sqrt{r_0r_1} \quad \text{and} \quad |v-ic^*| \ge \sqrt{r_0r_1} \;.$$

This proves (3.2).

<u>Case II</u>. $(V_1,V_0) = (D_1,D_0)$ or (cD_1,cD_0). Here $\sigma_0\sigma_1 = 1$. By (3.10), $\sigma_0\sigma_2 = 1$ so that $\sigma_0 = \sigma_1 = \sigma_2$. From (2.27c) we must have $\sigma_0 = \sigma_1 = \sigma_2 = -1$, that is, $(V_1,V_0) = (-1 \in \mathrm{Int}(cD_1), cD_0)$, and since $\sigma_0\sigma_1\sigma_2 = -1$, by (2.36b), $v^2 \in E$ if and only if (3.12) holds with $\sigma_0\sigma_1 = 1$, that is,

(3.17) $$Q_{-1}(v)Q_{+1}(v) \le 0 \quad ,$$

while (2.30) holds in its specific form

(3.18) $$|v|^2 \geq \frac{r_0}{r_1}(|1+c_1|^2-r_1^2) = |c^*|^2 - r_0r_1 \geq 0 ,$$

recalling (3.4). We will show that $v^2 \in E$ if and only if v satisfies (3.18) and either

(3.19a) $$Q_{-1}(v) \leq 0 \quad \text{or}$$

(3.19b) $$Q_{+1}(v) \leq 0 .$$

Clearly (3.17) implies (3.19a) or (3.19b). Conversely we will show that given (3.18) and either (3.19a or b), then (3.17) holds. Assume that $Q_{-1}(v) \leq 0$. From (3.13a) for $Q_{-1}(v)$,

$$|v|^2 + |c^*|^2 - r_0r_1 \leq 2\mathrm{Re}[v\overline{(ic^*)}]$$

and consequently from (3.13a) for $Q_{+1}(v)$,

$$Q_1(v) = |v|^2 + |c^*|^2 - r_0r_1 + 2\mathrm{Re}[v\overline{(ic^*)}]$$

$$\geq 2[|v|^2+|c^*|^2-r_0r_1] \geq 4[|c^*|^2-r_0r_1] \geq 0$$

The last two inequalities following from (3.18). Hence (3.17) holds. A similar argument shows that if $Q_1(v) \leq 0$ then $Q_{-1}(v) \geq 0$, and again (3.17) holds. Thus $v^2 \in E$ if and only if (3.18) and either (3.19a) or (3.19b) hold, that is, if and only if

$$|v| \geq \sqrt{|c^*|^2-r_0r_1}$$

and from (3.13b), either

$$|v+ic^*| \leq \sqrt{r_0r_1} \quad \text{or} \quad |v-ic^*| \leq \sqrt{r_0r_1} .$$

Thus (3.3) is proved. This completes the proof of Theorem 2.

References

1. Jones, William B. and Walter M. Reid, Uniform twin-convergence regions for continued fractions $K(a_n/1)$, These Proceedings (1982).
2. Jones, William B. and R.I. Snell, Truncation error bounds for continued fractions, SIAM J. Numer. Anal. 6 (1969), 210-221.
3. Jones, William B. and R.I. Snell, Sequences of convergence regions for continued fractions $K(a_n/1)$, Trans. Amer. Math. Soc. 170 (1972), 483-497.
4. Jones, William B. and W.J. Thron, Convergence of continued fractions, Canad. J. Math. 20 (1968), 1037-1055.
5. Jones, William B. and W.J. Thron, Twin-Convergence regions for continued fractions $K(a_n/1)$, Trans. Amer. Math. Soc. 150 (1970), 93-119.
6. Jones, William B. and W.J. Thron, Continued Fractions: Analytic Theory and Applications, Encyclopedia of Mathematics and Its Applications, vol. 11, Addison-Wesley Publishing Co., Inc., Reading, Massachusetts, (1980).
7. Lane, R.E., The convergence and values of periodic continued fractions, Bull. Amer. Math. Soc. 51 (1945), 246-250.

8. Lange, L.J., On a family of twin convergence regions for continued fractions, Illinois J. Math 10 (1966), 97-108.

9. Lange, L.J. and W.J. Thron, A two parameter family of best twin convergence regions for continued fractions, Math. Zeitschr. 73 (1960), 295-311.

10. Reid, Walter McAllister, "Uniform Convergence and Truncation Error Estimates of Continued Fractions $K(a_n/1)$," Ph.D. Thesis, University of Colorado, Boulder, Colorado 80309, (1978).

11. Thron, W.J., Zwillingskonvergenzgebiete für Kettenbrüche $1 + K(a_n/1)$, deren eines die Kreisscheibe $|a_{2n-1}| \leq \rho^2$ ist, Math. Zeitschr. 70 (1959), 310-344.

12. Wall, H.S., Analytic Theory of Continued Fractions, D. Van Nostrand Co., New York, (1948).

Walter M. Reid

Department of Mathematics

University of Wisconsin - Eau Claire

Eau Claire, Wisconsin 54701

ON A CERTAIN TRANSFORMATION
OF CONTINUED FRACTIONS

W.J. Thron

Haakon Waadeland

1. Introduction. In another place in these proceedings we have surveyed the history of modifications $\{S_n(w_n)\}$ of the sequence of approximants $\{S_n(0)\}$ of a continued fraction $b_0 + K(a_n/b_n)$. There we pointed out the use of modifications for accleration of convergence, analytic continuation, and for obtaining necessary and sufficient conditions for the uniqueness of the solution of the Hamburger moment problem, among others.

In this article a transformed continued fraction

$$\tilde{b}_0 + \operatorname*{K}_{n=1}^{\infty} \left(\frac{\tilde{a}_n}{\tilde{b}_n}\right) ,$$

whose approximants $\tilde{S}_n(0)$ are equal to $S_n(w_n)$ for all $n \geq 0$, will be derived. One advantage of having such a transformation is that it allows us to bring known results of continued fraction theory to bear on the modified sequence $\{S_n(w_n)\}$. Another reason for considering the transformation is that in this way repeated modifications of the initial sequence become possible. That is one can study $\tilde{S}_n(\tilde{w}_n)$ as well as continue the process a finite or even an infinite number of times.

In Section 2 we first derive the transformed continued fraction for the most general case and then specialize it to a more restricted case namely $a_n = a + \delta_n$, $b_n = 1$, $w_n = x$, where x is a solution of the quadratic equation $a - w - w^2 = 0$. In Section 3 the very intractable case $a = -1/4$, $\delta_n \to 0$ is studied. By using Theorem 2.2 a little more light is thrown on the convergence behavior of this class of limit periodic continued fractions. In Section 4 repeated transformations of continued fractions of the form $K((a+ck^n)/1)$, $|k| < 1$, are investigated. Section 5 is devoted to numerical examples and Section 6 contains some speculations as to possible future developments.

2. Formulas for the transformed continued fraction.

For a given continued fraction

$$b_0 + \operatorname*{K}_{n=1}^{\infty} \left(\frac{a_n}{b_n}\right) \tag{2.1}$$

and a given sequence $\{w_n\}_{n=0}^{\infty}$ we wish to determine another continued fraction

$$\tilde{b}_0 + \operatorname*{K}_{n=1}^{\infty} \left(\frac{\tilde{a}_n}{\tilde{b}_n}\right) \tag{2.2}$$

in such a way that not only

$$S_n(w_n) = \frac{A_n + w_n A_{n-1}}{B_n + w_n B_{n-1}} = \tilde{S}_n(0) = \frac{\tilde{A}_n}{\tilde{B}_n} ,$$

but that $\tilde{A}_{-1} = A_{-1} = 1$, $\tilde{B}_{-1} = B_{-1} = 0$ and for $n \geq 0$

$$\text{(2.3)} \qquad \begin{aligned} \tilde{A}_n &= A_n + w_n A_{n-1} , \\ \tilde{B}_n &= B_n + w_n B_{n-1} . \end{aligned}$$

Here A_n , B_n , $\tilde{A}_n$, $\tilde{B}_n$ are the nth numerators and denominators of their respective continued fractions and hence satisfy the recursion relations (DN6) with coefficients a_n , b_n and $\tilde{a}_n$, $\tilde{b}_n$, respectively.

A well known theorem (see for example [7, Th. 2.2]) allows us to accomplish this. Slightly rephrased for our purpose the theorem can be stated as follows. Let $\{\tilde{A}_n\}$, $\{\tilde{B}_n\}$ be sequences of complex numbers, such that

$$\text{(2.4)} \qquad \tilde{A}_{-1} = 1 , \quad \tilde{A}_0 = \tilde{b}_0 , \quad \tilde{B}_{-1} = 0 , \quad \tilde{B}_0 = 1 ,$$

and

$$\text{(2.5)} \qquad \tilde{A}_n \tilde{B}_{n-1} - \tilde{A}_{n-1} \tilde{B}_n \neq 0 , \quad n \geq 0 .$$

Then there exists a uniquely determined continued fraction (2.2) with nth numerator $\tilde{A}_n$ and denominator $\tilde{B}_n$ for all $n \geq 1$. Moreover

$$\text{(2.6)} \qquad \begin{aligned} \tilde{a}_n &= \frac{\tilde{A}_{n-1}\tilde{B}_n - \tilde{A}_n \tilde{B}_{n-1}}{\tilde{A}_{n-1}\tilde{B}_{n-2} - \tilde{A}_{n-2}\tilde{B}_{n-1}} , \\ \tilde{b}_n &= \frac{\tilde{A}_n \tilde{B}_{n-2} - \tilde{A}_{n-2}\tilde{B}_n}{\tilde{A}_{n-1}\tilde{B}_{n-2} - \tilde{A}_{n-2}\tilde{B}_{n-1}} , \end{aligned}$$

for $n \geq 1$.

We want to apply this theorem to $\tilde{A}_n$, $\tilde{B}_n$ determined by (2.3) . Since $A_{-1} = 1$, $A_0 = B_0$, $B_{-1} = 0$, $B_0 = 1$ the relations (2.4) are satisfied with

$$\tilde{b}_0 = b_0 + w_0 .$$

Condition (2.5) is equivalent to

$$S_{n-1}(w_{n-1}) \neq S_n(w_n) = S_{n-1}(s_n(w_n)) ,$$

which in turn is the same as

$$w_{n-1} \neq s_n(w_n) = \frac{a_n}{b_n + w_n} ,$$

or

$$\text{(2.7)} \qquad a_n - b_n w_{n-1} - w_{n-1} w_n \neq 0 .$$

From (2.3) and the determinant formula [7, Th. 2.1]

$$A_n B_{n-1} - B_n A_{n-1} = (-1)^{n-1} \prod_{k=1}^{n} a_k , \quad n \geq 0$$

it follows by a simple, but somewhat lengthy computation, that

(2.8)
$$\tilde{A}_n\tilde{B}_{n-1} - \tilde{A}_{n-1}\tilde{B}_n$$
$$= (-1)^{n-1}\left(\prod_{k=1}^{n-1} a_k\right)(a_n - b_n w_{n-1} - w_{n-1}w_n) \quad , \quad n \geq 1$$

(2.9)
$$\tilde{A}_0\tilde{B}_{-1} - \tilde{A}_{-1}\tilde{B}_0 = -1$$

(2.10)
$$\tilde{A}_n\tilde{B}_{n-2} - \tilde{A}_{n-2}\tilde{B}_n$$
$$= (-1)^{n-2}\left(\prod_{k=1}^{n-2} a_k\right)(b_n a_{n-1} + a_{n-1}w_n - (a_n + b_n b_{n-1})w_{n-2} - b_{n-1}w_n w_{n-2}) \quad , \quad n \geq 2$$

(2.11)
$$\tilde{A}_1\tilde{B}_{-1} - \tilde{A}_{-1}\tilde{B}_1 = -(b_1 + w) \quad .$$

By using these formulas in (2.6) we easily find the formulas for $\tilde{a}_n$ and $\tilde{b}_n$, and are thus led to the result below:

<u>Theorem</u> 2.1. <u>Let</u> (2.1) <u>be a given continued fraction and let</u> $\{w_n\}$ <u>be a sequence of complex numbers satisfying</u>

(2.7)
$$a_n - b_n w_{n-1} - w_{n-1}w_n \neq 0 \quad .$$

<u>Then the continued fraction</u> $\tilde{b}_0 + K(\tilde{a}_n/\tilde{b}_n)$ <u>where</u>

(2.12)
$$\tilde{b}_0 = b_0 + w_0 \quad ,$$
$$\tilde{a}_1 = a_1 - b_1 w_0 - w_0 w_1 \quad ,$$
$$\tilde{b}_1 = b_1 + w_1$$
$$\tilde{a}_n = a_{n-1}\frac{a_n - b_n w_{n-1} - w_{n-1}w_n}{a_{n-1} - b_{n-1}w_{n-2} - w_{n-2}w_{n-1}} \quad , \quad n \geq 2 \quad ,$$
$$\tilde{b}_n = \frac{b_n a_{n-1} + a_{n-1}w_n - (a_n + b_n b_{n-1})w_{n-2} - b_{n-1}w_n w_{n-2}}{a_{n-1} - b_{n-1}w_{n-2} - w_{n-2}w_{n-1}} \quad , \quad n \geq 2 \quad ,$$

<u>is such that for all</u> $n \geq 0$ <u>its</u> nth <u>approximant</u> $\tilde{S}_n(0)$ <u>is equal to</u> $S_n(w_n)$.

The simplest case is when all w_n are equal to a fixed complex number w. We shall make this assumption for the remainder of this paper. Condition (2.7) then becomes

(2.13)
$$a_n - b_n w - w^2 \neq 0 \quad , \text{ for all } n \geq 1 \quad .$$

Any number w, except those on an at most denumerable set, is admissible in that it satisfies (2.13).

We shall now further assume that the given continued fraction can be written in the form $K(a_n/1)$. This will be the case if for all $n \geq 1$ $b_n \neq 0$ (see [7, p. 34]). The transformed continued fraction can be expressed in the same form (except for a term in front) by applying to $\tilde{b}_0 + K(\tilde{a}_n/\tilde{b}_n)$ the equivalence transformation

$$a_1^* = \frac{\tilde{a}_1}{\tilde{b}_1} \quad , \quad a_n^* = \frac{\tilde{a}_n}{\tilde{b}_{n-1}\tilde{b}_n} \quad , \quad n \geq 2 \quad .$$

Since the equivalence transformation requires $\tilde{b}_n \neq 0$, $n \geq 1$, we must avoid an additional set of w-values. This set however is also countable.

From now on it will be convenient to write

$$a_n = a + \delta_n \quad , \quad n \geq 1 \ ,$$

for some $a \neq 0$. With this notation the elements in $w + K(a_n^*/1)$ are given by the formulas

$$a_1^* = \frac{(a-w-w^2) + \delta_1}{1 + w} \ ,$$

$$a_2^* = (a+\delta_1)\frac{(a-w-w^2) + \delta_2}{(1+w)(\delta_1+(\delta_1-\delta_2)w + (a-w-w^2))} \ ,$$

$$a_n^* = (a+\delta_{n-1})\frac{(\delta_n+a-w-w^2)(\delta_{n-2}+a-w-w^2)}{(\delta_{n-1}+(\delta_{n-1}-\delta_n)w+(a-w-w^2))(\delta_{n-2}+(\delta_{n-2}-\delta_{n-1})w+(a-w-w^2))} \ .$$

These formulas suggest a choice for a and w such that

$$a - w - w^2 = 0 \ .$$

In this case they become much simpler. The condition (2.13) now reduces to

$$\delta_n \neq 0 \quad , \quad n \geq 1 \ ,$$

and we have the following result.

<u>Theorem</u> 2.2. *Let* $K((a+\delta_n)/1)$ *be a given continued fraction with*

$$\delta_n \neq 0 \quad , \quad n \geq 1 \ .$$

Further let x *be one of the roots of the quadratic equation*

$$a - w - w^2 = 0 \ .$$

Then the continued fraction

$$x + \operatorname*{K}_{n=1}^{\infty}\left(\frac{a_n^*}{1}\right) \ ,$$

with

$$a_1^* = \frac{\delta_1}{1+x}$$

(2.14)
$$a_2^* = (a+\delta_1)\frac{\delta_2}{(1+x)(\delta_1+(\delta_1-\delta_2)x)}$$

$$a_n^* = (a+\delta_{n-1})\frac{\delta_n\delta_{n-2}}{(\delta_{n-1}+(\delta_{n-1}-\delta_n)x)(\delta_{n-2}+(\delta_{n-2}-\delta_{n-1})x)} \ , \quad n \geq 3 \ ,$$

is such that for all $n \geq 1$ *its approximants* $S_n^*(0)$ *are equal to* $S_n(x)$.

Theorem 2.2 is valid even if the δ_n do not tend to zero. In the applications to be made of the theorem in the next two sections the continued fraction

(2.15) $$\overset{\infty}{\underset{n=1}{K}}\left(\frac{a_n}{1}\right)$$

however will be assumed to be limit periodic. It then will be natural to choose

(2.16) $$\lim_{n\to\infty} a_n = a$$

and hence have

$$\lim_{n\to\infty} \delta_n = 0 \ .$$

We recall from [9, p. 95] that if a is not on the ray $t \leq -1/4$ the limit periodic continued fraction (2.15) converges (possibly to ∞). For $a = -1/4$ it converges if $\delta_n \to 0$ sufficiently fast (see for instance [12, p. 162]). If a is located on the ray $t < -1/4$ we know that (2.15) diverges at least if $\delta_n \to 0$ sufficiently fast.

From now on we shall assume that $w = x$, where x is one of the solutions of the equation

$$a - w - w^2 = 0 \ .$$

If x is one solution then $-(1+x)$ is the other. Hence it could be used in (2.14) in place of x . Observe that x and $-(1+x)$ are the fixed points of the linear fractional transformation

(2.17) $$w = \frac{a}{1+w} \ .$$

Observe also that for $\delta_n \to 0$ the numerator and denominator of a_n^* in (2.14) both tend to zero, but if the ratio δ_n/δ_{n-1} behaves sufficiently nicely, the transformed continued fraction will still be useful.

We further recall that if in the limit periodic continued fraction (2.15) a (as defined in (2.16)) is different from zero and not on the ray $s \leq -1/4$ of the negative real axis, then the two fixed points x and $-(1+x)$ of (2.17) are both different from zero and of different absolute values. If

$$|x| < |1+x|$$

then x is the <u>attractive</u> fixed point and $\{S_n(x)\}$ tends to the value of the continued fraction faster than $\{S_n(0)\}$ in the following sense:

$$\frac{f-S_n(x)}{f-S_n(0)} \to 0 \quad \text{as} \quad n \to \infty \ .$$

This holds regardless of how slowly $\delta_n \to 0$. For $a = -1/4$ both fixed points coincide, that is

$$x = -(1+x) = -1/2 \quad ,$$

and the result above holds under certain conditions on the speed at which $\delta_n \to 0$ [12] . Replacing $\{S_n(0)\}$ by $\{S_n(x)\}$ in these cases has been called the <u>"right" modification</u>. If

$$|x| > |1+x| \quad ,$$

then x is the <u>repulsive</u> fixed point. Under additional conditions on the rate at which $\delta_n \to 0$ ($\delta_n \leq Ck^n$ for $0 < k < |(1+x)/x|$ is known to be sufficient) the sequence $\{S_n(x)\}$ will converge but in general to a value different from $\lim S_n(0)$. Replacing $\{S_n(0)\}$ by $\{S_n(x)\}$ in this case has been called the <u>"wrong" modifiction</u> since it leads to a value different from the value to which the continued fraction converges. The result on the wrong modification is proved in two different ways in the articles [13] and [14]. In the first paper it is also shown how the wrong modification can be used for analytic continuation of functions defined by regular C-fractions or general T-fractions. A crucial point is that the modification works also on the ray $s < -1/4$ under the condition mentioned above. On this ray we have $|1+x| = |x|$, $x \neq -(1+x)$ so that the fixed points are distinct but are neither attractive nor repulsive, since the transformation (2.17) is elliptic in this case.

3. <u>An application to limit periodic continued fractions with</u> $a = -1/4$. In this case the two fixed points coincide $x = -(1+x) = -1/2$ and the formulas (2.14) take the form

$$a_1^* = 2\delta_1 ,$$

(3.1)
$$a_2^* = a_1 \frac{4\delta_2}{\delta_1+\delta_2} ,$$

$$a_n^* = a_{n-1}\frac{4\delta_n\delta_{n-2}}{(\delta_n+\delta_{n-1})(\delta_{n-1}+\delta_{n-2})} , \quad n \geq 3 .$$

It follows from a result of Pringsheim [10] that

$$|\delta_n| \leqslant \frac{1}{4(4n^2-1)}$$

insures convergence of $K((-1/4+\delta_n)/1)$. In this formula n can be replaced by $n+k$; if in addition one sets $k - 1/2 = \theta$ then one is led to consider the bound

$$\frac{1}{16(n+\theta)(n+\theta+1)} .$$

An attractive but as yet unproved conjecture is that if

$$\delta_n < \frac{-1}{16(n+\theta)(n+\theta+1)} ,$$

and possibly is subject to additional conditions, then $K((-1/4+\delta_n)/1)$ can be made to diverge.

Now it turns out that, if in (3.1) δ_n is chosen to be

(3.2)
$$\delta_n = -\frac{C}{16(n+\theta)(n+\theta+1)} ,$$

where C is a complex number and θ is a real number $\geq -1/2$, then (3.1) takes on the very simple form

$$a_1^* = -\frac{C}{8(1+\theta)(2+\theta)} ,$$

(3.3)
$$a_2^* = -\frac{C}{8(2+\theta)^2} - \frac{1+\theta}{2(2+\theta)} = 2a_1 \frac{1+\theta}{2+\theta} ,$$

$$a_n^* = a_{n-1} , \quad n \geq 3 .$$

The transformed continued fraction $x + K(a_n^*/1)$ then can be written as

(3.4)
$$-\frac{1}{2} - \cfrac{\frac{C}{8(1+\theta)(2+\theta)}}{1 - \cfrac{\frac{C}{8(2+\theta)^2} + \frac{1+\theta}{2(2+\theta)}}{1 + \mathop{K}\limits_{n=2}^{\infty} (a_n/1)}} .$$

In the cases where the original continued fraction and (3.4) both converge and where they converge to the same value, this value f must satisfy the quadratic equation

(3.5)
$$f = -\frac{1}{2} - \cfrac{\frac{C}{8(1+\theta)(2+\theta)}}{1 + \cfrac{2a_1 \frac{1+\theta}{2+\theta}}{a_1/f}} ,$$

since

$$1 + \mathop{K}\limits_{n=2}^{\infty} (a_n/1) = a_1/f .$$

The quadratic equation (3.5) can be written as

(3.5')
$$f^2 + \frac{3+2\theta}{2(1+\theta)} f + \frac{2+\theta}{4(1+\theta)} + \frac{C}{16(1+\theta)^2} = 0 .$$

The solutions of this equation are

(3.6)
$$f = -\frac{3+2\theta \pm \sqrt{1-C}}{4(1+\theta)} .$$

If $|C| \leq 1$ we know from [12, p. 162] that both continued fractions converge to the same limit which is then one of the two values given in (3.6). It can be proved that if $\sqrt{1-C}$ is taken to be the root with non-negative real part then f is

(3.6')
$$f = -\frac{3+2\theta-\sqrt{1-C}}{4(1+\theta)} .$$

The case $C = 1$ is particularly simple since the square root vanishes. We have in this case

(3.7)
$$f = -\frac{3+2\theta}{4(1+\theta)}$$

and for the tails

$$f^{(n)} = -\frac{3+2n+2\theta}{4(1+n+\theta)} = -\frac{1}{2} - \frac{1}{4(1+n+\theta)} .$$

Since we know all of its tails, the continued fraction

$$(3.8) \qquad \underset{n=1}{\overset{\infty}{K}} \left(\frac{-\frac{1}{4} - \frac{1}{16(n+\theta)(n+\theta+1)}}{1} \right)$$

may then serve as an <u>auxiliary continued fraction</u> in the sense of L. Jacobsen ([3] and her article in these proceedings).

The case $\theta = 0$ was treated numerically in [12, Ex. 5.4]; the one millionth approximant was found to be $-.714$ and for $S_n(-1/2)$, $n = 10^6$, the computed value was $-.718$, so we conjectured that the limit would be approximately $-.72$. From (3.7) we now see that the correct value is $-3/4$. Thus we gravely under-estimated the slowness of the convergence, which is of order $1/\log n$. The convergence of the tails to $-1/2$ we now see to be of order $1/4n$.

The case $\theta = -1/2$ gives the continued fraction

$$\underset{n=1}{\overset{\infty}{K}} \left(\frac{-\frac{1}{4} - \frac{1}{4(4n^2-1)}}{1} \right) .$$

This is connected to a hypergeometric function in the following way

$$(3.9) \qquad \underset{n=1}{\overset{\infty}{K}} \left(\frac{\left(-\frac{1}{4} - \frac{1}{4(4n^2-1)}\right) z}{1} \right) = \frac{1}{{}_2F_1(\frac{1}{2},1;\frac{3}{2};z)} - 1$$

$$= \frac{1}{1+\sum_{n=1}^{\infty} \frac{z^n}{2n+1}} - 1 \quad \text{for } |z| < 1 .$$

For $z = t$, $0 < t < 1$ the function is a decreasing function of t, whose limit as $t \uparrow 1$ is -1. From (3.7) we have that the value of the continued fraction is also -1.

It is easily seen from (3.4) that if either the original or the transformed continued fraction converges (including convergence to ∞) then so does the other. However, it does not follow that they both converge to the same limit. For no $R > 1$ is it true that the continued fractions both converge to the same value for <u>all</u> C in $|C| \leq R$. In order to see this take C real and $C > 1$. If the continued fraction and its transform both converged to the same value, that value would have to be non-real by (3.6). However the elements of the continued fraction are all real. Hence we have a contradiction. Unfortunately we are not able to arrive at the stronger conclusion (which is probably true) that both continued fractions diverge in this case.

4. <u>Some results of repeated transformation.</u> Next to be investigated is the case when $a_n \to a$ geometrically. In its simplest form - the one to be studied here - the continued fraction is

$$(4.1) \qquad \underset{n=1}{\overset{\infty}{K}} \left(\frac{a+ck^n}{1} \right) , \quad 0 < |k| < 1 , \quad a \neq 0 .$$

Let x and $-(1+x)$ denote the fixed points of the linear fractional transformation (2.17) . Here x need not be the attractive fixed point. It follows directly from the formulas (2.14) that for $n \geq 3$ the transformation with $w_n = x$ gives

$$a_n^* = \frac{(a+ck^{n-1})k}{(1+(1-k)x)^2} ,$$

and hence, if $k \neq 1 - 1/x$, the transformed continued fraction is limit periodic with

$$\lim_{n\to\infty} a_n^* = \frac{ak}{(1+(1-k)x)^2} = \frac{\frac{kx}{1+x}}{\left(1 - \frac{kx}{1+x}\right)^2} . \tag{4.2}$$

From well known properties of the Koebe function it follows that for

$$\left|\frac{x}{1+x}\right| < \frac{1}{|k|} \tag{4.3}$$

$\lim a_n^*$ is not on the ray $(-\infty,-1/4]$ of the negative real axis, so that the convergence of the transformed continued fraction may be substantially improved by the method of [12] .

For later convenience we now introduce superscripts in (4.1) and shall write it as

$$\operatorname*{K}_{n=1}^{\infty} \left(\frac{a^{(0)} + c^{(0)}k^n}{1}\right) , \tag{4.4}$$

where $0 < |k| < 1$, $c^{(0)} \neq 0$, $a^{(0)} \neq 0$. The approximants of (4.4) shall be denoted by $S_{n,0}(0)$. We shall also assume that (4.3) is satisfied with x replaced by $x^{(0)}$. This is always possible by a proper choice of $x^{(0)}$. If $a^{(0)}$ is on the ray $a^{(0)} \leq -1/4$ of the negative real axis then both choices work, since in this case the left hand side is < 1 . It is easy to see that in the angular opening

$$\pi - \beta < \arg(a^{(0)} + 1/4) < \pi + \beta , \tag{4.5}$$

where

$$\beta = \pi - 4\arctan|k| \tag{4.6}$$

$x^{(0)}$ may be taken to be either the attractive or the repulsive fixed point.

We use the $x^{(0)}$-modification on (4.4) and get the transformed continued fraction

$$x^{(0)} + \operatorname*{K}_{n=1}^{\infty} \left(\frac{\gamma_n^{(0)}}{1}\right) , \tag{4.7}$$

where (from (2.14))

$$\gamma_1^{(0)} = \frac{c^{(0)}k}{1+x^{(0)}} ,$$

(4.8) $$\gamma_2^{(0)} = a_1^{(0)}\frac{k}{(1+x^{(0)})(1+(1-k)x^{(0)})} ,$$

$$\gamma_n^{(0)} = a_{n-1}^{(0)}\frac{k}{(1+(1-k)x^{(0)})^2} .$$

Its approximants shall be denoted by $S_{n,1}(0)$ so that we have

$$S_{n,0}(x^{(0)}) = S_{n,1}(0) \quad , n \geq 1 .$$

For $n \geq 1$ we define

$$a_n^{(1)} = a^{(1)} + c^{(1)}k^n ,$$

where

$$a^{(1)} = a^{(0)}\frac{k}{(1+(1-k)x^{(0)})^2} ,$$

$$c^{(1)} = c^{(0)}\frac{k^2}{(1+(1-k)x^{(0)})^2} .$$

It follows from the assumption $k \neq 1 - 1/x^{(0)}$ that the denominators do not vanish. We now set

$$\gamma_{n+2}^{(0)} = a_n^{(1)} \quad , n \geq 1 .$$

The transformed continued fraction may thus be written in the form

(4.9) $$x^{(0)} + \frac{\gamma_1^{(0)}}{1} + \frac{\gamma_2^{(0)}}{1 + \operatorname*{K}_{n=1}^{\infty}\left(\frac{a_n^{(1)}}{1}\right)} .$$

Before proceeding with the iteration we make a comment on the case where $x^{(0)}$ is the repulsive fixed point. In this case the direct computation of $\lim_{n\to\infty} S_n(x^{(0)})$ by using the wrong modification on the continued fraction (4.4) is numerically unstable. The reason for this is that

$$\lim_{n\to\infty} S_n(w) = \lim_{n\to\infty} S_n(0)$$

for all $w \neq x_0$. Hence a roundoff error in $x^{(0)}$ may cause the process to tend to the value $\lim S_n(0)$ instead of $\lim S_n(x^{(0)})$. On the other hand, the continued fraction (4.9) converges to $\lim S_n(x^{(0)})$ and can be computed by using the backward recurrence algorithm in the usual way. As was shown in [6] this process is numerically stable. Examples 5.2 and 5.3 in Section 5 illustrate this difference in stability.

We shall now apply a transformation to the continued fraction $\operatorname*{K}_{n=1}^{\infty}(a_n^{(1)}/1)$. For it we have $a_n^{(1)} \to a^{(1)}$. The two roots of the quadratic equation

$w^2 + w = a^{(1)}$ are $x^{(1)}$ and $-(1+x^{(1)})$. We shall choose them so that

$$x^{(1)} = \frac{kx^{(0)}}{1+(1-k)x^{(0)}} \quad , \quad -(1+x^{(1)}) = \frac{-(1+x^{(0)})}{1+(1-k)x^{(0)}} \quad . \tag{4.10}$$

This choice insures that

$$\left|\frac{x^{(1)}}{1+x^{(1)}}\right| = |k|\left|\frac{x^{(0)}}{1+x^{(0)}}\right| < 1 \tag{4.11}$$

so that $x^{(1)}$ is the attractive fixed point of the transformation $w = a^{(1)}/(1+w)$ and so that $\{S_{n,1}(x^{(1)})\}$ is the right modification of (4.9) regardless of what type of modification $x^{(0)}$ induced. The next step leads to a continued fraction of the form

$$x^{(0)} + \frac{\gamma_1^{(0)}}{1} + \frac{\gamma_2^{(0)}}{1+x^{(1)}} + \frac{\gamma_1^{(1)}}{1} + \frac{\gamma_2^{(1)}}{1 + \overset{\infty}{\underset{n=1}{K}}\left(\frac{a_n^{(2)}}{1}\right)} \quad .$$

The process can be repeated indefinitely and we are led to the continued fraction

$$x^{(0)} + \frac{\gamma_1^{(0)}}{1} + \frac{\gamma_2^{(0)}}{1+x^{(1)}} + \frac{\gamma_1^{(1)}}{1} + \frac{\gamma_2^{(1)}}{1+x^{(2)}} + \frac{\gamma_2^{(2)}}{1} + \frac{\gamma_2^{(2)}}{1+x^{(3)}} + \cdots \tag{4.12}$$

or the equivalent form

$$x^{(0)} + \frac{\gamma_1^{(0)}}{1} + \frac{\frac{\gamma_2^{(0)}}{1+x^{(1)}}}{1} + \frac{\frac{\gamma_1^{(1)}}{1+x^{(1)}}}{1} + \frac{\frac{\gamma_2^{(1)}}{1+x^{(2)}}}{1} + \frac{\frac{\gamma_1^{(2)}}{1+x^{(2)}}}{1} + \cdots \quad . \tag{4.13}$$

Before proceeding to the formulas for the elements we shall describe what the continued fraction (4.13) does. It is a "diagonal process" in the sequence of modified sequences. We have already defined $S_{n,0}(0)$, $S_{n,1}(0)$. We now extend this definition as follows

$$S_{n,k}(0) = S_{n,k-1}(x^{(k-1)}) \quad , \quad k \geq 2 \quad .$$

The sequence of approximants of (4.13) will be the staircase illustrated below:

$S_{0,1}(0)$,	$S_{1,1}(0)$,	$S_{2,1}(0)$,	$S_{3,1}(0)$,	$S_{4,1}(0)$,	$S_{5,1}(0)$,	$S_{6,1}(0)$,	$S_{7,1}(0)$
$S_{0,2}(0)$,	$S_{1,2}(0)$,	$S_{2,2}(0)$,	$S_{3,2}(0)$,	$S_{4,2}(0)$,	$S_{5,2}(0)$,	$S_{6,2}(0)$,	$S_{7,2}(0)$
$S_{0,3}(0)$,	$S_{1,3}(0)$,	$S_{2,3}(0)$,	$S_{3,3}(0)$,	$S_{4,3}(0)$,	$S_{5,3}(0)$,	$S_{6,3}(0)$,	$S_{7,3}(0)$
$S_{0,4}(0)$,	$S_{1,4}(0)$,	$S_{2,4}(0)$,	$S_{3,4}(0)$,	$S_{4,4}(0)$,	$S_{5,4}(0)$,	$S_{6,4}(0)$,	$S_{7,4}(0)$
$S_{0,5}(0)$,	$S_{1,5}(0)$,	$S_{2,5}(0)$,	$S_{3,5}(0)$,	$S_{4,5}(0)$,	$S_{5,5}(0)$,	$S_{6,5}(0)$,	$S_{7,5}(0)$

Hence for $n = 2q$, $q \geq 0$ the nth approximant of (4.13) is $S_{2q,q+1}(0)$ and for $n = 2q + 1$, $q \geq 0$, the nth approximant of (4.13) is $S_{2q+1,q+1}(0)$.

The formulas for $\gamma_1^{(n)}$, $\gamma_2^{(n)}$ are now easily obtained by induction. We omit the details and state the result below, having replaced $x^{(0)}$ by x again to simplify the statement.

Theorem 4.1. *Let* $K\left(\frac{a+ck^n}{1}\right)$ *be a continued fraction where* $0 < |k| < 1$, $c \neq 0$, $a \neq 0$. *Let* x *be one of the solutions of the quadratic equation*

$$w^2 + w - a = 0$$

chosen such that

$$\left|\frac{x}{1+x}\right| < \frac{1}{|k|} .$$

Then the sequence $\{S_n(x)\}$ *converges and the continued fraction*

$$(4.14) \qquad x + \frac{\frac{ck}{1+x}}{1} + \frac{\frac{k}{(1+x)^2}(a+ck)}{1} + \frac{\frac{ck^3}{(1+x)^2}}{1} + \frac{\frac{k^2}{(1+x)^2}(a+ck^2)}{1} + \cdots + \frac{\frac{ck^{2p-1}}{(1+x)^2}}{1} + \frac{\frac{k^p}{(1+x)^2}(a+ck^p)}{1} + \cdots$$

converges to the same value, but substantially faster. If, in particular, $|x| < |1+x|$ *then the continued fraction* (4.14) *converges to the same value as* $K\left(\frac{a+ck^2}{1}\right)$.

5. *Numerical examples*. The purpose of the present section is to illustrate the acceleration of convergence by *numerical examples*. A proper analysis of the acceleration, for instance by giving estimates for ratios of different types of truncation errors, is beyond the scope of the present paper. Let it be mentioned, though, that Tor Leknes in a not yet completed investigation [8] has found upper estimates for the truncation errors

$$|f - S_n(0)| , \quad |f - S_n(x)| , \quad |f - S_{n/x}(0)| ,$$

where the continued fraction in question is (4.1) with a *off* the ray $a \leq -\frac{1}{4}$ and where f is the value of the continued fraction. x is chosen such that $|x| < |x+1|$, $S_n(0)$ is the nth approximant and $S_n(x)$ the nth x-modified approximant, whereas $S_{n/x}(0)$ is the nth approximant of the continued fraction (4.14) . His estimates are, under certain mild conditions, in the three cases of the following orders of magnitude

$$O\left(\left|\frac{x}{1+x}\right|^n\right) , \quad O\left(\left|\frac{kx}{1+x}\right|^n\right) , \quad o\left(|k|^{\frac{3}{8}n^2}\right)$$

The numerical examples to be studied here are all computed on the NORD - 10/1000 - computer at the University of Trondheim, the real cases with double precision (15 decimals), the complex one with ordinary precision (8 decimals), in most of the cases by Kent Holing. Example 5.1 explains how the results will be presented also in the other examples, as far as the acceleration of convergence to the right value of the continued fraction is concerned. In Ex. 5.1 all 15 decimals are listed, whereas in the other real examples we shall restrict ourselves to 10 (even if the computation is done with 15). In two of the examples we shall also illustrate the wrong modification.

Example 5.1. For

$$a = 56 \quad , \quad c = 1 \quad , \quad k = \frac{1}{2}$$

the value of the continued fraction (4.4) , as given by the computer, rounded off to 15 decimals, is

$$7.0433816741\ 38279 \quad .$$

The table below gives in each case the smallest n-value n_0 , such that the desired accuracy is obtained for all $n \geq n_0$. Approximants of (4.14) are called "super-approximants".

	4 decimals	8 decimals	15 decimals
Ordinary approx. $S_n(0)$	97	176	281
Modified approx. $S_n(7)$	9	22	39
Superapprox. $S_{n/7}(0)$	2	4	8

Example 5.2. For

$$a = 2 \quad , \quad c = 1 \quad , \quad k = \frac{1}{4}$$

the value of the continued fraction (4.4) , rounded in the 10th decimal place, is

$$1.1096400019 \quad .$$

A table, to be understood as in Ex. 5.1., is here:

	4 decimals	10 decimals
Ord. approx. $S_n(0)$	18	36
Mod. approx. $S_n(1)$	4	11
Superapprox. $S_{n/1}(0)$	2	5

The "wrong" value of the continued fraction (4.4) , $\lim_{n\to\infty} S_n(-2)$, rounded in the 10th decimal place, is

$$-\ 2.1607903165 \quad .$$

This is obtained from $n = 6$ on by using the superapproximants $S_{n/-2}(0)$.

(Here $\left|\frac{x}{x+1}\right| = \left|\frac{-2}{-1}\right| = 2$, whereas $\frac{1}{|k|} = 4$, and the condition (4.3) is thus satisfied.)

An attempt to compute the sequence $\{S_n(-2)\}$ directly from the formulas

$$S_n(-2) = \frac{2+4^{-1}}{1} + \frac{2+4^{-2}}{1} + \cdots + \frac{2+4^{-n}}{1+(-2)}$$

(by the backwards recurrence algorithm, starting with -2 instead of 0) leads to the following adventure: The values produced by the computer first seem to approach the value we want, and for $n = 26$ it gives the value

$$-\,2.1607903112 \quad .$$

This is the closest value to the correct one. Further computation produces values approaching $\lim_{n\to\infty} S_n(0)$, which is taken on with 10 decimals from $n = 92$ on. With four decimals instead of ten $S_{n/-2}(0)$ give the value $-\,2.1608$ from $n = 4$ on, whereas the "direct" method gives $-\,2.1608$ for $13 \leq n \leq 38$ and the value $\lim_{n\to\infty} S_n(0) = 1.1096$ from $n = 72$ on.

Example 5.3. For

$$a = 6 \;, \quad c = 1 \;, \quad k = \frac{1}{2}$$

the value of the continued fraction (4.5) , rounded in the 10th decimal place, is

$$2.1228302993 \quad .$$

A table as in Ex. 5.1 and Ex. 5.2 is here:

	4 decimals	10 decimals
Ord. approx. $S_n(0)$	30	62
Mod. approx. $S_n(2)$	8	20
Superapprox. $S_{n/2}(0)$	4	6

The "wrong" value, $\lim_{n\to\infty} S_n(-3)$, is in this case

$$-\,3.1393059686 \quad ,$$

obtained from $n = 7$ on by using the superapproximants $S_{n/-3}(0)$. Direct computation from (4.5) shows the same pattern as in the previous example. The value closest to the correct one is

$$-\,3.1393058294 \quad ,$$

obtained for $n = 50$. Fron $n = 157$ the attempted computation of $\lim_{n\to\infty} S_n(-3)$

gives us back the value of (4.5), rounded in the 10th decimal place.

Example 5.4. For

$$a = -\frac{1}{4}, \quad c = -1, \quad k = \frac{1}{2}$$

the value of the continued fraction (4.4), rounded in the 10th decimal place, is

$$-0.6303611673 \quad .$$

Table:

	4 decimals	10 decimals
Mod approx. $S_n(-\frac{1}{2})$	22	39
Superapprox. $S_{n/-1/2}(0)$	8	11

As for the ordinary approximants, the convergence is so slow that even for $n = 9999$ the computed value of $S_n(0)$,

$$-0.6299770019$$

only gives the value of the continued fraction rounded in the 2nd place.

Example 5.5. For

$$a = -1 + i, \quad c = 2, \quad k = \frac{i}{2}$$

the value of the continued fraction (4.4) rounded in the 5th decimal place, is

$$0.78568 + 1.07662\ i$$

Table:

	3 decimals	5 decimals
Ord. approx. $S_n(0)$	28	43
Mod. approx. $S_n(i)$	9	14
Superapprox. $S_{n/i}(0)$	5	6

6. Final remarks. The special results of Sections 3 and 4 may prove to be of more interest than merely to serve as illustrations to Theorem 2.2. Partly the same methods may turn out to be useful and lead to manageable continued fractions in more general cases, for instance

$$a_n = a + ck^n + o(k^n) \tag{6.1}$$

in Section 4. More interesting is, that the continued fractions of Section 4 may serve as an auxiliary continued fraction in acceleration of the continued fractions by the method of L. Jacobsen [3], [4]. It is likely that from a numerical and practical point of view for continued fractions $K(\frac{a_n}{1})$, where the a_n's are as in (6.1), this method will be superior to a method with direct application of repeated transformation on (6.1). As for continued fractions "near" the ones we studied in Section 3 L. Jacobsen's method is not applicable, but there is no doubt that it will be extended also to such cases.

References.

1. John Gill, The use of attractive fixed points in accelerating the convergence of limit-periodic continued fractions, Proc. Amer. Math. Soc. 47 (1975), 119-126.

2. John Gill, Enhancing the convergence region of a sequence of bilinear tranformations, Math. Scand. 43 (1978), 74-80.

3. Lisa Jacobsen, Convergence Acceleration for Continued Fractions $K(a_n/1)$, Trans. Amer. Math. Soc., to appear.

4. Lisa Jacobsen, A Method for Convergence Acceleration for Continued Fraction $K(a_n/1)$, these Lecture Notes.

5. W.B. Jones and R.I. Snell, Truncation error bounds for continued fractions $K(a_n/1)$, SIAM J. Numer. Anal. 6 (1969), 210-221.

6. W.B. Jones and W.J. Thron, Numerical stability in evaluating continued fractions, Math. of Comput. 28 (1974), 795-810.

7. William B. Jones and Wolfgang J. Thron, Continued fractions: Analytic theory and applications, ENCYCLOPEDIA OF MATHEMATICS AND ITS APPLICATIONS, No. 11, Addison-Wesley, Reading, Mass. 1980.

8. Tor Leknes, Truncation errors for certain continued fractions, Thesis for the degree cand. real. at the University of Trondheim. In preparation.

9. O. Perron, Die Lehre von den Kettenbrüchen, Band 2, B.G. Teubner, Stuttgart 1957.

10. A. Pringsheim, Über die Konvergenz unendlicher Kettenbrüche. Bayer. Akad. nat. wiss. Kl. Sitzber 28 (1899), 295-324.

11. W.J. Thron, On parabolic convergence regions for continued fractions, Math. Z. 69 (1958), 173-182.

12. W.J. Thron and Haakon Waadeland, Accelerating convergence of limit periodic continued fractions $K(a_n/1)$, Numer. Math. 34 (1980), 155-170.

13. W.J. Thron and Haakon Waadeland, Analytic continuation of functions defined by means of continued fractions, Math. Scand. 47 (1980), 72-90.

14. W.J. Thron and Haakon Waadeland, Convergence questions for limit periodic continued fractions, Rocky Mountain J. Math., 11 (1981), 641-657.

15. W.J. Thron and Haakon Waadeland, Survey of Modifications of Continued Fractions, these Lecture Notes.

W.J. Thron
Department of Mathematics
Campus Box 426
University of Colorado
Boulder, Colorado 80309
U.S.A.

Haakon Waadeland
Department of Mathematics
and Statistics
University of Trondheim
N-7055 Dragvoll
Norway

Vol. 787: Potential Theory, Copenhagen 1979. Proceedings, 1979. Edited by C. Berg, G. Forst and B. Fuglede. VIII, 319 pages. 1980.

Vol. 788: Topology Symposium, Siegen 1979. Proceedings, 1979. Edited by U. Koschorke and W. D. Neumann. VIII, 495 pages. 1980.

Vol. 789: J. E. Humphreys, Arithmetic Groups. VII, 158 pages. 1980.

Vol. 790: W. Dicks, Groups, Trees and Projective Modules. IX, 127 pages. 1980.

Vol. 791: K. W. Bauer and S. Ruscheweyh, Differential Operators for Partial Differential Equations and Function Theoretic Applications. V, 258 pages. 1980.

Vol. 792: Geometry and Differential Geometry. Proceedings, 1979. Edited by R. Artzy and I. Vaisman. VI, 443 pages. 1980.

Vol. 793: J. Renault, A Groupoid Approach to C*-Algebras. III, 160 pages. 1980.

Vol. 794: Measure Theory, Oberwolfach 1979. Proceedings 1979. Edited by D. Kölzow. XV, 573 pages. 1980.

Vol. 795: Séminaire d'Algèbre Paul Dubreil et Marie-Paule Malliavin. Proceedings 1979. Edited by M. P. Malliavin. V, 433 pages. 1980.

Vol. 796: C. Constantinescu, Duality in Measure Theory. IV, 197 pages. 1980.

Vol. 797: S. Mäki, The Determination of Units in Real Cyclic Sextic Fields. III, 198 pages. 1980.

Vol. 798: Analytic Functions, Kozubnik 1979. Proceedings. Edited by J. Ławrynowicz. X, 476 pages. 1980.

Vol. 799: Functional Differential Equations and Bifurcation. Proceedings 1979. Edited by A. F. Izé. XXII, 409 pages. 1980.

Vol. 800: M.-F. Vignéras, Arithmétique des Algèbres de Quaternions. VII, 169 pages. 1980.

Vol. 801: K. Floret, Weakly Compact Sets. VII, 123 pages. 1980.

Vol. 802: J. Bair, R. Fourneau, Etude Géometrique des Espaces Vectoriels II. VII, 283 pages. 1980.

Vol. 803: F.-Y. Maeda, Dirichlet Integrals on Harmonic Spaces. X, 180 pages. 1980.

Vol. 804: M. Matsuda, First Order Algebraic Differential Equations. VII, 111 pages. 1980.

Vol. 805: O. Kowalski, Generalized Symmetric Spaces. XII, 187 pages. 1980.

Vol. 806: Burnside Groups. Proceedings, 1977. Edited by J. L. Mennicke. V, 274 pages. 1980.

Vol. 807: Fonctions de Plusieurs Variables Complexes IV. Proceedings, 1979. Edited by F. Norguet. IX, 198 pages. 1980.

Vol. 808: G. Maury et J. Raynaud, Ordres Maximaux au Sens de K. Asano. VIII, 192 pages. 1980.

Vol. 809: I. Gumowski and Ch. Mira, Recurences and Discrete Dynamic Systems. VI, 272 pages. 1980.

Vol. 810: Geometrical Approaches to Differential Equations. Proceedings 1979. Edited by R. Martini. VII, 339 pages. 1980.

Vol. 811: D. Normann, Recursion on the Countable Functionals. VIII, 191 pages. 1980.

Vol. 812: Y. Namikawa, Toroidal Compactification of Siegel Spaces. VIII, 162 pages. 1980.

Vol. 813: A. Campillo, Algebroid Curves in Positive Characteristic. V, 168 pages. 1980.

Vol. 814: Séminaire de Théorie du Potentiel, Paris, No. 5. Proceedings. Edited by F. Hirsch et G. Mokobodzki. IV, 239 pages. 1980.

Vol. 815: P. J. Slodowy, Simple Singularities and Simple Algebraic Groups. XI, 175 pages. 1980.

Vol. 816: L. Stoica, Local Operators and Markov Processes. VIII, 104 pages. 1980.

Vol. 817: L. Gerritzen, M. van der Put, Schottky Groups and Mumford Curves. VIII, 317 pages. 1980.

Vol. 818: S. Montgomery, Fixed Rings of Finite Automorphism Groups of Associative Rings. VII, 126 pages. 1980.

Vol. 819: Global Theory of Dynamical Systems. Proceedings, 1979. Edited by Z. Nitecki and C. Robinson. IX, 499 pages. 1980.

Vol. 820: W. Abikoff, The Real Analytic Theory of Teichmüller Space. VII, 144 pages. 1980.

Vol. 821: Statistique non Paramétrique Asymptotique. Proceedings, 1979. Edited by J.-P. Raoult. VII, 175 pages. 1980.

Vol. 822: Séminaire Pierre Lelong–Henri Skoda, (Analyse) Années 1978/79. Proceedings. Edited by P. Lelong et H. Skoda. VIII, 356 pages, 1980.

Vol. 823: J. Král, Integral Operators in Potential Theory. III, 171 pages. 1980.

Vol. 824: D. Frank Hsu, Cyclic Neofields and Combinatorial Designs. VI, 230 pages. 1980.

Vol. 825: Ring Theory, Antwerp 1980. Proceedings. Edited by F. van Oystaeyen. VII, 209 pages. 1980.

Vol. 826: Ph. G. Ciarlet et P. Rabier, Les Equations de von Kármán. VI, 181 pages. 1980.

Vol. 827: Ordinary and Partial Differential Equations. Proceedings, 1978. Edited by W. N. Everitt. XVI, 271 pages. 1980.

Vol. 828: Probability Theory on Vector Spaces II. Proceedings, 1979. Edited by A. Weron. XIII, 324 pages. 1980.

Vol. 829: Combinatorial Mathematics VII. Proceedings, 1979. Edited by R. W. Robinson et al.. X, 256 pages. 1980.

Vol. 830: J. A. Green, Polynomial Representations of GL_n. VI, 118 pages. 1980.

Vol. 831: Representation Theory I. Proceedings, 1979. Edited by V. Dlab and P. Gabriel. XIV, 373 pages. 1980.

Vol. 832: Representation Theory II. Proceedings, 1979. Edited by V. Dlab and P. Gabriel. XIV, 673 pages. 1980.

Vol. 833: Th. Jeulin, Semi-Martingales et Grossissement d'une Filtration. IX, 142 Seiten. 1980.

Vol. 834: Model Theory of Algebra and Arithmetic. Proceedings, 1979. Edited by L. Pacholski, J. Wierzejewski, and A. J. Wilkie. VI, 410 pages. 1980.

Vol. 835: H. Zieschang, E. Vogt and H.-D. Coldewey, Surfaces and Planar Discontinuous Groups. X, 334 pages. 1980.

Vol. 836: Differential Geometrical Methods in Mathematical Physics. Proceedings, 1979. Edited by P. L. García, A. Pérez-Rendón, and J. M. Souriau. XII, 538 pages. 1980.

Vol. 837: J. Meixner, F. W. Schäfke and G. Wolf, Mathieu Functions and Spheroidal Functions and their Mathematical Foundations Further Studies. VII, 126 pages. 1980.

Vol. 838: Global Differential Geometry and Global Analysis. Proceedings 1979. Edited by D. Ferus et al. XI, 299 pages. 1981.

Vol. 839: Cabal Seminar 77 – 79. Proceedings. Edited by A. S. Kechris, D. A. Martin and Y. N. Moschovakis. V, 274 pages. 1981.

Vol. 840: D. Henry, Geometric Theory of Semilinear Parabolic Equations. IV, 348 pages. 1981.

Vol. 841: A. Haraux, Nonlinear Evolution Equations- Global Behaviour of Solutions. XII, 313 pages. 1981.

Vol. 842: Séminaire Bourbaki vol. 1979/80. Exposés 543–560. IV, 317 pages. 1981.

Vol. 843: Functional Analysis, Holomorphy, and Approximation Theory. Proceedings. Edited by S. Machado. VI, 636 pages. 1981.

Vol. 844: Groupe de Brauer. Proceedings. Edited by M. Kervaire and M. Ojanguren. VII, 274 pages. 1981.

Vol. 845: A. Tannenbaum, Invariance and System Theory: Algebraic and Geometric Aspects. X, 161 pages. 1981.

Vol. 846: Ordinary and Partial Differential Equations, Proceedings. Edited by W. N. Everitt and B. D. Sleeman. XIV, 384 pages. 1981.

Vol. 847: U. Koschorke, Vector Fields and Other Vector Bundle Morphisms – A Singularity Approach. IV, 304 pages. 1981.

Vol. 848: Algebra, Carbondale 1980. Proceedings. Ed. by R. K. Amayo. VI, 298 pages. 1981.

Vol. 849: P. Major, Multiple Wiener-Itô Integrals. VII, 127 pages. 1981.

Vol. 850: Séminaire de Probabilités XV. 1979/80. Avec table générale des exposés de 1966/67 à 1978/79. Edited by J. Azéma and M. Yor. IV, 704 pages. 1981.

Vol. 851: Stochastic Integrals. Proceedings, 1980. Edited by D. Williams. IX, 540 pages. 1981.

Vol. 852: L. Schwartz, Geometry and Probability in Banach Spaces. X, 101 pages. 1981.

Vol. 853: N. Boboc, G. Bucur, A. Cornea, Order and Convexity in Potential Theory: H-Cones. IV, 286 pages. 1981.

Vol. 854: Algebraic K-Theory. Evanston 1980. Proceedings. Edited by E. M. Friedlander and M. R. Stein. V, 517 pages. 1981.

Vol. 855: Semigroups. Proceedings 1978. Edited by H. Jürgensen, M. Petrich and H. J. Weinert. V, 221 pages. 1981.

Vol. 856: R. Lascar, Propagation des Singularités des Solutions d'Equations Pseudo-Différentielles à Caractéristiques de Multiplicités Variables. VIII, 237 pages. 1981.

Vol. 857: M. Miyanishi. Non-complete Algebraic Surfaces. XVIII, 244 pages. 1981.

Vol. 858: E. A. Coddington, H. S. V. de Snoo: Regular Boundary Value Problems Associated with Pairs of Ordinary Differential Expressions. V, 225 pages. 1981.

Vol. 859: Logic Year 1979–80. Proceedings. Edited by M. Lerman, J. Schmerl and R. Soare. VIII, 326 pages. 1981.

Vol. 860: Probability in Banach Spaces III. Proceedings, 1980. Edited by A. Beck. VI, 329 pages. 1981.

Vol. 861: Analytical Methods in Probability Theory. Proceedings 1980. Edited by D. Dugué, E. Lukacs, V. K. Rohatgi. X, 183 pages. 1981.

Vol. 862: Algebraic Geometry. Proceedings 1980. Edited by A. Libgober and P. Wagreich. V, 281 pages. 1981.

Vol. 863: Processus Aléatoires à Deux Indices. Proceedings, 1980. Edited by H. Korezlioglu, G. Mazziotto and J. Szpirglas. V, 274 pages. 1981.

Vol. 864: Complex Analysis and Spectral Theory. Proceedings, 1979/80. Edited by V. P. Havin and N. K. Nikol'skii, VI, 480 pages. 1981.

Vol. 865: R. W. Bruggeman, Fourier Coefficients of Automorphic Forms. III, 201 pages. 1981.

Vol. 866: J.-M. Bismut, Mécanique Aléatoire. XVI, 563 pages. 1981.

Vol. 867: Séminaire d'Algèbre Paul Dubreil et Marie-Paule Malliavin. Proceedings, 1980. Edited by M.-P. Malliavin. V, 476 pages. 1981.

Vol. 868: Surfaces Algébriques. Proceedings 1976–78. Edited by J. Giraud, L. Illusie et M. Raynaud. V, 314 pages. 1981.

Vol. 869: A. V. Zelevinsky, Representations of Finite Classical Groups. IV, 184 pages. 1981.

Vol. 870: Shape Theory and Geometric Topology. Proceedings, 1981. Edited by S. Mardešić and J. Segal. V, 265 pages. 1981.

Vol. 871: Continuous Lattices. Proceedings, 1979. Edited by B. Banaschewski and R.-E. Hoffmann. X, 413 pages. 1981.

Vol. 872: Set Theory and Model Theory. Proceedings, 1979. Edited by R. B. Jensen and A. Prestel. V, 174 pages. 1981.

Vol. 873: Constructive Mathematics, Proceedings, 1980. Edited by F. Richman. VII, 347 pages. 1981.

Vol. 874: Abelian Group Theory. Proceedings, 1981. Edited by R. Göbel and E. Walker. XXI, 447 pages. 1981.

Vol. 875: H. Zieschang, Finite Groups of Mapping Classes of Surfaces. VIII, 340 pages. 1981.

Vol. 876: J. P. Bickel, N. El Karoui and M. Yor. Ecole d'Eté de Probabilités de Saint-Flour IX – 1979. Edited by P. L. Hennequin. XI, 280 pages. 1981.

Vol. 877: J. Erven, B.-J. Falkowski, Low Order Cohomology and Applications. VI, 126 pages. 1981.

Vol. 878: Numerical Solution of Nonlinear Equations. Proceedings, 1980. Edited by E. L. Allgower, K. Glashoff, and H.-O. Peitgen. XIV, 440 pages. 1981.

Vol. 879: V. V. Sazonov, Normal Approximation – Some Recent Advances. VII, 105 pages. 1981.

Vol. 880: Non Commutative Harmonic Analysis and Lie Groups. Proceedings, 1980. Edited by J. Carmona and M. Vergne. IV, 553 pages. 1981.

Vol. 881: R. Lutz, M. Goze, Nonstandard Analysis. XIV, 261 pages. 1981.

Vol. 882: Integral Representations and Applications. Proceedings, 1980. Edited by K. Roggenkamp. XII, 479 pages. 1981.

Vol. 883: Cylindric Set Algebras. By L. Henkin, J. D. Monk, A. Tarski, H. Andréka, and I. Németi. VII, 323 pages. 1981.

Vol. 884: Combinatorial Mathematics VIII. Proceedings, 1980. Edited by K. L. McAvaney. XIII, 359 pages. 1981.

Vol. 885: Combinatorics and Graph Theory. Edited by S. B. Rao. Proceedings, 1980. VII, 500 pages. 1981.

Vol. 886: Fixed Point Theory. Proceedings, 1980. Edited by E. Fadell and G. Fournier. XII, 511 pages. 1981.

Vol. 887: F. van Oystaeyen, A. Verschoren, Non-commutative Algebraic Geometry, VI, 404 pages. 1981.

Vol. 888: Padé Approximation and its Applications. Proceedings, 1980. Edited by M. G. de Bruin and H. van Rossum. VI, 383 pages. 1981.

Vol. 889: J. Bourgain, New Classes of $\mathcal{L}^p$-Spaces. V, 143 pages. 1981.

Vol. 890: Model Theory and Arithmetic. Proceedings, 1979/80. Edited by C. Berline, K. McAloon, and J.-P. Ressayre. VI, 306 pages. 1981.

Vol. 891: Logic Symposia, Hakone, 1979, 1980. Proceedings, 1979, 1980. Edited by G. H. Müller, G. Takeuti, and T. Tugué. XI, 394 pages. 1981.

Vol. 892: H. Cajar, Billingsley Dimension in Probability Spaces. III, 106 pages. 1981.

Vol. 893: Geometries and Groups. Proceedings. Edited by M. Aigner and D. Jungnickel. X, 250 pages. 1981.

Vol. 894: Geometry Symposium. Utrecht 1980, Proceedings. Edited by E. Looijenga, D. Siersma, and F. Takens. V, 153 pages. 1981.

Vol. 895: J.A. Hillman, Alexander Ideals of Links. V, 178 pages. 1981.

Vol. 896: B. Angéniol, Familles de Cycles Algébriques – Schéma de Chow. VI, 140 pages. 1981.

Vol. 897: W. Buchholz, S. Feferman, W. Pohlers, W. Sieg, Iterated Inductive Definitions and Subsystems of Analysis: Recent Proof-Theoretical Studies. V, 383 pages. 1981.

Vol. 898: Dynamical Systems and Turbulence, Warwick, 1980. Proceedings. Edited by D. Rand and L.-S. Young. VI, 390 pages. 1981.

Vol. 899: Analytic Number Theory. Proceedings, 1980. Edited by M.I. Knopp. X, 478 pages. 1981.